WETLANDS

Revised Edition

WETLANDS

Revised Edition

Peter D. Moore

Illustrations by Richard Garratt

An imprint of Infobase Publishing

WETLANDS, Revised Edition

Copyright © 2008, 2000 by Peter D. Moore

Facts On File, Inc.
An imprint of Infobase Publishing
132 West 31st Street
New York NY 10001

ISBN-10: 0-8160-5931-4
ISBN-13: 978-0-8160-5931-7

Library of Congress Cataloging-in-Publication Data
Moore, Peter D.
    Wetlands / Peter D. Moore ; illustrations by Richard Garratt—Rev. ed.
      p. cm.—(Ecosystem)
    Includes bibliographical references and index.
    ISBN 0-8160-5931-4
    1. Wetland ecology—Juvenile literature. 2. Wetlands—Juvenile literature. I. Garratt, Richard, ill. II. Title.
    QH541.5.M3M664 2007
    577.68—dc22   2006037399

You can find Facts On File on the World Wide Web at http://www.factsonfile.com

Text design by Erika K. Arroyo
Illustrations by Richard Garratt
Photo research by Elizabeth H. Oakes

Printed in the United States of America

Bang Hermitage 10 9 8 7 6 5 4 3 2

This book is printed on acid-free paper.

To Amelia, Amanda, Madeleine, and Michael.
Your hands may be very small but in them lies the future.

✦ ✦ ✦ ✦

Contents

Preface

Increasingly, scientists, environmentalists, engineers, and land-use planners are coming to understand the living planet in a more interdisciplinary way. The boundaries between traditional disciplines have become blurred as ideas, methods, and findings from one discipline inform and influence those in another. This cross-fertilization is vital if professionals are going to evaluate and tackle the environmental challenges the world faces at the beginning of the 21st century.

There is also a need for the new generation of adults, currently students in high schools and colleges, to appreciate the interconnections between human actions and environmental responses if they are going to make informed decisions later, whether as concerned citizens or as interested professionals. Providing this balanced interdisciplinary overview—for students and for general readers as well as professionals requiring an introduction to Earth's major environments—is the main aim of the Ecosystem set of volumes.

The Earth is a patchwork of environments. The equatorial regions have warm seas with rich assemblages of corals and marine life, while the land is covered by tall forests, humid and fecund, and contain perhaps half of all Earth's living species. Beyond are the dry tropical woodlands and grassland and then the deserts, where plants and animals face the rigors of heat and drought. The grasslands and forests of the temperate zone grow because of the increasing moisture in these higher latitudes but grade into coniferous forests and eventually scrub tundra as the colder conditions of the polar regions become increasingly severe. The complexity of diverse landscapes and seascapes can, nevertheless, be simplified by considering them as the great global ecosystems that make up our patchwork planet. Each global ecosystem, or biome, is an assemblage of plants, animals, and microbes adapted to the prevailing climate and the associated physical, chemical, and biological conditions.

The six volumes in the set—*Deserts, Revised Edition; Tundra; Oceans, Revised Edition; Tropical Forests; Temperate Forests, Revised Edition;* and *Wetlands, Revised Edition*—

between them span the breadth of land-based and aquatic ecosystems on Earth. Each volume considers a specific global ecosystem from many viewpoints: geographical, geological, climatic, biological, historical, and economic. Such broad coverage is vital if people are to move closer to understanding how the various ecosystems came to be, how they are changing, and, if they are being modified in ways that seem detrimental to humankind and the wider world, what might be done about it.

Many factors are responsible for the creation of Earth's living mosaic. Climate varies greatly between Tropics and poles, depending on the input of solar energy and the movements of atmospheric air masses and ocean currents. The general trend of climate from equator to poles has resulted in a zoned pattern of vegetation types, together with their associated animals. Climate is also strongly affected by the interaction between oceans and landmasses, resulting in ecosystem patterns from east to west across continents. During the course of geological time even the distribution of the continents has altered so that the patterns of life currently found on Earth are the outcome of dynamic processes and constant change. The Ecosystem set examines the great ecosystems of the world as they have developed during this long history of climatic change, continental wandering, and the recent meteoric growth of human populations.

Each of the great global ecosystems has its own story to tell: its characteristic geographical distribution; its pattern of energy flow and nutrient cycling; its distinctive soils or bottom sediments, vegetation cover, and animal inhabitants; and its own history of interaction with humanity. The books in the Ecosystem set are structured so that the different global ecosystems can be analyzed and compared and so that the relevant information relating to any specific topic can be quickly located and extracted.

The study of global ecosystems involves an examination of the conditions that support the planet's diversity, but environmental conditions are currently changing rapidly. Human beings have eroded many of the great global ecosystems as they have reclaimed land for agriculture and urban settlement and built roads that cut ecosystems into ever smaller units. The fragmentation of Earth's ecosystems is proving to be a serious problem, especially during times of rapid climate change, itself the outcome of intensive industrial activities on the surface of the planet. The next generation of ecologists will have to deal with the control of global climate and also the conservation and protection of the residue of Earth's biodiversity. The starting point in approaching these problems is to understand how the great ecosystems of the world function, and how the species of animals and plants within them interact to form stable and productive assemblages. If these great natural systems are to survive, then humanity needs to develop greater respect and concern for them, and this can best be achieved by understanding better the remarkable properties of our patchwork planet. Such is the aim of the Ecosystem set.

Acknowledgments

I should like to record my gratitude to the editorial staff at Facts On File for their untiring support, assistance, and encouragement during the preparation of this book. Frank K. Darmstadt, executive editor, has been a constant source of advice and information and has been meticulous in his checking of the text and coordinating the final assembling of materials. My gratitude also extends to Ms. Alana Braithwaite, his editorial assistant, and the production department. I should also like to thank Richard Garratt for his excellent illustrations and Elizabeth Oakes for her perceptive selection of photographs. Particular thanks are due to my wife, who has displayed a remarkable degree of patience and support during the writing of this book, together with much needed critical appraisal. I must also acknowledge the contribution of many generations of students in the Life Sciences Department of the University of London, King's College, who have been a constant source of stimulation, critical comment, and new ideas. I also acknowledge a considerable debt to my colleagues in teaching and research at King's College, especially those who have accompanied me on field courses and research visits to many parts of the world. Their work, together with that of countless other dedicated ecologists, underlies the science presented in this book.

Introduction

The world is changing rapidly. Scientific knowledge of the Earth's ecosystems is increasing exponentially as ecological research develops, and the ecosystems themselves are in a constant state of flux as human populations expand and place them under increasing threat of drainage and development. Much has changed since the first edition of this book several years ago both in the scientific understanding of wetland ecosystems and in the state of the wetlands themselves. A revised edition of the book is therefore urgently needed.

Wetlands, Revised Edition contains much additional information on the global cycling of water, which will become an increasingly scarce resource in coming decades, and also concerning the relationship among the water cycle, rock weathering, and wetland formation. More is included regarding the development of wetlands and the techniques used in their study, such as radiocarbon dating and pollen analysis. The plants and animals that inhabit wetlands are remarkable in their adaptations, and in this new edition it has been possible to expand accounts of individual species with more information on their ways of coping with life in the wet. Finally, the section dealing with the future of wetlands in a world of rapid environmental change has been brought up to date, particularly with respect to the interaction between wetlands and the global carbon cycle, which in turn has major global climatic implications. The expansion and revision of the entire set of books dealing with the world's ecosystems has also offered an opportunity to restructure the books' contents, which will aid the student in extracting information regarding any specific topic. The revised edition is thus an up-to-date and accessible resource for all who wish to study this rich and varied habitat.

Wetlands may not be as spectacular as towering mountain ranges or as diverse in their inhabitants as the tropical rain forests, but they have a captivating appeal to all who study them and come to know them well. Wetlands are among the wildest places on Earth; they are an untamed wilderness. There can be few sounds more stirring to the human ear than that of a flock of wild geese circling at sunset over an expanse of marshland and few sights more impressive than their descent as the birds tumble from the sky, collecting in groups to spend the night feeding on the wetland's rich and productive resources.

Explorers and pioneers have often been inspired by extensive wetlands and their rich abundance of wildlife. They have seen them as sources of food and have exploited them for fish and fowl. They have found them valuable as a means of transport and communication, allowing pioneers to penetrate into the heart of continents where overland travel would have been impossible. There have also been times when wetlands have presented obstacles to settlement, development, agriculture, and progress. Farming and wetlands do not fit together comfortably, and wetlands have invariably suffered in the conflict. For an agriculturalist, the best kind of wetland is a drained wetland. In recent years, however, the tide of opinion is beginning to change, and wetlands are being appreciated for the advantages they can bring to human cultures. Drainage and destruction are not the only way to deal with an intractable habitat that does not lend itself to conventional cultivation. The complexity and the fragility of wetlands need to be appreciated and understood if the mistakes of the past are to be avoided or even rectified. This book aims to examine the diversity of wetlands, to analyze the way they work, and to set out ways in which they could be conserved.

■ WHAT ARE WETLANDS?

Many diverse habitats fall within the term *wetland,* so defining the concept of wetland has proved difficult. The one obvious feature that all wetlands have in common is their wetness, but oceans are wet and so are tropical rain forests, yet these habitats do not qualify as wetlands in the generally accepted sense of the term. Shallow lakes with aquatic plants emerging from them; extensive marshes

Forested raised bog in Alaska *(Vera Bogaerts)*

of reeds and sedges; swamps in which trees grow with their roots surrounded by water; peat lands that blanket the landscape; saline shallow water, where salt marshes or mangrove forests line the shore; all of these habitats could be included within the wetland definition. In 1971 an international conference was held in Ramsar on the southern coast of the Caspian Sea in Iran. Delegates gathered from many countries to discuss the global problems facing wetlands and how these problems could be solved. One of the tasks facing the meeting was to define the habitats to be covered by the conference so that the final agreement and recommendations of the meeting, known as the Ramsar Convention, could be clearly interpreted. The Ramsar Convention defines wetlands as "all areas of marsh, fen, peatland, or water, whether natural or artificial, permanent or temporary, with water that is static or flowing, fresh, brackish, or salt." The convention set a depth of 20 feet (6 m) as the limit for any water body to be included in the term *wetland*.

On the basis of this widely accepted definition, it is clear that wetlands are very diverse in their nature, ranging from open water to forested ecosystems, as shown in the illustration. Most of the other great ecosystems, or *biomes,* of the world can be defined according to the structure, or architecture, of their vegetation. Tropical forest is characterized

by tall trees, tropical savanna consists of tall grassland with scattered trees, chaparral has a low cover of shrubs, and so on. Wetlands cannot be defined according to vegetation structure, only by the abundance of water.

Most biomes are also controlled in their development and distribution by climate. Rain forests require high levels of precipitation; prairies are treeless because of their summer drought; deserts experience constantly dry conditions. Each of these biomes is therefore restricted in its distribution to particular zones of the Earth, defined broadly by latitude, but wetlands are found in almost all parts of the world, and this fact adds to their great diversity of form and the wide range of plants and animals that occupy them. It is possible to construct a definition of the word *wetland,* but this cannot do justice to the great range of wetland types found around the world, from the frozen wastes of the Arctic to the steamy swamps of the tropical jungle. Wetlands occupy about 6 percent of the world's surface, but not in a single block. They are scattered, often in small patches, over a very wide area, and their fragmentation and wide dispersion add to their vulnerability. It is easy for wetland sites to be lost through drainage and development in different parts of the world without the overall impact on the global ecosystem being appreciated. This was one of the reasons why the Ramsar Convention was formulated—to ensure that wetlands and their conservation received attention at an international rather than a purely local level.

Not only are wetlands found in many different climatic zones of the Earth, but they are also present in a very wide range of geological settings. The rocks that underlie wetlands have a strong impact on the chemistry of their waters, and, as will become apparent later in this book, water chemistry is a major factor in determining the plants and animals that live in a wetland. The acidity of a wetland depends to a large extent on the nature of the rocks through which the water has soaked during its passage. Rocks with high lime content may neutralize any acids present, and the wetlands receiving such water can support a very different set of organisms from more acidic sites. People affect water chemistry by industrial pollution or by releasing fertilizers from agricultural land into drainage ditches, streams, and eventually wetland habitats. Some species require waters rich in nutrients, but many others fail to compete under such fertile conditions and survive only in waters that have low concentrations of nutrient elements. Wetlands with water supplies of different chemical composition thus support very different sets of plants and animals.

Wetlands are never static ecosystems but are forever changing. Lakes become infilled with sediment; reeds invade shallow water; trees take root in marshes and turn them into swamps; bogs form mounds of peat and raise their surfaces above the surrounding lands. All wetlands are dynamic systems, and the plants and animals they contain remain only for a limited period while the conditions suit them and then move on to other parts of the developing ecosystem. In most wetlands, therefore, there exists a complex mosaic of different stages in development, and this adds to the diversity of the landscape and the range of species found. Wetland scientists need to understand the pattern of development so they can predict the next stage. Sometimes conservationists may wish to deflect or hold back the course of wetland development in order to preserve a particular species, and this can only be achieved if the process of wetland development is fully understood. The fact that wetlands change in this way can sometimes be helpful in the repair and rehabilitation of damaged sites.

A wetland develops over the course of time because its living components interact with one another. With its abundant supply of water, a wetland can be a very productive ecosystem, especially in tropical regions. Aquatic vegetation grows quickly in tropical marshes and swamps, providing a lush supply of food for herbivorous animals, from caterpillars to antelope. Predators, ranging in size from dragonflies to lions, feed on the grazing animals and build complex interactions that support high levels of diversity. The wetness that encourages plant productivity also affects other processes within the ecosystem, sometimes causing them to slow down. The bacteria and fungi responsible for consuming any residual plant material left by the grazers do not function as efficiently in very wet environments as they do in the air, so, in a wetland, they may not be able to decompose plant detritus as quickly as it is deposited. When this happens the wetland accumulates excess organic matter in the sediments as peat. Peat-forming wetlands are thus exceptional ecosystems in being unbalanced, accumulating dead materials within their soils, and steadily growing in their overall mass.

Just as standing water slows the rate of decomposition, it also creates problems for other organisms. Animals living in mud at the bottom of a wetland may find it difficult to obtain enough oxygen. Oxygen is needed for respiration, and it is available in water in a dissolved form, so fish and many other aquatic creatures, including plant roots, take their oxygen directly from the water. Oxygen molecules dissolved in water, however, diffuse much more slowly than in the atmosphere, so when it is used up, oxygen is slow in being replaced in the surrounding water and living creatures can experience shortages. Overcoming this difficulty has led to the development of a wide range of structures and biochemical pathways in the plants and animals that inhabit wetlands. These adaptations will be examined in some detail later in this book.

This brief survey of some of the features and characteristics of wetlands illustrates their great range and diversity and illustrates why it is so difficult to establish a simple definition for the term. The Ramsar Convention definition provides only a simple starting point from which to approach the rich assembly of the Earth's wetlands.

■ WHY ARE WETLANDS IMPORTANT?

Does it really matter if the world's wetlands are lost? Are they of any value to humans or to the global environment in general? It is always very difficult to place a value on a particular habitat, and wetlands are no exception. Wetlands have existed on Earth for many hundreds of millions of years, so one could argue that it would be irresponsible for people to lose a feature of such great antiquity that would be almost impossible to replace. Destroying wetlands is rather like burning an art gallery in that it involves the loss of things that cannot be re-created, but, unlike works of art, it is impossible to place a monetary value on a wetland.

Ironically, it is easier to evaluate a really ancient wetland in financial terms than a modern one. The great swamps of the past have left behind them deposits of peat that have now been compressed and altered to form coal, and this energy resource has had a very great economic impact on the history of humanity. It was the ancient wetlands, through their coal, that fueled the Industrial Revolution and took human cultures to new planes of development. Wetlands have indeed modified the course of history, and deposits of coal are still valuable as energy reserves to fuel modern industry, even though oil and gas, also products of wetland biological processes, have largely replaced coal as a primary energy resource. More recently formed wetland sediments, such as the peat deposits of the northern bogs, have also supplied some human energy needs, as well as other functions such as soil additives, so these peat deposits can also be evaluated financially. All of these wetland products, ancient and modern, are essentially nonrenewable; once they have been used up they cannot be replaced. Even in the case of peat-forming wetlands, the rate at which new peat is being laid down is much too slow to form the basis for sustainable production in the long term.

Wetlands serve other functions besides their production of energy-rich materials, however. By definition, they contain reserves of water, and the supply of freshwater for a growing global human population will become increasingly important in the future. Without wetlands, rainwater would soak through to streams and go rushing toward the ocean with no brake on its progress. In wetlands the rate of water movement is slowed down, and water is conserved in reservoirs within their basins. Wetlands can act like giant sponges, absorbing excess water and releasing it slowly into the rivers from which it can be extracted for use in agricultural irrigation, industrial processes, or for domestic consumption. Draining wetlands is equivalent to punching holes in a water storage tank. By storing water, wetlands can also act as flood controls, retaining some of the water that would otherwise spread over regions occupied by people or their crops. Along shorelines, mangrove wetlands can absorb some of the energy of storms at sea and prevent coastal flooding.

For many people, wetlands are also a source of food. Fish provide a staple source of protein to many of the world's people. The people around Lake Victoria in East Africa, for example, rely heavily upon the fish stocks of abundant wetlands of that region. The mangrove habitats of many tropical coastal areas are also an important source of fish. Even in developed parts of the world where fish are not a major component of the diet, the use of wetlands for angling represents a valued source of recreation. Wildfowl can also provide a sustainable supply of food from wetlands, and some wetland sites have a long history of management for hunting.

Perhaps the greatest value of wetlands, however, lies in the animals and plants that inhabit them. Many wetland species are found in no other type of habitat, so the loss of the world's wetlands would entail the extinction of many highly specialized organisms. It is impossible to place a monetary value on this biodiversity, but it is not difficult to appreciate that financial loss could result from extinctions. The biochemistry of those plants that tolerate waterlogging and flood, or the high salt content of saline lakes, may help agriculturalists to develop new types of crops that can deal with these conditions. Many of the drugs used to combat disease are derived from the poisons contained in plants or in the venom of snakes and spiders. It is impossible to predict what unlikely creature lurking somewhere in a remote wetland may one day supply the answer to a medical problem. Biodiversity is one of the Earth's most precious resources, and the wetlands contain an important and distinctive component of the world's biological richness.

The wide variety of the world's wetland types and the many possible uses that wetlands provide for human life on the planet should now be apparent. Wetland is an ecosystem that is worthy of study; its complexity requires careful analysis so that the habitat can be managed with wisdom. In this book, the many aspects of wetland ecology and geography will be considered. Wetlands are part of the human support system of this planet, and we would do well to examine each component with care. Those who care for the future of the planet need to look closely at wetlands, one of the Earth's richest yet most threatened ecosystems.

1

Geography of Wetlands

There are many different kinds of wetlands, but the one thing they all have in common is water. Water is a remarkable material that is essential for all living things. It is no exaggeration to say that without water life itself would be impossible. The operations of all biochemical systems within the living cell are determined by, and dependent on, the physical and chemical properties of water. The abundance of water on the surface of the Earth is a major contributory factor to the richness of life on this planet. The climatic conditions found on the Earth mean that global temperatures fall within a range where water can exist in liquid, gaseous, and solid forms, and this ensures that water is available to living organisms in most parts of the planet. The Earth is rich in water and is therefore rich in wetlands.

■ THE WATER CYCLE

Water is constantly on the move, and its movement ensures that it is distributed around the world's surface. By far the biggest reservoirs of water on Earth are the oceans, which contain more than 96 percent of the world's water resources. About 0.32 billion cubic miles (1.34 billion cu km) of water resides in the oceans. The freshwater resources of the world therefore account for less than 4 percent of the total water, and all land plants and animals, including humans, are totally dependent on this limited freshwater resource. Of this 4 percent, roughly half is locked up in the ice of ice sheets and glaciers, and much of the remainder is present below ground, soaked into the rocks of the Earth's mantle, where it lies in deep *aquifers,* or storage reservoirs, gradually moving toward the sea. The water that is readily available to the organisms living on land is only about 0.004 percent of the Earth's total water resources. This available resource includes water in soils, rivers, lakes, and wetlands, together with water present in living organisms and the water vapor in the atmosphere. Water is constantly on the move, but the quantity actually moving around landscapes is thus a very small proportion of the total resource.

Another way of looking at this cycle is to consider how long a water molecule spends in any of the various locations available to it. The oceans have the longest residence time; any water molecule entering the ocean can expect on average to spend the next 2,600 years there. When a water molecule happens to come near the surface of the oceans, it may enter a vapor state and move out into the atmosphere, beginning its cycle once more. In the atmosphere the water molecule on average will only stay for eight days before being deposited as precipitation. It is entirely possible (the chances are about 90 percent) that it will fall right back into the ocean, in which case it can expect an additional long term of imprisonment there, but there is a 10 percent chance that it will fall over a continental landmass and begin its slower return to the sea. If the molecule should fall upon one of the world's great ice sheets, in Greenland or Antarctica, its residence time in a frozen form will be about 9,000 years before it is likely to melt and start on its way to the oceans again. If it enters the soil, it will on average remain there only 50 days and then may be taken up by a plant, but if so it will stay as part of the biomass for only about six days before being evaporated back into the atmosphere. If it is not taken up by a plant, it may soak down into the ground under the influence of gravity and become part of a deep aquifer. Here it will stay for an average of 500 years before reaching the sea or being drawn up by human activities to supply the water requirements of farms and cities. The final option as the water molecule continues on its way back to the ocean is that it will enter a stream, river, lake, or other type of wetland, in which case it can expect to remain there for about 4.5 years before passing on its way. Wetlands contain only about 0.014 percent of the world's water, but the residence time there for a water molecule is long enough to ensure that the water is readily available to living things. Wetlands thus act as a kind of brake, holding back the water that is on is way to long-term incarceration in the oceans and ensuring that living creatures can avail themselves of this precious resource.

There are thus long cycles and short cycles involved in the overall water cycle of the Earth. Cycles of water molecules

ice 2.25%   freshwater 0.75%   saltwater oceans 97%

P precipitation
E evaporation
E/T evapotranspiration

© Infobase Publishing

Global hydrological cycle. The figures indicate what proportion of the world's water is present as ice, freshwater, and salt water.

involving residence in the oceans or in ice sheets will take many thousands of years to complete, but cycles in which water molecules move simply between soil, vegetation, and atmosphere are considerably shorter.

Some wetlands occur along the edge of the ocean, and they have a ready and reliable source of water. Other wetlands have developed inland and are dependent on the sup-

ply of water from precipitation and catchment drainage. The existence of these wetlands is determined by the efficiency of the water cycle and in particular the rate and the seasonal timing of water supply. The diagram shows a simplified water cycle for a terrestrial landscape. The figures are the total global rates of movement of water among the major storage reservoirs (oceans, groundwater, and atmosphere). As can be seen from the diagram, more than 100 cubic miles of water evaporates (see sidebar "Evaporation and Precipitation") from the oceans each year. Of this, more than 90 percent returns directly to the ocean as precipita-

Evaporation and Precipitation

Water molecules are relatively simple in purely chemical terms. They consist of two hydrogen atoms joined to an oxygen atom to form the stable water molecule H_2O. Water molecules, however, tend to stick together because the hydrogen atoms of one molecule form weak bonds (called *hydrogen bonds*) with the oxygen component of neighboring molecules. This holds the molecules together and gives them a cohesive force. When energy is withdrawn from water by cooling it, the molecules move more slowly and pack together more tightly, causing them to contract in volume, but if water is heated the energetic molecules move faster and farther apart, leading to expansion. As a molecule takes up additional energy and becomes more active, it may reach a point where it breaks the hydrogen bonds linking it to other molecules and detaches itself from the rest of the water body. This is called *evaporation,* and the molecule has now entered the vapor state from the liquid state. The amount of heat required to liberate a molecule from its liquid neighbors is called the *latent heat of evaporation.*

Evaporation is also affected by the density of water molecules already present as vapor in the atmosphere. If there is a high density of water vapor molecules in the

layer of air above a water surface, the pressure exerted by these molecules makes it more difficult for liquid water to evaporate. There comes a point where the air is saturated with water vapor and no more vapor molecules can be accepted. The pressure exerted by the water vapor at this point is called *saturated vapor pressure,* and it is dependent on the temperature of the atmosphere. At higher temperature a packet of air can hold more water vapor molecules than at lower temperature. Air at 104°F (40°C) can hold six times the amount of water that can be held by air at 50°F (10°C). When air saturated with water vapor cools, some of the water contained within it must revert to a liquid state. Molecules lose their latent heat and condense into water droplets, which may initially remain suspended in the atmosphere as mist or cloud or may become sufficiently large and heavy to fall as raindrops. If the air temperature is below the freezing point of water, the droplets solidify into crystals or snowflakes. It is even possible for water vapor to change directly into a solid form without passing through a liquid phase. This is known as *sublimation.* The deposition of water from the atmosphere in either a liquid or a solid form is called *precipitation.*

Transpiration and Evapotranspiration

All plant and animal cells require water as a medium in which they carry out their metabolism. Plants are static, so they have to obtain water from the soil through their roots. Most plants are also photosynthetic, which means that they have to take in carbon dioxide from the atmosphere in order to build the energy-rich sugars that they need. Plants collect this atmospheric gas through pores, or *stomata,* on their leaves and stems, and they are able to control the aperture of the pores according to their needs. The presence of pores, however, results in water loss because these apertures open into small chambers within the structure of the leaf, where thin-walled cells easily lose their water by evaporation to the internal atmosphere. The water molecules in the saturated atmosphere within these chambers quickly diffuse through the pore, and water is thus lost to the plant in the process known as *transpiration.* Transpiration can be regarded as an inevitable and unfortunate consequence of a plant's need to take in carbon dioxide, but it does also have some positive value. Water evaporating from a leaf takes up latent heat of evaporation (see sidebar "Evaporation and Precipitation," page 2), and this has the effect of cooling the leaf. Transpiration thus serves a purpose similar to perspiration in some animals. Transpiration also creates a tension on the water columns within the stem of a plant. As water vapor is lost from the leaf, a pull is exerted on the liquid water within the plant, resulting from the bonding and adhesion between water molecules. Tubular cells, called *vessels,* within the veins of the leaf and the conducting tissues of the stem contain an unbroken column leading all the way down to the root, and as water is lost at the top of the column, water is drawn up through the plant, creating the pull that also brings water into the root from the soil. There is thus a transpiration stream running from soil to atmosphere through the plant, and this is an important pathway by which water molecules can move through this part of the water cycle.

Vegetation varies in its architecture and therefore in its efficiency as a conduit for water moving to the atmosphere from the soil. The transpiration also varies between day and night and seasonally, especially in the case of deciduous plants. Evaporation from soil and water surfaces also varies with diurnal and seasonal factors, and any calculations regarding the water cycle need to take into account the combined processes of transpiration and evaporation. Climatologists usually consider the two together as *evapotranspiration.* This is an important concept for wetland scientists because the state of water in any ground reservoir is determined by the balance between water input from precipitation and groundwater flooding and the output from drainage and evapotranspiration.

tion. The remaining water vapor is carried by winds over the land surface, where it is joined by additional water vapor produced by evaporation and transpiration (see sidebar "Transpiration and Evapotranspiration") from soil and water surfaces and from the leaves of vegetation respectively. This accumulated water vapor in the atmosphere condenses and precipitates over the land, where it may find its way back to the ocean or may take a shorter route back into the atmosphere to be recycled there. Ultimately, the budget is balanced, with the same amount of water returning to the oceans each year as the quantity received by the land from oceanic evaporation.

Most of the major habitats of the Earth—tropical rain forest, desert, tundra, savanna, and so on—occur only in very restricted parts of the globe, normally determined by climate. Wetlands are different. As the accompanying map of the world illustrates, wetlands give the impression of being scattered all over the Earth's surface, not confined to any particular area. The reason for this is fairly obvious; wetlands occur wherever conditions are wet, and this may be a consequence of a rainy climate, or because the climate is so cold that water does not evaporate quickly, or because a region is supplied with water by river systems, or because the area is adjacent to the sea and is regularly flooded, or for many other reasons. So although climate plays an important part in determining where wetlands occur, there are other important geographical factors that also come into play, especially those relating to the form of the landscape and its drainage patterns.

A closer look at the global map reveals that there is a general pattern underlying the scatter of wetlands: There are some latitudinal zones where wetlands are more abundant and others where they are less so. In the Northern Hemisphere, for example, the land between approximately 45°N and 75°N is richer in wetlands than the zone immediately to the south of this band. In the Southern Hemisphere, there is very little land that lies between 45°S and 75°S—only the tip of South America—but what land there is has an abundance of wetlands. The zone enclosed by the latitudes 15°N and 15°S of the equator is again relatively rich

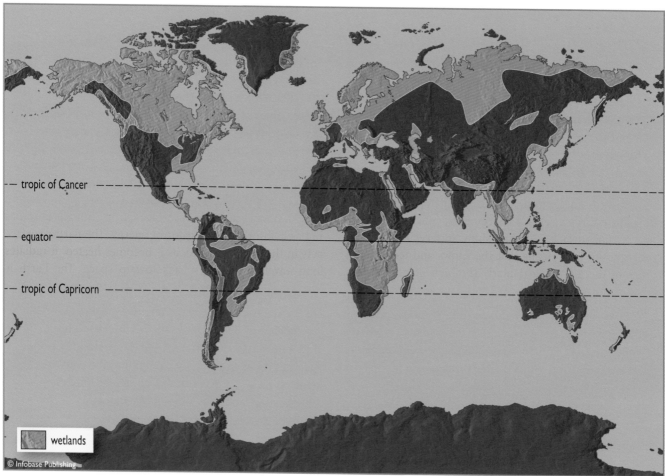

Distribution of wetlands around the world. Pale green areas show where wetland ecosystems are most abundant.

in wetlands when compared with the bands of land that lie along the tropics of Cancer and Capricorn and immediately outside them. Using global maps to provide an overall picture of wetland abundance has its dangers, however, because it is easy to forget that the map is a flat projection of what is really the surface of a sphere. This results in a tendency to exaggerate the land area at higher latitudes (nearer the poles) and to underestimate the land area near the equator. Even taking this into account, there is clearly some latitudinal, zonal pattern in the global distribution of wetlands.

CLIMATE AND THE GLOBAL DISTRIBUTION OF WETLANDS

In order to explain the zonal distribution of wetlands around the world it is necessary to consider the way in which global climates are also arranged in latitudinal

bands. The Earth's climate is quite evidently not uniform; conditions in the Tropics are generally hotter than those at the North and South Poles, and conditions in Texas and Arizona are more arid than those in Washington or Maine. The variation of climate over the surface of the Earth is the most important factor in determining the vegetation of a region, and wetlands are particularly sensitive to climatic conditions, especially those that affect the balance of the water cycle in a region.

The first reason why the Earth's climate is not uniform is because the planet is spherical, so that incident energy from the Sun is received at different angles according to the location on its surface. In the Tropics the Sun rises to a position close to overhead at noon each day. There is some variation with season, when the Sun is directly overhead at the tropic of Cancer at the height of the Northern Hemisphere summer and at the tropic of Capricorn at the height of the Southern Hemisphere summer. At the equator itself the midday Sun is overhead twice a year, at the equinoxes, but even in midwinter in the Tropics, the Sun is not far from being overhead in the middle of the day. When a surface receives energy, how efficiently it absorbs that energy depends in part upon the angle of incidence. A surface at right angles to the direction of the energy source is the most efficient for absorption, and this efficiency decreases as the angle of incidence becomes more oblique. The spherical nature of the Earth's surface causes the angle of incidence of the Sun's rays to increase as one moves away from the equator and toward the poles, as shown in the diagram. In the polar latitudes a larger proportion of the incoming energy is reflected rather than absorbed, so there is less heating of the Earth's surface.

The absorption of solar energy by the Earth is affected not only by the angle at which the energy is received but also by the nature of the surface, in particular its color. A dark object absorbs energy more efficiently than a light object, and a shiny surface reflects energy more efficiently than a matte or dull surface. So white shiny surfaces reflect a lot of energy, and this means that the snow- and ice-covered regions of the globe, particularly the high polar latitudes where the Sun is always low, reflect a greater proportion of the incident energy than the darker canopy of a tropical forest. The reflectivity of a surface is called its *albedo* (see sidebar on page 6). Oceans and land, forest and desert, all have their own characteristics of albedo, and these influence the total quantity of the available energy that is absorbed.

The high latitudes (so called because the number assigned to a line of latitude reflects the angle it subtends to the equator, giving the South and North Poles a latitude 90°S and 90°N, respectively, and the equator 0°) also receive sunlight that has passed through a greater thickness of atmosphere before it reaches the ground. There is a greater chance that some of the energy has been absorbed or reflected by dust, water droplets, or ice on its passage. This further reduces the energy reaching the Earth's surface in these polar regions.

Light from the Sun arrives at the Earth as parallel beams, but the light heats different areas of the Earth's surface depending upon the angle at which they arrive. Tropical areas receive more intense energy than the polar regions. The tilt of the Earth upon its axis means that the position of the overhead noonday Sun varies with season.

One final consideration is the seasonal variation in day length. Beyond the Arctic and Antarctic Circles there are certain days when the Sun never rises and also some days when the Sun never sets. Close to the poles the day and the night each last for about six months, and this has its effect on the energy balance of the region.

The Earth receives its energy as short-wavelength radiation that humans perceive as light. Very short-wavelength radiation, such as ultraviolet radiation, is absorbed high in the atmosphere so that does not penetrate to the Earth's surface. When short-wavelength light radiation reaches a surface and is absorbed, it is transformed into long-wavelength radiation, or heat. A dark surface left in sunlight becomes hot as it absorbs and transforms energy. The Earth's surface acts in the same way and, as it becomes heated, it radiates long-wavelength energy back toward space. The Earth is surrounded by an atmosphere containing some gases that absorb energy in the long-wavelength range. These include water vapor, carbon dioxide, ozone, methane, and nitrous oxide, which are transparent to short-wavelength radiation, so they allow the passage of light but take up heat energy on its way back into space. The atmosphere thus acts as a thermal blanket around the Earth, retaining some of the solar heat and preventing some radiation loss. This characteristic of the atmosphere is known as the *greenhouse effect* because it resembles the function of the glass in a greenhouse, allowing light in but preventing the escape of heat. The atmospheric gases that absorb long-wavelength radiation are called *greenhouse gases* (see "Wetlands and the Carbon Cycle," pages 73–77).

When considering the pattern of climate over the Earth's surface, therefore, the first factor to take into account is the balance between incoming energy and radiated heat loss. The diagram shows the Earth's radiation balance at different latitudes. Although the total energy budget for the Earth is approximately in balance (if it were otherwise, the Earth would be rapidly getting hotter or colder), gains and losses are not evenly distributed with latitude. In the equatorial regions more energy is being received than is being radiated, while

Northern Hemisphere summer
North Pole
North Pole
South Pole
South Pole
Southern Hemisphere summer
© Infobase Publishing

Albedo

When light falls upon a surface, some of the energy is reflected and some is absorbed. The absorbed energy is converted into heat and results in a rise in the temperature of the absorbing surface. What proportion of the light is absorbed and what proportion is reflected varies with the nature of the surface. A dark-colored surface absorbs a greater proportion of the light energy and reflects less than a light-colored surface; it therefore becomes warmer more rapidly. A dull, matte-textured surface absorbs energy more effectively than a shiny one.

A mirror reflects most of the light that falls upon it, which is why mirrors are usually cold to the touch. The angle at which light falls upon a surface also affects the proportion of light absorbed and reflected; light arriving vertically is reflected less than light arriving from a low angle.

Scientists express the degree of reflectivity of a surface as its *albedo*. It is expressed as the amount of light reflected divided by the total incident light. In order of efficiency of reflectivity, the following list gives examples of the albedo of various surfaces.

SURFACE	PROPORTION REFLECTED	ALBEDO
Snow	75–95%	0.75–0.95
Sand	35–45%	0.35–0.45
Concrete	17–27%	0.17–0.27
Deciduous forest canopy	10–20%	0.10–0.20
Road surface	5–17%	0.05–0.17
Evergreen forest canopy	5–15%	0.05–0.15
Water (overhead sun)	5%	0.05
Overall average for Earth	39%	0.39

in the higher latitudes (above 30°N and 30°S) the radiation losses are greater than the energy gains from sunlight. If this graph portrayed the whole story, then the low latitudes would be constantly getting hotter and the high latitudes colder. Although there are overall differences in temperature between high and low latitudes, the contrast is not as great as might be expected from this simple model. There are mechanisms in operation that redistribute some of this energy imbalance, and these are found in the movements of the atmosphere and the oceans.

The strong heating of the Earth's surface in the equatorial regions leads to an instability in the atmosphere. Molecules of air become more active when they are heated, and this results in expansion. As air expands it becomes less dense so that it is easily displaced by colder, denser air that pushes the heated air upward. This happens at a grand scale over the equatorial zone, creating an updraft of heated air and converging winds over the Earth's surface that replace the displaced air masses, as shown in the diagram on page 7. The equatorial air may be forced up to 60,000 feet (18,000 m) when it spreads out north and south toward the poles. The air cools at these high altitudes, and its contraction and increasing density causes it to sink, creating zones of descending air between 25° and 30° north and south of the equator. So the equator is generally a region of low pressure, while the tropics of Cancer and Capricorn are zones of high pressure. Low pressure and rising air usually create conditions of high precipitation, and this is true of the equatorial regions, where tropical rain forests are the characteristic type of vegetation. In high-pressure systems, however, precipitation is low, and deserts predominate in these regions of descending air. The descending air masses may be deflected back toward the equator or may move poleward, in which case they meet the cold, high-pressure air moving out from the poles. The meeting of these two air masses, tropical and polar, creates an unstable set of conditions, with warm, moist air being forced up over cold air. The temperate latitudes where the two masses collide are characterized by variable weather patterns with frequent rain, especially in regions close to the oceans. The Poles are sites of cold, dense air falling to the surface and bearing little snow or rain, so that the polar regions often have no more precipitation than the hot deserts of the world.

The global circulation pattern of the atmosphere can be simply described as a set of circulation cells, as shown in the illustration. At the equator the surface winds converge from the north and the south to form an *Intertropical Convergence Zone*. At the point of convergence, which varies a little with

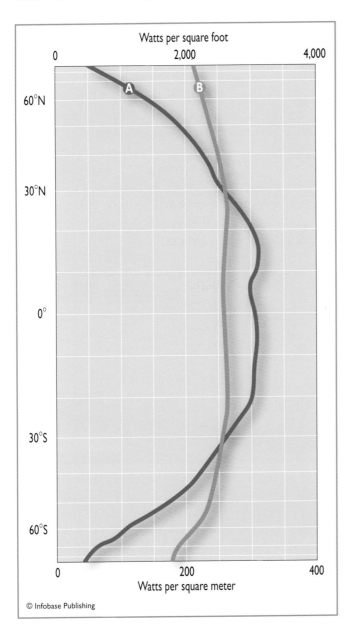

© Infobase Publishing

Curve A represents the distribution of solar energy absorbed by the Earth's surface. The polar regions absorb considerably less energy than the equatorial regions. Curve B shows the radiation of the energy from the Earth's surface at different latitudes. The high latitudes radiate more energy than they receive, while the Tropics absorb more energy than they radiate. Stability can only be maintained by the redistribution of energy by the atmosphere and the oceans.

season but is located over the equator at times of equinox, there is very little wind at all. Sailors named this zone the doldrums because in times of sailing ships there was a very real likelihood of becoming becalmed in these waters, waiting for a breath of wind. The converging winds outside this zone do not blow directly from the north and the south but from the northeast and the southeast, as shown in the diagram. The reason for this is the spin of the Earth on its axis. Any

free-moving object in the Northern Hemisphere, whether in the oceans or in the atmosphere, tends to be deflected to the right by the Earth's spin, hence motions become clockwise. In the Southern Hemisphere free objects are deflected to the left, so the motions become anticlockwise in direction. This tendency was first described by a French engineer, Gaspard Gustave de Coriolis (1792–1843) in 1835 and has come to be called the *Coriolis effect.*

The winds heading for the poles to the north and the south of latitudes 30°N and 30°S are also displaced in the same way, leading to the formation of a strong general air movement from west to east in these latitudes, especially at high altitude, where it is known as the *jet stream.* The jet stream often reaches speeds of 150 miles per hour (240 km per hour) and has a strong effect on aircraft; journeys from west to east are invariably faster than the same journey from east to west. The jet stream runs just to the south of the

The circulation of the atmosphere occurs in a series of cells, creating areas of low pressure near the equator and in the region of the polar fronts and areas of high pressure close to the tropics of Cancer and Capricorn and at the North and South Poles.

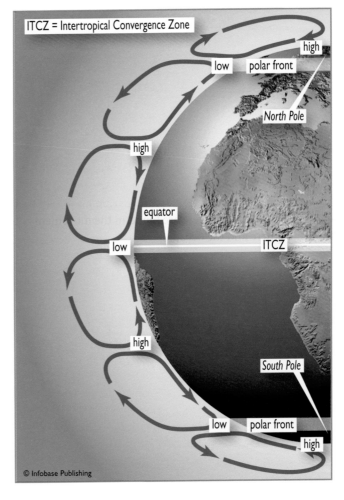

© Infobase Publishing

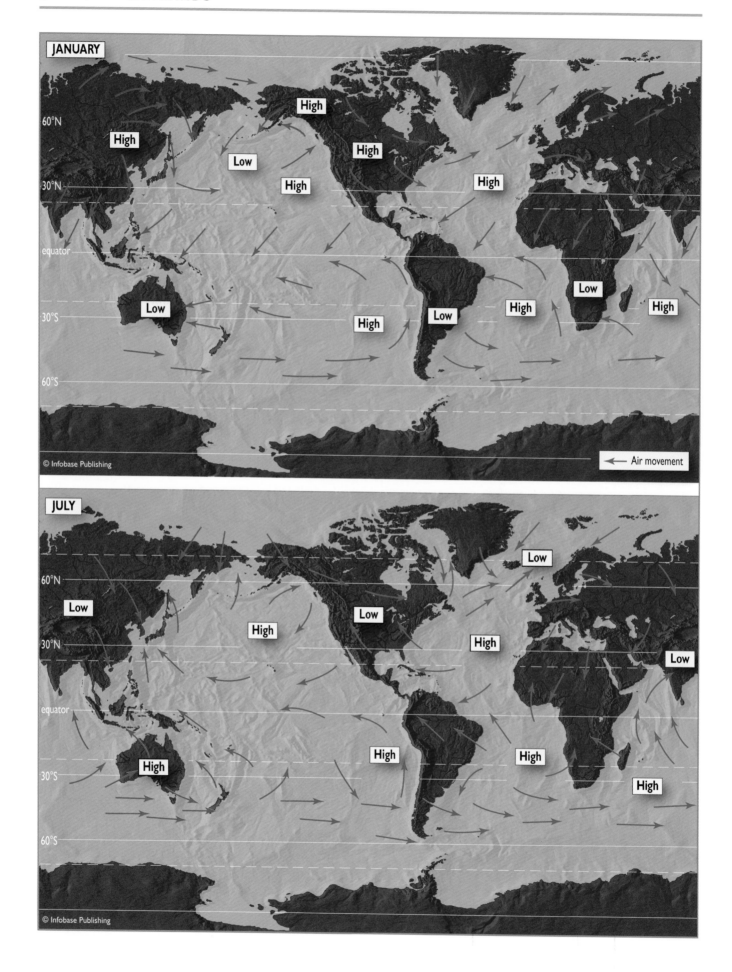

JANUARY

60°N

High

High

Low

High

High

30°N

High

High

equator

Low

Low

30°S

Low

High

Low

High

High

60°S

© Infobase Publishing

← Air movement

JULY

Low

60°N

Low

Low

High

High

30°N

Low

equator

High

High

30°S

High

High

High

60°S

© Infobase Publishing

(opposite page) The pattern of air movement over the Earth's surface and the distribution of high- and low-pressure systems in January and July. In the Northern Hemisphere, the northward movement of the Intertropical Convergence Zone in summer causes a reversal of the winds in the Indian Ocean resulting in monsoon conditions over India.

boundary zone between the polar air and the tropical air masses, a boundary that is known as the *polar front.* The jet stream causes cyclones (low-pressure systems) and anticyclones (high-pressure systems) to move around the world in an easterly direction. The latitude of the polar front and the jet stream varies with the season. In summer in North America the polar front and jet stream lie roughly along the Canadian border, dipping south to Washington, D.C., in the east. In winter it runs approximately from Los Angeles through northern Texas to South Carolina.

In the Southern Hemisphere, the polar front encircles Antarctica and lies within the Southern Ocean. For much of its length, where the Southern Ocean meets the South Atlantic Ocean and the Indian Ocean, it lies approximately along the 50°S latitude. In the region where the Southern Ocean meets the South Pacific, however, the southern polar front is deflected farther south and is located along latitude 60°S.

The overall outcome of the atmospheric circulation pattern is a series of bands of high and low pressure over the globe. High pressure occurs along the tropics of Capricorn and Cancer and also at the North and South Poles, resulting in low precipitation at these latitudes. Low-pressure systems, on the other hand, are concentrated at the equator and along the polar fronts, leading to high precipitation in these regions. This banding of precipitation provides part of the explanation for the distribution of wetlands shown in the diagram on page 4. The equatorial and temperate zones have a high concentration of wetlands because of this high level of precipitation, while the high-pressure regions of the desert latitudes do not. The northern regions that border on and extend into the Arctic do not exactly follow this pattern. Although precipitation is low in these regions, the energy input from the Sun is also weak and the resulting average temperature is low, so evaporation is slow and water remains in and on the ground, favoring the development of wetlands.

■ THE OCEANS AND CLIMATE

The atmosphere is not the only determinant of global patterns of climate; the oceans also have a profound influence. Water has certain physical properties that impact on climate. A sample of water both gains and loses heat more slowly than an equivalent weight of rock. Areas of land surrounded by water are therefore protected to some extent from extreme variations in temperature; land close to the ocean is kept cool in summer and relatively warm in winter. The proximity of the sea therefore has a considerable influence on the climate of a region. A site close to the sea is said to have an *oceanic climate,* whereas one situated far from the ocean's influence has a *continental climate.* For example, the city of Archangel at a latitude of 65°N in western Russia lies on the edge of the Barents Sea and has an average annual temperature of 33°F (0.4°C) and an average temperature for its coldest month of 5°F (-15°C). In continental east Russia, Verkhoyansk in Siberia, which lies on the same latitude as Archangel, has an average annual temperature of 3°F (-16°C), and the average for its coldest month is as low as -58°F (-50°C). The oceanic regions of Russia are kept warmer by the proximity of the sea. Oceanic regions also receive more precipitation because the sea is a source of atmospheric moisture. Thus Archangel receives 18 inches (466 mm), while Verkhoyansk has only 5 inches (128 mm) of precipitation each year.

The oceans also act as channels for the redistribution of energy around the world. Surface water moves around and between the great oceans in the form of currents, as shown in the diagram on page 10. In general, the pattern follows that of the winds and is driven by them. The spin of the Earth on its axis creates a deflection of freely moving water due to the Coriolis effect (see page 7), just as in the case of wind movements. In the Southern Hemisphere the main surface circulatory pattern of the oceans is in the form of anticlockwise motions, whereas in the Northern Hemisphere the movements are clockwise. In the North Pacific Ocean and the North Atlantic Ocean there are west–east currents bringing warm tropical water into the higher latitudes, but in the Atlantic Ocean there is a gap between Iceland and the British Isles that allows the continued movement of warm water (called the Gulf Stream) north past Scandinavia and into the Arctic Ocean. The penetration of warm water into the Arctic Ocean is far less from the Pacific basin. The northeast-directed waters in the Pacific either circulate around southern Alaska and return south or pass down the west coast of California and Mexico and back into the tropics.

Oceanic circulation occurs in deep water as well as at the surface. There is a global movement of water around the Earth that is driven by changes in the density of seawater, itself a product of salinity and temperature. Warm water with relatively low levels of salt is less dense than cold water with higher salt concentrations. As water moves around the world it changes in its temperature and its salinity. This circulation pattern is called *thermohaline circulation (thermo-* refers to the water temperature, and *-haline* relates to its saltiness), or the *oceanic conveyor belt.* The overall pattern of global thermohaline circulation is shown in the diagram and is explained in the sidebar on page 10.

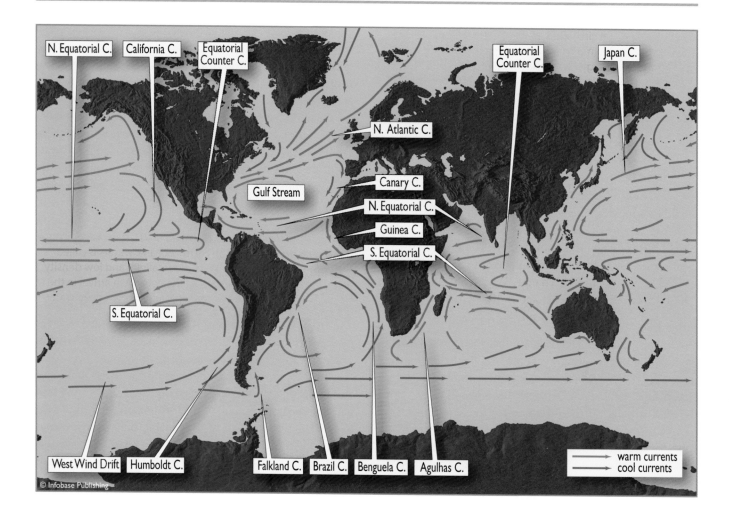

The Oceanic Conveyor Belt

In addition to the surface currents in the oceans, which are driven mainly by winds in the atmosphere (see the diagram above), there is a global circulation of waters driven by the changing density of seawater, as shown in the accompanying diagram. The surface water of the oceans tends to be warmer, less salty, and hence less dense than the deep water. When these warm surface waters move into the high latitudes, they give up some of their heat to their surroundings and consequently cool, become denser, and sink. This chilling of surface water occurs most strongly in the North Atlantic Ocean, where the warm tropical surface waters of the Gulf Stream pass northward between Iceland and the British Isles and meet the cold waters of the Arctic Ocean. As the warm waters lose their heat to the waters of the Arctic Ocean, they cool, sink, and begin to make their way back south through the Atlantic and eventually into the Southern Ocean surrounding Antarctica. Passing eastward into the Pacific Ocean, these cold deep waters eventually surface either in the Indian Ocean or in the North Pacific. There they become warm and drift westward once more around the Cape of Good Hope in South Africa and into the Atlantic to begin their circulation over again.

This thermohaline circulation of water through the oceans of the world plays an important part in the redistribution of energy between the Tropics and the high latitudes, particularly in the North Atlantic. These movements of water have strongly influenced world climate in the past. A failure in the North Atlantic system 12,000 years ago temporarily plunged the Earth into an ice age, so oceanographers are currently expending much research effort in understanding the movements and the changes that occur in this oceanic conveyor belt.

(opposite page) The circulation pattern of the Earth's oceans. The currents are an important means of redistributing the Earth's incident energy. The passage of the warm Gulf Stream waters across the Atlantic and into the Arctic Ocean has a warming effect on northern Europe.

The thermohaline circulation of water in the oceans is an important means of exchanging energy between different parts of the world. Without this energy exchange there would be a much stronger gradient of temperature between the Tropics and the poles. There have been times in the Earth's history when this conveyor belt has slowed down or even ceased altogether in certain parts of its cycle, and this has led to severe climatic change. When the conveyor belt faltered around 12,000 years ago in the North Atlantic, for example, the Earth began to move into a new ice age, but the normal circulation pattern was restored within a few centuries and the climate began to warm once again. This illustrates the critical role played by oceanic circulation in controlling the climate of the entire Earth.

Oceanic circulation thus carries some of the tropical warmth to the high latitudes and helps to redistribute energy around the world. The high specific heat of water also assists in the retention of heat during the high latitude winter, preventing excessive cooling. The specific heat of water also prevents overheating in summer in those regions

The global circulation of the Earth's oceanic waters is called the oceanic conveyor belt. Warm, low salinity and low density water moves along the surface of the ocean. On arrival in the North Atlantic Ocean the water cools and forms a deep water current that flows in the opposite direction.

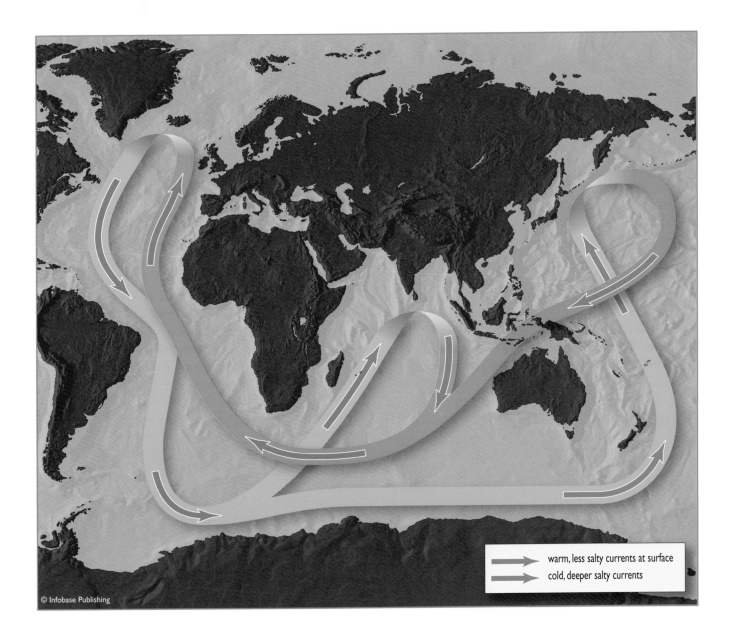

© Infobase Publishing

→	warm, less salty currents at surface
→	cold, deeper salty currents

bordering the oceans. Seawater is also a major source of atmospheric moisture and therefore precipitation (see the diagram on page 2). So, areas close to the oceans are likely to receive more rainfall or snowfall than areas far from the oceans, deep within continental interiors. Climates can thus be regarded as oceanic or continental, depending on their proximity to the oceans. The degree of climatic oceanicity is a strong influence on the development of wetlands (see "Factors Affecting Wetland Types," pages 12–14).

The pattern of wetlands over the surface of the Earth can therefore be related in general terms to climate patterns. The banded, zoned pattern of precipitation is reflected in the distribution of wetlands, and local variations in this zonation also has its impact on the distribution pattern of wetlands.

There are, of course, breaks in the overall pattern. In Southeast Asia, for example, the regions around Bangladesh and Burma lie on the tropic of Cancer, which is the latitudinal position of the high-pressure belt, and yet the regions are rich in wetlands. These areas are fed by heavy monsoon rains from the Indian Ocean and lie at the estuaries of major river systems. North of these, but on the same longitude, the continental regions of Central Asia lie far from any oceans, and the monsoonal rains are blocked by the Himalaya mountain chain, so dry climates penetrate far to the north, and wetland systems are limited to regions supplied by the monsoon and the rivers draining from the mountains.

The pattern of wetlands over the surface of the Earth can, therefore, be related in general terms to climate patterns, but the relationship is complicated by the movements of water over the surface of the Earth, leading to the development of wetlands in some situations where one might not expect to find them. Even the arid continental interior of Australia has its wetlands, such as Lake Eyre, so the geographical conditions underlying the distribution of wetlands are evidently more complicated than is the case for almost all other major biomes. These complications also lead to a great diversity of wetland types, depending in part on the source of their water supply.

■ FACTORS AFFECTING WETLAND TYPES

Water availability is the key to wetland development. Water arriving at a site directly by precipitation is the most important factor in determining the frequency of wetland habitats in an area, as shown in the diagram, but the movement of water once on and in the ground can also influence the pattern of wetland development. Precipitation and groundwater are the two main determinants of wetland distribution patterns.

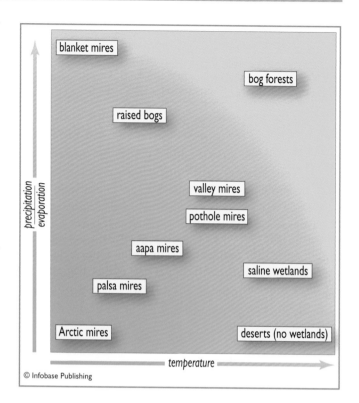

The distribution of different types of wetlands in relation to temperature and the ratio of precipitation to evaporation. Wetlands are generally more abundant in regions with high precipitation, or low temperature, or both.

In oceanic situations where precipitation is extremely high and evaporation is low, wetlands can form over entire landscapes, almost irrespective of topography. On the western coast of Norway, for example, or in parts of Newfoundland, the combination of high rainfall, distributed throughout the year, together with a generally cool climate and consequent low evaporation rates, leads to a permanent excess of water over the entire landscape. Under such conditions, bog vegetation develops not only in hollows but over flat surfaces and even slopes. Obviously, because of fast drainage and instability, there is a limit to the steepness of slope over which bog can develop, but the peat of these so-called blanket bogs (see "Blanket Bogs," pages 107–111) can accumulate on quite steep slopes, in excess of 15°. They are largely confined to wet temperate regions of the Earth, but they can occur in the wet tropics on high mountains, where evaporation is low.

These wetlands are not dependent on groundwater for their saturation. Some other peat-forming wetlands (given the general term *mires*) are precipitation-dependent but begin their development in groundwater hollows, either in lake basins or estuaries. They become independent of groundwater as their peat mass builds up and elevates their surfaces beyond its influence. These are the raised bogs (see

"Temperate Raised Bogs," pages 103–105), which occur in those temperate areas with high enough precipitation to support elevated peat surfaces, and also in some lowland, coastal regions of the wet tropics, such as New Guinea (see "Tropical Raised Mires," pages 106–107).

All other wetlands depend, at least in part, on groundwater for maintaining their saturation, so they are less immediately restricted by precipitation in their geographical distribution. The diagram illustrates the relationship between the different wetland types and the two main climatic factors of precipitation/evaporation ratio and temperature. No sharp boundaries can be drawn around the wetland types because they often grade into one another and because they frequently occur in the same areas when the landscape is diverse. The upper part of this diagram contains the mire systems that are entirely rain-fed, or *ombrotrophic*, while the lower part of the diagram contains those that require a groundwater supply for their maintenance. These occur in regions of the world where precipitation is lower and may be nonuniformly distributed, with wet and dry seasons. In the lower right-hand corner of the diagram are the hotter, drier regions where mires become infrequent,

are always dependent on groundwater, and are often saline because of high evaporation rates.

Some wetland types are not included in this diagram. Shallow lakes can occur in almost any part of the diagram. The same is true of the coastal wetland systems—salt marshes—which are limited only by extreme cold and are replaced by mangroves in the tropics. Reed beds, marshes, and swamps are also omitted from the diagram because they tend to be found throughout most of the conditions covered, apart from the most cold conditions, where low temperature and short growing season limit the productivity, and ultimately the survival, of wetland plants.

Groundwater-fed, or *rheotrophic*, wetlands are less restricted in their global distribution than rainwater-fed ones. The reason for this is not difficult to understand. Suppose a region has a precipitation of just 8 inches (200 mm) per year. This is unlikely to be sufficient to support a rain-fed peatland, but if a valley has a water catchment that is, say, 10 times the area of the valley floor itself, then the water supply is effectively increased tenfold (apart from incidental losses, such as evaporation and transpiration by catchment vegetation). The water supply of valleys can be sufficient to support wetland ecosystems even under conditions of relatively low rainfall. It follows that, as one moves into regions of lower precipitation and higher temperature, groundwater wetlands become increasingly important compared with rain-fed wetlands.

The source of water supply in wetlands has its impact on the chemistry, the vegetation, and the nature of the sediments that accumulate within them. In rain-fed wetlands, the supply of mineral elements is poor, an acid-loving vegetation develops, and the sediments are almost purely organic (peat). In wetlands fed by groundwater, a range of chemical elements are present in the water entering the ecosystem, vegetation may be nutrient-demanding, and the sediments that are laid down have a high proportion of inorganic materials such as silts and clays. This is illustrated in the diagram above, where wetland distribution is shown in relation to precipitation: evaporation (P:E) ratio and the increasing influence of groundwater.

Ombrotrophic (purely rain-fed) wetlands, which produce peat sediments, are found only where P:E is high and the influence of groundwater is low. Valley mires, fens, swamp forest, and mangroves are more strongly influenced by groundwater drainage and accumulate sediments with a mixture of peat and inorganic materials. At the other end of the spectrum are rheotrophic marshes, reed swamps, and salt marshes with their strong groundwater influence and inorganic sediments. All wetlands in the warm and dry climate regions will be of the latter type.

From this brief overview it can be seen that the type of wetland that develops in any particular location is dependent on many different factors. Climate, geology, and topography

The distribution of different types of wetlands in relation to groundwater influence and the ratio of precipitation to evaporation. In regions of high precipitation and low evaporation, raised and blanket ombrotrophic bogs develop, while in dryer areas rheotrophic fens and marshes are more frequent.

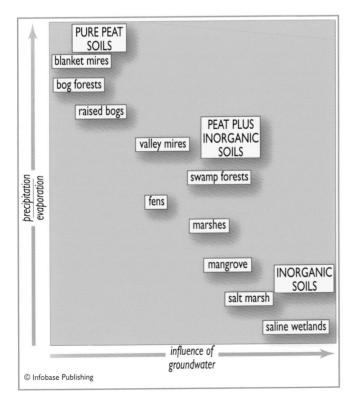

all play a part in determining where wetlands can form and what type of wetlands is possible. Climate, water chemistry, and the biogeographical position of a site will then determine what kinds of plants and animals are present and appropriately equipped to live in that wetland. Finally, the growth of the plants themselves causes such profound alterations in the local environment that they change the very nature of the wetland in the course of time. One wetland type can replace another in a successional sequence.

■ SUCCESSION: WETLAND CHANGES IN TIME

All habitats on Earth are constantly changing, but some change faster than others. In a forest, trees are constantly dying and being replaced by young individuals that sprout up from below. If there is a forest fire or a strong wind that devastates an area, then plant and animal life soon reinvade and the wounds begin to heal. The changes that occur in a habitat are often reasonably predictable. It may not be possible to say exactly what plants and animals will be next in a sequence, but there is a general pattern and order to changes in a habitat in the course of time. When ecologists first recognized this fact, they called the process *succession*.

One of the characteristics of succession is that it is directional, which means that the final outcome is predictable. An abandoned field in New England will change its character over the course of time, but its eventual conclusion is likely to be deciduous forest. A forest fire in northern Manitoba, on the other hand, will undergo a series of changes over many decades, eventually stabilizing with a forest of coniferous trees. The predictability of the outcome of succession is due largely to the influence of climate. The severe winter cold and the high level of snowfall in Manitoba give evergreen conifers a competitive advantage over deciduous trees, such as oak, but in New England, where winters are milder, the deciduous way of life can be advantageous, so oaks tend to predominate over conifers. In general, therefore, it is climate that determines the outcome of successions and that gives them their predictable character.

The same is true for wetland habitats; climate ultimately limits and controls the type of wetland that can develop in any particular region (see "Factors Affecting Wetland Types," pages 12–14). Blanket mires develop only in the cool wet oceanic regions of the world, while the great elevated bog forests are found only in the hot wet Tropics. In the case of wetlands there are complicating factors. Chemistry can change the course of succession, as in the case of the saline wetlands, and landscape form, or *topography,* influences patterns of water movement and accumulation. An additional complication is that many wetland types are linked together in a successional series.

The details of the processes of wetland change will be discussed later (see chapter 4, "Different Kinds of Wetlands," pages 81–128), but, in essence, the development of wetlands is driven by two types of forces, one from outside the habitat and the other from within. The outside factors that can drive succession are called *allogenic* forces. Many wetland ecosystems receive an input of material from outside their own boundaries. Silt, soil, and plant-derived materials (such as leaf litter) are carried down streams and rivers and can become deposited within the wetland. The supply of these materials is entirely dependent on conditions outside the wetland itself. In particular, the supply varies with the shape of the basin in which the wetland lies, the nature of the rocks in the catchment, and the quantity of water that erodes those rocks and bears the products of erosion into the wetland. Imported materials, therefore, provide the means by which the wetland can change. Sedimentation of the derived materials makes a lake shallower and creates new conditions for invasive plants and animals, thus driving the process of succession forward. This input of material is thus an allogenic driving force.

Other driving forces are internal. When a plant grows in a shallow lake, it slows the rate of water flow and encourages suspended particles to settle out on the bottom. Thus the presence of a plant changes the local conditions in favor of sedimentation and infilling of the lake. In addition, the roots of the plant in the basal mud create more stable conditions, preventing erosion by the flowing water, and the constant production of organic matter as leaves, stems, and roots die add to the process of deposition. These are internal, or *autogenic,* driving forces in the course of succession. Ultimately, the development and growth of the ecosystem is encouraged by both allogenic and autogenic forces.

Living organisms clearly play an important part in succession. The presence of any plant or animal modifies its immediate surroundings, as is illustrated by the case of a submerged aquatic plant slowing down the movement of water and causing additional sedimentation. By changing its surroundings the organism may make it easier for other species to invade and survive. Most aquatic plants, for example, are limited in how deep in the water they can grow, so water lilies can grow in deeper water than reeds and cattails. When water lilies grow in a lake for many years, they may cause sedimentation that makes the lake shallow enough for the invasion of reeds, and in this way they push the course of succession onward. When one organism makes conditions more suitable for the invasion of another species, it is called *facilitation.* Facilitation is not an altruistic process; there is no suggestion that the first plant is seeking advantage for the new invader. Indeed, the reverse is often true; the invader may prove a stronger competitor than the facilitating species and may eliminate its inadvertent benefactor. The process may then continue as the second plant, in this case a reed, continues to accumulate sediment around it, eventually leading to the invasion of a willow tree, and the reed becomes a victim of its own activities. So facilitation is followed by *competition,* and between them these two autogenic forces drive the succession along its predictable course.

Because successions are often very predictable, their progress is said to be *deterministic.* Each step inevitably leads to the next so that the entire course of succession can be mapped out and modeled. This is an ideal that is rarely achieved in practice. More often some random event takes place that changes the course of succession and sends it off in a less predictable direction. A river may change its course and erode a developing area; a beaver may take up residence and build a dam; or there may be a landslip that will block the drainage and change the water level. Such events cannot be predicted in advance; they are caused by chance and are said to be *stochastic* events. Ecologists argue about the relative importance of deterministic processes and stochastic events in molding the course of a succession. Sometimes it is difficult to distinguish between these two factors because what may appear as a random catastrophe, such as drought, may be predictable to a certain extent if one considers the long-term climatic conditions of an area. Even seemingly random events can have a certain pattern underlying them.

Wetlands such as the inland Okavango Delta in southern Africa, for example, or Lake Eyre in central Australia are subjected to both flooding and drought. The Okavango is flooded seasonally, so the water level is reasonably predictable, though it does vary from year to year. Lake Eyre, on the other hand, is very much less predictable in its floods, and the animals and plants that inhabit the area need to be adapted to cope with less regular provision of water. Only species that can cope with catastrophic, stochastic events can survive.

Whether strictly predictable or not, wetlands are ecosystems that are constantly developing from one state to another, which is one reason why their classification is difficult. They never sit still. Many wetland types are linked to one another in a successional series, and a knowledge of these relationships is important in understanding wetland processes. It is particularly important that wetland conservationists and managers should be aware of these links between wetland types as this is the key to manipulating and conserving them.

Successions follow a number of basic rules, even though there are exceptions to most of them. On the whole, the total amount of living matter in the ecosystem (the *biomass*) increases in the course of succession. Early stages in succession have few organisms and those present tend to be small in stature, while later stages have increasingly large amounts of vegetation and animal matter. This is not always the case in wetlands. In some types of bog the total living biomass actually decreases in the late stages of the succession. In general, the total number of species present per unit area of ground (*species richness*) increases through succession, but again there are exceptions in some wetland successions.

When ecologists first developed the idea of succession, they postulated that all successions eventually end in a state of equilibrium and stability when the community of living organisms is perfectly in balance with the prevailing climate. They called this ideal state the *climax.* Recent research has called this concept into question for a number of reasons. The final stages of a succession often exhibit a kind of cyclic turnover of species. An aged, fallen oak tree, for example, may be replaced by birch rather than oak, or by sugar maple. It may take several generations of trees before the oak regains dominance on that spot. So the final stage in succession is actually a mosaic of patches, each patch at a different stage in the cyclic turnover of species. Perhaps it is this patchwork nature of the climax that makes it so rich in species.

The other question that needs to be answered about the supposedly stable climax state is whether the climate itself remains stable long enough for the living communities of plants and animals to attain any kind of equilibrium. Over the past few centuries there have been some very considerable changes in global climate. Undoubtedly, this is partly due to human activity in burning fossil fuels and creating an enhanced greenhouse effect, but even before people had developed industrial processes the climate fluctuated quite strongly from one century to the next. Successions take two or more centuries to reach their final stages, and by that time it is quite possible that conditions will have changed and a new shuffling of species begins. Succession may well prove to be a never-ending process, and this is particularly true of wetlands.

Whether wetland succession ever reaches a conclusion may be open to doubt, but the fact that wetlands undergo succession is very evident. Indeed, wetlands are perhaps the most accessible of all habitats for the study of succession because they leave a record of their past history in their sediments. The infilling of lakes, the accretion of marsh sediments, the growth of trees in swamps, and the accumulation of peat in bogs—all leave behind, in a stratified, layered arrangement, a complete and chronologically arranged account of all that has gone before. From fossils in the sediments it is possible to piece together the precise course of succession, and this is a unique feature of wetlands (see "Wetland Stratigraphy," pages 32–34). As a result of the study of stratigraphy in wetlands, it is possible to trace their development over thousands of years. Indeed, it is even possible to study the wetlands of many millions of years ago.

■ CONCLUSIONS

Wetlands, by definition, can only develop in the presence of water, so their presence on continental landmasses is dependent on the global hydrological cycle. The oceans account for about 96 percent of the world's water, and an additional 2 percent is locked up as ice. Only 2 percent is actually in circulation in the atmosphere, the soils, the rocks, and the wetlands. The oceans occupy much of the surface area of the planet, but they provide less than 40 percent of the atmospheric moisture that leads to precipitation; the rest comes from evaporation and transpiration over land surfaces and freshwater bodies. When a water molecule enters the ocean, it takes an average 2,600 years before it enters circulation once again.

Unlike most other terrestrial biomes, wetlands are not confined to a single latitudinal zone but are distributed widely through the Earth. They are most abundant, however, in the equatorial zone, where rising air masses produce abundant rainfall, and in the northern temperate zone, where the jet stream, coupled with a generally cool climate, results in wet prevailing conditions. Oceanic climates are milder than continental ones, with cooler summers and warmer winters, and have high levels of precipitation. They are particularly conducive to the development of wetlands. Wetlands occur in more arid parts of the world, but they are then dependent on topographic conditions that assist in the collection of the limited supplies of water. Wetlands in these drier lands are therefore almost exclusively fed by flowing water (rheotrophic), while those of wetter climates may be either rheotrophic or rain-fed (ombrotrophic).

Wetlands are dynamic ecosystems that are constantly changing. The development of wetlands often follows a directional and predictable sequence, called succession. Succession is driven by both external (allogenic) and internal (autogenic) forces, chief among the latter being facilitation and competition. Sometimes the presence of an organism, typically a plant, modifies the environment in such a way that it becomes easier for another species to invade; this is facilitation. The new arrival may then prove more efficient at acquiring resources than the original species, so competition comes into play and the invader eventually eliminates the pioneer. The repetition of these two processes drives the course of succession. Eventually, a climax or equilibrium state may be attained, but this often takes the form of a mosaic of different disturbance stages rather than a uniform and stable cover of vegetation.

Wetlands accumulate beneath them stratified sediments that record in fossil form the stages they have passed through in their development. The study of the geology of wetland sediments thus supplies information about their successional processes and their history.

2

Geology and Chemistry of Wetlands

Geology is the study of rocks, their formation, and their breakdown. Geologists define rock as any mass of mineral matter found in the Earth's crust that has formed naturally. Most rocks are composed of hard materials, solid stony masses built up by the fusion of different minerals, but the geologist does not limit the word *rock* to hard materials; rocks can also be soft. Deep desert sands, the layers of clay sediments in a lake bed, and the detritus left behind when a glacier melts are all rocks according to the geological definition. Only those materials constructed by human activity, such as concrete and plastics, lie beyond the confines of the word *rock*.

Neither the Earth nor the rocks that compose its crust are static—they are constantly changing. Rocks are forming even today as sand, silt, clay, and the bodies of microscopic animals and plants sediment to the floor of the ocean and become compacted. Rocks formed in this way are called *sedimentary rocks*. Rocks are being modified by the intense pressures and high temperatures they experience within the Earth's crust. These are called *metamorphic rocks*. Slate is an example of a metamorphic rock that began its existence as a sedimentary deposit but has been subjected to pressures and temperatures that have compressed and changed its physical constitution. Metamorphic rocks are often produced in the vicinity of volcanic activity or at great depth within the crust. *Igneous rocks* are the third major rock type, and these come into being when molten rock, or magma, from the Earth's core breaks up through the crust and solidifies as it cools.

Wetlands are agents in the geological processes of the Earth. The sediments that accumulate within a lake basin are technically rocks, even though they happen to be soft. If they are not eroded by water or by glacial advance, then they may in time become compressed and hardened to form a hard rock. Peat is an organic rock that can also under the right circumstances be transformed into a hard rock, namely coal. So wetlands are contributors to the Earth's rock cycle.

■ THE ROCK CYCLE

The interior of the Earth consists of molten rock, or magma, and convection currents are set up within this fluid core that push upward into the solid crust at the Earth's surface. Sometimes these rising currents in the magma force their way into the crust but fail to reach the surface before the rock cools and solidifies. Occasionally the magma bursts through the crust and erupts in the form of a volcano, and the molten rock solidifies on the surface of the Earth as masses of larva. Solidification of the magma involves many chemical changes in its composition, depending in part on the location (above or below ground) where it finally cools. Magma contains about 14 percent of dissolved gases, mainly water vapor and carbon dioxide, and these may be discharged into the atmosphere. The chemical constituents of the magma, called *minerals*, have different melting points, so they may separate out during the cooling process, depending on the rate at which the temperature falls. Consequently, there are different kinds of igneous rock that vary chemically according to the conditions under which they were formed.

Igneous rocks that have failed to reach the surface and are buried deep in the crust are protected from the process of rock breakdown, or *weathering*, suffered by surface rocks. Eventually even these may be exposed to the atmosphere as the crust above them is worn away. Weathering involves the chemical and physical breakdown of the rocks into their component minerals, or even into smaller chemical units. Living organisms also play a part in this decomposition of the rocks, and the process is the basis of soil formation (see "Geology and Rock Weathering," pages 21–22). Solid rock is converted by weathering into smaller particles, some of which may be soluble in water, and they are transported by water draining over and through the soils in which they lie. This transport of particles is called *erosion*, and the movement of dissolved materials is called *leaching*. As a result of these two processes, the rocks exposed at the Earth's surface are constantly being worn away, and their components are transported down to the oceans, where they may remain suspended for a while but eventually sediment to the bottom.

The sedimentation of particles derived from land-based rocks is accompanied in the oceans or in wetlands by the deposition of tiny remains of planktonic organisms that have spent their lives in the surface layers of the water. Creatures such as foraminifera and coccolithophorids, single-celled organisms belonging to the Kingdom Protoctista, accumulate calcium carbonate (lime) from the seawater and build cases for themselves that survive and sink into the ocean sediments after the death of the living creature. The diatoms (also members of the Kingdom Protoctista) are found in both saline and freshwater and build cases out of silicon dioxide, and this relatively inert material, with the same composition as sand, also joins the other particles in the steady rain of fine materials that constantly accumulate in the sediments of the ocean or wetland floor. These sediments build up in

the course of time and are converted into harder rocks by the pressure from above. The sedimentary rocks produced in this way may eventually be forced beneath the crust to enter the magma once more. Alternatively, they may be crushed, folded, and buckled by movements in the Earth's crust, raised above the surface of the ocean, and exposed to the forces of weathering and erosion once again. In this way the rocks of the Earth are involved in a constant cycle that takes many millions of years to complete (see diagram below).

The rock cycle. Volcanic activity creates igneous rocks derived from the molten magma of the Earth's interior and also modifies the physical and chemical nature of adjacent materials, forming metamorphic rocks. Uplifted landmasses are weathered and eroded by the atmosphere and climate, and the transported materials are carried to the oceans where they are sedimented, in time creating the sedimentary rocks.

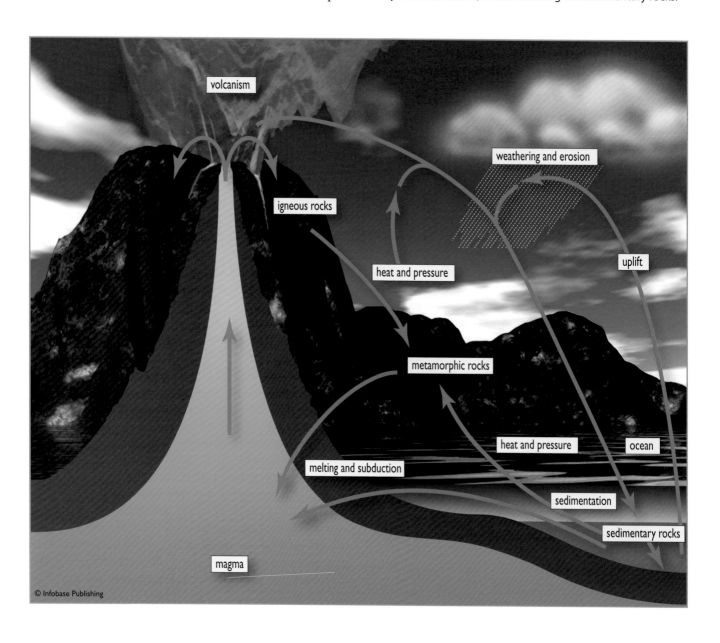

Wetlands are therefore involved in the creation of rocks, but they are also strongly affected by the older rocks that surround them. Wetlands interact with their catchment geology because water constantly moves in and out of wetlands, bringing particles and dissolved chemicals from the surrounding rocks. The nature of the bedrock also determines whether water can accumulate at a site or whether it simply sinks into deep aquifers and is lost to the surface landscape. Water, therefore, is the key to understanding the interaction between wetlands and geology.

THE NATURE OF WATER

Water is an important component of the Earth's surface, residing largely in the oceans and the ice sheets but with a small proportion, about 2 percent of the total, in active circulation through the atmosphere and terrestrial landscapes (see "The Water Cycle," pages 1–4). This small mobile component of the Earth's water resources is vital for the maintenance of wetlands. Water is a vital component of the living organisms of wetlands, sometimes accounting for more than 95 percent of their fresh weight, and all wetland creatures are dependent on its presence for their survival. Water can be regarded as a critical material in the construction of wetlands and wetland inhabitants, but water can also be a destructive agent, degrading rocks, eroding mineral particles in its flow, and dissolving the chemicals out of the rocks, leaving them impoverished and weakened. This erosive activity provides many of the chemical resources of wetlands, so the losses in one habitat are converted into gains for another.

Water has many remarkable properties. Most of the water on Earth is at a temperature between 32° and 212°F (0 and 100°C), so it is in liquid state. Many other molecules of similar size and constitution are gases at these temperatures, for example methane (CH_4), ammonia (NH_3), and nitrogen dioxide (NO_2), so the liquid state of water is an unusual and important feature of its chemistry. There is a bonding between water molecules (*hydrogen bonding*) that causes them to cohere so that columns of water are not easily broken (see sidebar "Transpiration and Evapotranspiration," page 3). The bonding results in a high surface tension force that enables water to be drawn into fine tubes, or capillaries. This is important for water movement in plants but also affects the behavior of water in rocks and soils. It is held tightly in fine tubes and is not easily withdrawn from such locations.

Water is also an excellent *solvent,* which means that many different materials can dissolve in water to form a *solution.* A salt, such as sodium chloride (NaCl), that is dissolved in water separates into two charged atoms, or *ions,* that can be represented as Na^+ and Cl^-. One might expect these two units to be attracted to one another too strongly

to be able to separate, or *dissociate* in this way, but water has the property to reduce the attraction so that they can remain apart and maintain themselves in solution. Their attraction in water is reduced by about 40 times compared with their attraction in an organic solvent. So, water has this remarkable capacity to keep ions in solution, and this affects wetlands in many ways. It means that ions removed from rocks in the catchment can be carried for long distances in aqueous solution and delivered to a wetland far from the original source. The solution of ions in water also ensures that they are available to wetland plants, which absorb them and carry them from one part to another in a water matrix. The transpiration stream in terrestrial plants provides a major transport system for the movement of ions. Plants in turn supply ions in solution to animals, which can move them around in their blood, or excrete any excess in their water-based urine. The solvent properties of water, therefore, are vital to life in general and wetlands in particular.

Water molecules themselves dissociate to a limited extent, producing the ions H^+ (hydrogen ions, or protons) and OH^- (hydroxyl ions). The presence of these ions in water means that even pure water that contains no dissolved materials will conduct electricity to a certain extent. The production of an excess of hydrogen ions is a characteristic of acids, but pure water produces an equal concentration of hydrogen and hydroxyl ions, so it is neutral, neither acidic nor alkaline. When other materials are dissolved in water, however, an excess of hydrogen ions or hydroxyl ions may result, and this renders the solution respectively acidic or basic in reaction. Chemists have devised an expression for the degree of acidity in a solution that they call pH (see sidebar on page 20). A low pH value on the scale indicates a high degree of acidity, and a high value indicates alkaline conditions. The diagram on page 20 illustrates the range of the pH scale and the values for some familiar materials.

Very many materials can dissolve in water, especially if they are capable of dissociating into charged ions. This includes gases as well as solids. Oxygen, carbon dioxide, and the oxides of nitrogen and sulfur are all capable of dissolving in water. Oxygen gas dissolves in water, but it is not altered in its chemical state as a result, it remains O_2. It is then able to move through the water matrix by diffusion, just as it moves through the air, but its movement is slower. Water is denser than air, which is why it is more difficult to walk or run in waist-deep water than in air. Oxygen has precisely the same problem, and its rate of diffusion in water is about 10,000 times slower in water than in air. This can give rise to problems for plants, animals, and microbes that require oxygen for respiration. The solubility of oxygen and other gases in water depends on temperature. Unlike solids, gases are more soluble in cold water than in hot. Water at 104°F (40°C) can hold only half the oxygen that can be dissolved in the same volume of water at 32°F (0°C).

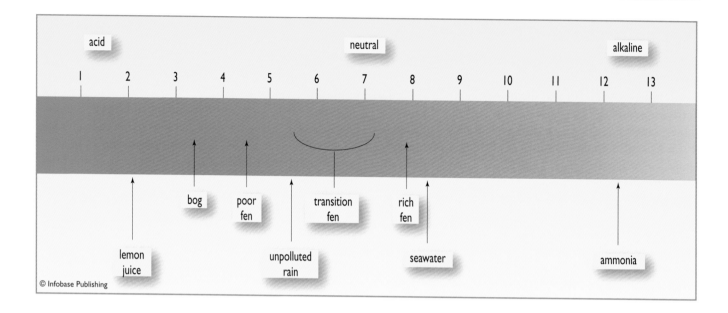

© Infobase Publishing

The pH scale of acidity and alkalinity. Neutrality is marked by pH 7. The scale is logarithmic and negative, so pH 4 is 10 times as acid as pH 5. The pH of a range of wetland waters is shown together with some familiar materials.

Carbon dioxide also dissolves in water, and its rate of diffusion is very similar to that of oxygen. Submerged aquatic plants that require carbon dioxide for photosynthesis can find the gas in short supply, especially on a warm summer day when plants are very active, the high temperature leads to low gas solubility, and the slow diffusion rate means that replenishment of used resources takes a long time. The solu-bility of carbon dioxide in water is even more sensitive to temperature than oxygen. At 104°F (40°C) a given volume of water can hold only 30 percent of its capacity at freezing point. Some dissolved carbon dioxide remains chemically unchanged in water, but some interacts with the water to form a new compound, carbonic acid,

$$H_2O + CO_2 \rightarrow H_2CO_3$$

Carbonic acid then dissociates as follows,

$$H_2CO_3 \rightarrow HCO_3^- + H^+$$

The liberation of hydrogen ions is the characteristic of an acid, and the presence of extra hydrogen ions in solution

Acidity and pH

An acid can be defined as a source of hydrogen ions, or protons, H^+. The more hydrogen ions present in a solution, the greater the acidity. Chemists therefore measure the acidity of a solution in terms of its hydrogen ion concentration, but this varies greatly and the numbers involved would be extremely large, so a scale, called the pH scale, has been devised in order to express this hydrogen-ion concentration in terms that can be more easily handled. As in the case of any very long scale of numbers, the simplest way of reducing it is to express the numbers as logarithms. Using a base 10 logarithm, the difference between one unit and the next is a factor of 10 times. In the case of pH, the scale is also expressed in a negative fashion so that high numbers represent low hydrogen-ion concentra-tions and low numbers indicate high concentrations, in other words, greater acidity.

The pH scale is shown in the diagram above. The scale runs from 0 to 14 and the figure of pH 7.0 is taken as an indication of neutrality, neither being acidic nor basic. At this pH, hydrogen ions have a concentration of 10^{-7} molar, or 0.0000001 gm per liter. This is the pH of pure water, where hydrogen ions and hydroxyl ions are in perfect balance. Above the figure of 7.0, the pH indicates basic or alkaline conditions, and below 7.0, the pH denotes acidity. The diagram shows the pH of certain common solutions to demonstrate how the scale works out in practice. It is important to remember that the scale is logarithmic, so pH 4.0, for example, indicates 10 times the acidity of pH 5.0 and 100 times the acidity of pH 6.0.

reduces the pH. Water containing dissolved carbon dioxide and carbonic acid thus has an acid reaction. Since carbon dioxide is a constituent of the atmosphere, raindrops invariably capture some of this gas and dissolve it as they fall to the ground, which means that all rainfall tends to be acidic. Carbonic acid is a relatively weak acid, which means that its dissociation to produce hydrogen ions is very incomplete. Other acids are produced from pollutant molecules in the air, such as oxides of nitrogen and sulfur, and the nitric and sulfuric acids generated in this way are strong acids that are more fully dissociated in water and are more active in their interactions with other chemical compounds.

Some rocks and their component chemicals are more soluble in acidic solution, so they are dissolved more rapidly. Lime, calcium carbonate, for example, is more readily dissolved when exposed to acidic water. So the acidity of rainfall affects the efficiency with which it attacks the rocks of a wetland catchment and carries dissolved materials down into the wetland. Rainfall pH can also have a direct impact on the vegetation of wetlands, especially those wetlands that receive their water input solely from precipitation. The acidity of water draining from a catchment and through a wetland is thus an important factor in determining the plant and animal life that can be supported.

The properties of water are thus important because they influence the degradation of rocks and the mobilization of essential elements in a catchment and also because they affect the transport of those elements to wetlands and their availability to living things within the wetland.

■ GEOLOGY AND ROCK WEATHERING

The solid rocks of a landscape, especially if unprotected by a layer of soil, lie exposed to the elements and the outcome is a process of destruction. In some locations rocks are subjected to the crushing and grinding of ice as they lie beneath ice sheets and glaciers. Rock is fragmented in this manner and emerges from the ice in a variety of particles of different sizes, from boulders right down to the finest clay, sometimes referred to as "rock flour." Geologists and soil scientists have developed a classification scheme for different sized particles (see sidebar at right), which makes it easier to describe the breakdown of rocks and the development of soils. Even when the glacier has gone, rocks face a whole range of factors that result in their continued degradation. This decay of rock to produce ever smaller components is called *weathering*. Weathering takes place as a result of the activity of physical, chemical, and biological factors.

Physical weathering is particularly important. Rocks usually contain tiny fissures over their surfaces, and water

Particle Size

Soil scientists and geologists have devised a scheme for classifying the different particles of rock according to their size. In this way they can provide a precise definition for terms that could otherwise be used very loosely, such as sand and clay. Unfortunately, scientists in different parts of the world have failed to agree on a universally accepted set of definitions, but within the United States there is some measure of agreement. A rock fragment with a diameter greater than 10 inches (25.4 cm) is called a *boulder,* while smaller particles that exceed 2.5 inches (6.4 cm) are *cobbles.* Any particle greater than 0.08 inches (0.2 cm) but smaller than a cobble is called *gravel. Sand* consists of particles between 0.0008 to 0.08 inches (0.002 to 0.2 cm) in diameter, while *silt* is still smaller, having a diameter of 0.00008 to 0.0008 inches (0.0002 to 0.002 cm). *Clay* is the smallest of all particles, having a diameter of less than 0.00008 inches (0.0002 cm).

Clay is distinctive not only as a consequence of its small size but also because of its chemical structure. The tiny particles of clay are composed of chemical sandwiches in which layers of silicon oxides alternate with layers of aluminum oxides. The outcome of this complex crystalline structure is that the surface of the clay particle is negatively charged, and when clay particles are suspended in water, the effect of these negative charges is to repel any neighboring clay particle. Clay is thus suspended in water both by its small size, which makes sedimentation slower, but also by the repulsion between the fine particles. Clay remains in suspension in water, for considerably longer than all other particles and eventually settles only under very calm conditions.

penetrates into the minute cracks and weakened layers of the particles. When it freezes the water expands, increasing its volume by about 10 percent, splitting the rock into smaller pieces. The heating of rocks in sunlight can also lead to their physical breakdown. As the dark rocks absorb the light their temperatures rise, and this can cause cracking and breakage along planes of weakness. Heating splits layers from rocks, rather like the peeling of an onion, in a process called *exfoliation.* Rocks consist of mixtures of crystalline minerals, and these become separated from one another during the weathering process. Granite, for example, is broken down into three main minerals, feldspar, mica, and quartz. During the rock's formation these three minerals crystallize separately from the molten magma that is the raw material of granite. The physical weathering of granite breaks these mineral crystals apart and liberates them into the developing soil.

Chemical weathering acts on the fragmented rock particles. Water is an important medium for chemical weathering because it is such an excellent solvent and contains many dissolved substances. Rainwater contains a range of acidic compounds, including weak acids (such as carbonic acid) and strong acids (such as sulfuric and nitric acids) (see "The Nature of Water," pages 19–21). When this collection of acids reacts with the mineral components of rocks and soils, the result is chemical breakdown. The feldspar mineral from granite, for example, is broken down by carbonic acid to form potassium carbonate and a relatively inert material, kaolinite. Kaolinite, or china clay, is a clay mineral, and is not broken down further. Carbonic acid also acts on the mica mineral derived from granite, breaking it into potassium and magnesium carbonates along with oxides of iron and aluminum. The third mineral derived from granite, quartz, however, is relatively inert, being composed of silicon dioxide (silica), better known in the form of sand.

Granite thus weathers by physical and chemical processes that liberate potassium, magnesium, iron, and aluminum compounds, together with a clay mineral and silica. Although several of these elements are needed for plant growth, aluminum can be toxic, and many of the other important elements plants need are absent, such as phosphorus and calcium. The absence of calcium in particular is important because the acids that accumulate in the soil are not neutralized, and the soil becomes acidic in reaction. In other words, it has a low pH (see sidebar "Acidity and pH," page 20).

Living organisms also play a part in the weathering process. The rocks exposed by retreating ice or laid bare by landslips and erosion are quickly invaded by algae, lichens, bacteria, various protists, invertebrates, mosses, and even higher plants. All of these have their impact on the physical and chemical environment and, in particular, on the rock particles that form the skeleton of the developing soil. Photosynthetic organisms, such as some bacteria, lichens, and green plants, add organic matter to the soil in the form of dead litter, and other bacteria, together with the animal life in the soil, use this as an energy source, converting dead organic compounds to carbon dioxide. As has been described, carbon dioxide dissolves in water to produce carbonic acid, and this acid attacks the rocks, dissolving some of its component parts. Some lichens (see sidebar "Lichens," page 148) grow flat on the surface of rocks, and they secrete organic acids that attack their substrates, releasing elements for their continued growth. In the process they wear away the rocks they inhabit. Higher plants produce roots, some fine and some robust, all of which can penetrate into the fine

crevices of rock particles and expand to split them apart. The weathering process thus accelerates when plants invade a habitat as a result of these biological weathering activities.

The combination of physical, chemical, and biological weathering gradually wears the rock particles into smaller parts and releases increasing quantities of chemical elements into the local environment. In the European Alps, geologists have tried to measure the overall rate of rock weathering, and they have come to the conclusion that on average 0.004 inches (0.01 cm) of rock are being worn away each year. As the rocks are degraded, a soil is gradually born. As the soil forms, it may become eroded or stripped of its chemical content by the very water that has played such an important part in its formation by physical and chemical weathering.

■ EROSION AND LEACHING

Erosion is the physical movement of particles of rock and soil from one place to another. Water and wind can play parts in erosion as can slope because the movement is ultimately a consequence of the action of gravity on loose fragments. In wetlands, water is the most important medium for transport, of course, and the volume of flow and the speed of the water movement is the main determinant of the quantity of material and the size of the particles moved.

Fast-moving waters occur where large volumes of water pass down steep gradients, and these have the capacity to move large rocks and boulders. Such water is highly energetic. Large rocks are moved mainly by being rolled along the bottom of a river or stream, and their movement will take place during periods of intense water flow, as in the spring snow-melt period when the volume of water moving is at its greatest. The beds of glacial streams and rivers become littered with large rocks moved in this way.

Smaller stones, pebbles, rock fragments, and gravel can be moved by less energetic waters. Like larger rocks, they may roll along the bed of a stream, or they may bounce along the bottom, being periodically lifted by the motion of the water and then falling back to the streambed. This type of motion is called *saltation*. It also occurs when wind blows across a sandy beach and causes sand grains to bounce over the ground in a series of leaps. Fine particles of rock flour, the silt and clay fraction of glacial debris, or the finer fraction of eroded soils are small enough to remain suspended in the moving waters of a stream and may give the water a turbid or milky appearance. These particles can be carried many miles, eventually settling when the water movement becomes much slower or less energetic, often in lakes and wetlands farther down the course of the river.

Large particles are thus unlikely to be carried to most wetlands because these ecosystems usually occur in sites where the water is less energetic and moves only gently over a landscape. The main particle input to most wetland sediments, therefore, consists of silts and clays. Only in times of flood or major catastrophe in the catchment, such as fire or landslip, will large rock fragments probably enter the sediments of a wetland.

Apart from particles, water draining from a catchment into a wetland contains dissolved materials derived from the weathering of rocks or from the foraging activity of water as it percolates through soils. Weathering—physical, chemical, and particularly biological—continues once a soil has formed. Particles of rocks and their exposed mineral components are still subject to breakdown, especially as the soil bacteria work around them, generating carbon dioxide (and hence carbonic acid) during their respiration. As a result, chemical compounds are released into the soil water, where many of them dissociate into ions. Some of the ions, such as sodium and calcium, are positively charged (*cations*), and others (such as chloride and nitrate) are negatively charged (*anions*). These may be carried away by water draining through the soil in a process called *leaching*. If this happens, they are borne by the water into streams and rivers that may enter wetlands.

The leaching process is complicated by the fact that ions can become loosely trapped in the soil by certain of its components. Clay particles, for example, are very small, so that they have a very large surface area in relation to their volume, and they also have a crystalline structure, which results in their surfaces being negatively charged (see sidebar "Particle Size," page 21). The negative charges on clay particles attract the positively charged cations, which become loosely bound to the clay surface. The clay thus acts as a kind of reservoir for cations in the soil. Organic matter in the soil has a very complex chemistry, but some of its constituent compounds, such as organic acids, also become negatively charged and thus attract and retain cations. Some organic materials also bear a positive charge and retain anions, but these are less abundant in the soil, and most anions are more easily lost in drainage water than are cations.

As has been explained (see "The Nature of Water," pages 19–21), rainwater, and therefore the water that percolates through soils, is usually acidic in reaction. It thus contains an excess of hydrogen ions that are positively charged and that can replace cations attached to the clay particles or the organic matter in the soil. This type of leaching is an active stripping of the soil resources, taking away the ions held within its structure and removing them in solution in the drainage water. The soil is then left with a higher concentration of hydrogen ions attached to its clays and organic materials, hence it becomes more acid in its reaction. Meanwhile, the enriched leachate, the drainage water containing collected ions, continues on its way downstream toward the wetlands.

Since water plays such an important role in wetland ecology and chemistry, it is important to understand what controls the movement of water in wetland *catchments*. The study of water movement and the transport of sediments and chemicals is called *hydrology*.

■ HYDROLOGY OF WETLAND CATCHMENTS

Wetlands, as their name implies, depend on a supply of water to maintain their existence, and this water arrives from two main sources, out of the air as precipitation and out of the ground either as surface runoff or seepage water from underground sources. Wetland types can be classified according to the relative importance of these two water sources.

Precipitation falls directly onto the wetland surface and is of particular importance in those wetlands that have no groundwater input—the true bogs (see "Temperate Raised Bogs," pages 103–105). Precipitation may arrive as rainfall, or as snowfall in winter, in which case it may lie for some while on a peatland surface before it melts in the spring and floods the buried vegetation. The winter snow cover of the peatlands of high latitudes is of importance to the plant and animal life of these ecosystems as it insulates them and protects them from the intense cold that they would otherwise suffer. It also serves as a blanket that protects the peat itself from deep freezing, and, where snow cover is thin or absent, deep pockets and blocks of ice may develop in the peat itself and cause surface heaving.

Rainfall may arrive in the form of storms, in which case much of the water will rapidly drain away from the wetland. This happens in some of the central Australian wetlands, which develop rapidly following storms and then gradually dry up in the ensuing drought period, becoming increasingly saline (see "Inland Saline Wetlands," pages 116–118). Rainfall may, on the other hand, arrive in small, regular amounts, which usually proves more appropriate for the maintenance of high water tables and permanent standing water in the wetland. Some types of wetland can only exist under a consistent supply of rainfall, such as the *blanket mires* of oceanic regions (see "Blanket Bogs," pages 107–111). This being so, the actual quantity of rain received in a year is not always indicative of the type of wetland that can develop—its distribution through the year must also be known. In the case of blanket mires, for example, rain must fall on at least 200 days of each year to permit their development.

One other form of precipitation must be considered, and that is mist, or what is sometimes called *occult precipitation,* meaning literally unseen precipitation. It is unseen in the sense that it is more difficult to measure than normal rainfall; the latter can be measured in a rain gauge, but mist does not fall out of the sky, it condenses on projecting surfaces as the mist moves across a wetland. Mountain mires, especially if forested, may accumulate much of their water in this way as clouds move through an area and droplets form on the leaves of trees, eventually falling to the ground.

Water also arrives at a wetland by drainage from the land surface and by soaking through local soils and rocks. This type of water is not available to true bogs, which are purely rain-fed, but is often an important part of the water budget of other wetlands. How much water arrives in this way depends in part on the local precipitation, but also on how big the drainage catchment of the wetland happens to be. Some wetland catchments are truly enormous, such as that of the Sudd marshes of the Sudan in East Africa. These wetlands are fed by the Nile River, which drains much of the mountainous region of East Africa, including the mountains of Uganda and the Lake Victoria basin. Other wetland catchments are very small and can be measured in acres; these will have less opportunity to gather water to channel to the wetland.

Other aspects of climate, apart from precipitation, influence the amount of water draining from a catchment and its seasonal pattern of flow. Temperature and wind affect the rate of water's evaporation, so in hot, windy conditions even a large catchment area may not produce as much water as might be expected. This is true of the Sudan, where much of the water evaporates before the Nile reaches the Sudd marshes. If much of the precipitation arrives as snow, the spring snowmelt can result in sudden flooding of a wetland. Such flooding greatly affects the ecology of a wetland since it may be damaged and eroded by fast-moving waters. These floods can also produce chemical changes in the wetland.

The vegetation of the catchment can also affect the quantity of water gained and lost by catchments. A forest cover may collect more occult precipitation than a grassland, but it also loses more by interception of rainfall and then evaporation from the surfaces of the leaves. The tree acts like an umbrella, preventing some water from reaching the ground, but trees are more than umbrellas—they also act like pumps, drawing water from the ground, which then passes up through the trunks and eventually evaporates through pores in the leaf surfaces. This process is known as transpiration (see sidebar "Transpiration and Evapotranspiration," page 3). Why plants transpire is still debated by plant physiologists. Perhaps the loss of water helps to cool the leaves, rather like the perspiration of animals, but transpiration is also an inevitable consequence of the plant's need to open its pores (stomata) to take in carbon dioxide from the atmosphere in order to conduct photosynthesis. The outcome of transpiration, however, is that a leafy tree takes much water from the soil and loses it back into the atmosphere.

A forested catchment, therefore, intercepts and removes water more efficiently than one covered in grasses or dwarf shrubs and has an important influence on the amount of water that reaches the wetland served by the catchment. This is important to understand when managing and conserving wetlands. If trees are removed from a catchment, more precipitation will pass into the soil and eventually into streams feeding the wetland, although it must be borne in mind that tree removal can result in less shading for soils, raising the soil temperature and, hence, leading to greater evaporation from the soil. So the equation is complicated by local climatic conditions, but experimental removals of forest cover from watersheds in the eastern United States have demonstrated that stream flow increases by about 10–40 percent. In temperate forested regions, therefore, forest removal leads to increased runoff and can lead to an enhanced water supply to valley wetlands. Clearance of forest can also lead to soil instability, however, and this causes an influx of mineral sediments to wetlands. The influence of catchment vegetation on hydrology is illustrated in the diagram.

Wetlands can only exist where the geology prevents water soaking away rapidly, so an impervious bedrock (whether hard or soft) is needed for wetland formation. Surrounding areas of the catchment, however, may contain rocks of other types, such as pervious limestones, and water may percolate through these and eventually emerge when the water arrives at underlying water-resistant layers. Water may then come to the surface in the form of springs and can feed fens and marshes or even give rise to spring mires (see "Spring Mires," pages 91–92). Some wetlands are present in Minnesota, where springwater under pressure forces its way up through more acidic peatlands to create unusual spring features within these mires.

■ HYDROLOGY OF PEAT

Peat consists largely of vegetable debris, compacted together in an organic mass. The degree of compaction varies considerably, from the relatively uncompressed litter of dead plant leaves, fruits, flowers, and fragments on the surface to the dense, dark mass of older peat, which bears a weight of material that has accumulated above. An increase in compaction means that there is less space between plant fragments, and it is this space that can be filled either with water or with air, depending on the wetness of the site at any time.

The leaves and stems of many aquatic plants are supplied with spaces and channels through which air can penetrate to underwater parts, carrying the oxygen required for the respiration of stem bases, rhizomes, and roots (see "Submerged Plant Life," pages 134–137). Some plants, such as the bog

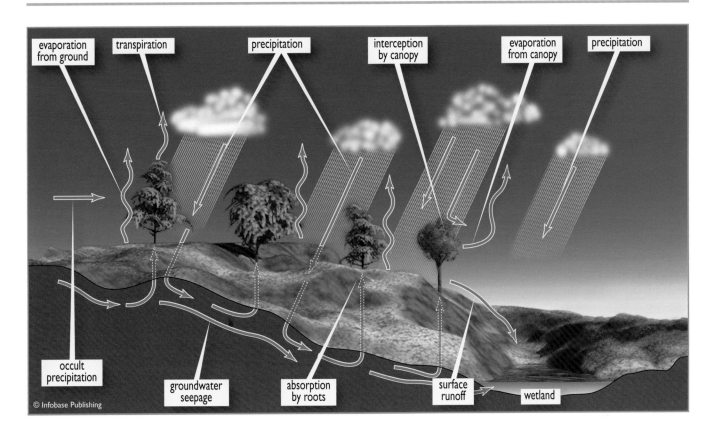

The movement of water within a watershed. Vegetation plays an important role in intercepting and evaporating precipitation, thus reducing the quantity of water in the soil and its rate of flow into wetlands.

moss, *Sphagnum,* have large storage cells that retain water even when conditions become dry, and these cells may also become partially filled with air as the water gradually evaporates during drought. So the plant detritus that accumulates in a wetland is generally of a spongy nature, rich in tubes and channels, spaces and pores, where water can be stored and where air may penetrate if the sediment surface dries out. The litter that first accumulates is generally complex in its structure, having many and varied spaces, while older, compressed sediments are generally more uniform and more dense in their structure.

In the case of permanently waterlogged wetlands, the water bodies, reed beds, marshes, and swamps, sediment drying may never occur. These are sometimes referred to as *hydric sediments,* which can, in turn, be divided into *limnic,* or underwater, *sediments* that accumulate in lakes, and *telmatic* (at the water surface) *sediments* that are deposited in reed beds and marshes. In wetlands such as fens and swamps, where the sediment surface is somewhat higher in relation to water level, seasonal drying takes place, and air penetration into the surface detritus is a regular event. These are called *terrestrial sediments.* True bogs have their peat surface

elevated above groundwater tables, and the saturation of the surface layers varies with the occurrence of rainfall. These sediments and their relative content of water and air are very variable, depending on current weather conditions.

Peat, because of its spongy texture when not severely compressed, is extremely absorptive of water. The *Sphagnum* peats of bogs may contain up to 95 percent water (by weight) when they are saturated. This percentage reflects the pore volume of the peat, for this space, if not occupied by water, would be occupied by air. The water-holding capacity of peat is one of the reasons it is prized as a soil conditioner in horticulture. As peat dries, its internal cavities gradually lose their water content, and the spaces become occupied by air. Some water is lost easily, simply draining under the force of gravity; some water is more tightly held within the smaller spaces and capillaries of the peat and is only lost after longer periods of drying. When peat becomes thoroughly dry, it may be very difficult to rewet. This is because the smallest capillaries and spaces are difficult for water to enter because of the surface tension that exists over the penetrating body of water. This can be observed experimentally with a desiccated houseplant, where added water may simply run off the surface of a dry peaty soil. The problem can be remedied by adding a few drops of liquid detergent to the water as this reduces its surface tension, allowing water to penetrate back into the peat capillaries.

Desiccation of this kind can occur in natural peatlands after severe drought, but the existence of a water table lower in the peat usually prevents complete drying, and even a small water residue in the peat allows a more rapid rewetting when the drought is ended. In this dry state, however, the surfaces of peatlands can be extremely prone to fire.

Water is not simply stagnant in the tubes and spaces of peat; it can move both vertically and horizontally. Just as the structure of the peat influences its drying and wetting properties, so it also affects water movements. In an uncompacted, loosely structured peat, like that found in the surface layers of mires, water moves relatively easily, while in the compressed and compacted layers of older peat, water movement is impeded. The ease with which water can move through peat is expressed by the term *hydraulic conductivity*. A peat with high hydraulic conductivity has a free flow of water, while one with a low conductivity has little water movement (see sidebar below).

The uncompacted peats of the surface layers have a high hydraulic conductivity, so they allow free passage of water, whereas the compacted peats of lower layers have a low hydraulic conductivity and are relatively impermeable to

Darcy's Law

The illustration shows how the hydraulic conductivity is calculated for a given type of peat. The rate at which water moves through a porous material depends on the nature of the material itself and how much resistance it offers to flow (just how porous it happens to be), how much force is exerted on the water, and how far the water has to pass through the material (L). The force in this case is the force of gravity, expressed as the head of water (H) applied to the peat. The resistance is a constant (K) for any particular material and this is the hydraulic conductivity. The relationship between conductivity and flow is called Darcy's law, after Henry Darcy, who first described it in 1856, and is expressed:

Volume of water flow = Hydraulic conductivity (K) ×
Head of water (H)
Length of passage (L)

The diagram illustrates the way in which the hydraulic conductivity for any material can be measured.

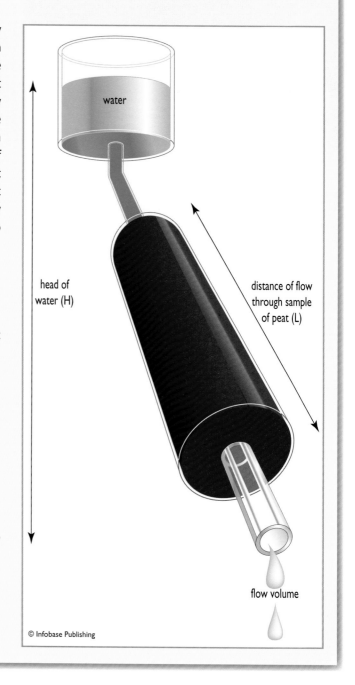

© Infobase Publishing

The rate at which water flows through peat is proportional to the pressure of the water (measured by its head) and is inversely related to the distance the water has to move in its descent. The permeability of the peat depends on its degree of compaction, and this determines the hydraulic conductivity. The relationship between these variables is given by Darcy's law.

Section of a peat deposit from a blanket mire in Norway. The loosely compacted, permeable layer of the acrotelm is clearly distinguished from the dark, compacted peat of the catotelm. *(Peter D. Moore)*

water. The layered structure of peat is well illustrated in the blanket bog profile shown in the illustration. These upper layers of peat are called the *acrotelm,* and within them the water table periodically rises and falls according to the conditions of rainfall and evaporation and also moves laterally through the mire. In the lower layers, the *catotelm,* water movement is very limited, both in a vertical and a horizontal direction. The water here is effectively a stagnant mass, except where particular circumstances, such as springwater under pressure, may disturb the system (as in some mires in Minnesota).

This understanding of the nature of water movement in peats has greatly influenced the interpretation of the hydrology of raised mires. It has long fascinated ecologists and hydrologists just how a waterlogged mass of peat can continue to grow in a dome-shaped form in which the surface is elevated several yards above the groundwater table.

The ecologists who first described these mires assumed that this could only be achieved by a system of capillarity, where the evaporation of water from the peat surface pulls water up from below, rather like the transpiration of a plant. This would require high hydraulic conductivity in the catotelm, however, and this is not the case. The alternative theory, now generally accepted, is that the saturated mass of peat in the catotelm effectively blocks downward drainage of water from the surface, so that all the rainwater seeps horizontally through the acrotelm, eventually to drain from the sloping sides of the raised bog. This maintains a surface water table perched above an impermeable elevated base.

An appreciation of how water behaves in peat is thus important in many areas. It helps us to understand the form and shape of peatlands, and it can be important in the horticultural use of peats, in peatland reclamation for agriculture and forestry, and also in the restoration of damaged peatlands for conservation (see "Bog Rehabilitation," pages 232–234).

■ GEOLOGY OF WETLAND CATCHMENTS

Wetlands can form only where an adequate supply of water can be maintained, and this condition is strongly influenced by the region's geology. If water is to be retained at a site long enough to permit the invasion of wetland plants and animals, then the input of water must be adequate to compensate for losses; in other words, drainage must be impeded by impervious underlying rocks, or there must be a generous water supply from rivers or aquifers. There may, of course, be a combination of both.

Impervious bedrock is typically formed by shales, slates, clays, or mudstones, which prevent the downward movement of water and permit only horizontal water movements on the surface. Instead of soaking away through the basal rocks, water is retained at the surface. Areas with such bedrocks are more likely to be rich in wetlands, depending on the supply of water by precipitation and water that flows in from rivers. A landscape of this type is likely to contain *topogenous* wetlands (see the diagram on page 28, panel A), that is, dependent simply on the local topography, and these will be found scattered through such an area. They may develop in hollows and valleys, wherever water is retained by the impervious bedrock. *Fens* will be found in such hollows, which may be rich or poor (see "Fens," pages 87–89) depending on the chemistry of the local rocks.

Landscapes that have experienced glaciation often have a geology that is complicated by the presence of overlying detritus, including morainic dams and boulder clays derived from the melting glaciers. These features can mask

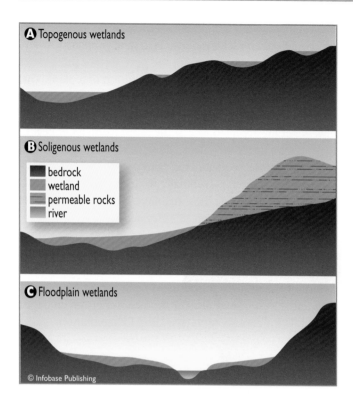

A Topogenous wetlands

B Soligenous wetlands

bedrock
wetland
permeable rocks
river

C Floodplain wetlands

© Infobase Publishing

Hydrological types of groundwater-fed wetlands. (A) Topogenous wetlands are formed where water collects in hollows in a landscape with impermeable bedrocks. (B) Soligenous wetlands occur where water drains through permeable rocks and then collects on impermeable strata. (C) Floodplain wetlands develop along river valleys where the volume of water flow in the river occasionally results in the spread of water over a floodplain.

the underlying geology and produce a land surface in which topogenous wetlands are liable to develop. Superficial glacial materials may even occur in landscapes of sandstone or limestone origin that are normally well drained and free of wetlands. Here, they create a veneer of impervious sediments leading to wetland formations that may seem out of keeping with the general nature of the area, as shown in the illustration of Malham Tarn in northern Britain.

In nonglaciated regions where the main bedrock is permeable, however, wetlands are generally likely to be less frequent. Sandstone and limestone regions are often poor in water bodies and mires, but even in such landscapes, wetlands may be found if the deep, underlying bodies of water—the aquifers—emerge at the ground surface in valleys. This may happen if there are deeper layers of rocks impervious to the downward movement of groundwater that result in its emergence at the surface in low-lying sites. Wetlands may develop where this groundwater reaches the surface; such wetlands are termed *soligenous* (see the diagram above, panel B). These wetlands are often richer in their chemistry than the topogenous types because the water has often

moved considerable distances through the ground and has become charged with a greater load of dissolved chemicals as a result.

Soligenous wetlands can be particularly apparent in landscapes affected by geological faulting. Here, pervious bedrocks may meet impervious ones with very sharp transitions and boundaries. An example is shown above a wetland system from a limestone region of northern Britain where older shale rocks have been brought into sharp contact with the limestone as a result of extensive faulting. Water draining through the limestone contacts the impervious shales and accumulates upon them as a shallow lake and associated peatlands. The site is further complicated by a deposition of glacial clays that helps to hold back the waters of the wetlands situated on top of the shales.

Similar to the soligenous wetlands are the *floodplain* wetlands created by the overflow of rivers during periods of flood (see the diagram at left, panel C). All of these types of wetlands, topogenous, soligenous, and floodplain, are rheotrophic, they are fed by a flow of groundwater through the ecosystem. Floodplains may create their own impervious strata by encouraging the deposition of eroded materials carried from the upper reaches of river catchments. Silts and clay are carried farther by rivers than sand particles because of their small particle size and slow sedimentation rates when in suspension. They may eventually be deposited when the river spreads across its floodplain and loses some of its energy of motion, thus creating a silty bed over which wetlands can develop. The Nile Delta is a classic example of such a floodplain mire that was once annually inundated by the floodwaters from the highlands of East Africa. As these waters approached the Mediterranean Sea, they became slower in their flow, and the suspension of fine materials was deposited through the lateral areas of the floodplain.

In some sites, the burning of vegetation in river catchments can lead to charcoal becoming deposited in valleys and floodplains, and this inert material can add to the water retention capacity of valley soils, as the fine particles impede drainage. Waterlogging, peat formation, and the development of valley mires may result from this process (see "Valley Mires," pages 95–97).

The geology of a wetland catchment has a strong influence on the chemistry of water draining from it. The presence of limestone in a catchment, for example, ensures a supply of calcium carbonate in the drainage waters, and this will result in rich fen vegetation with an abundance of mollusks (which need lime for their shells). In granitic and sandstone catchments, the supply of calcium is liable to be lower, and poor fens will develop in the wetlands. Ombrotrophic (rainfed) mires are independent of groundwater, so they are unaffected by the geology of their catchment. They may develop from either topogenous or soligenous types of wetland.

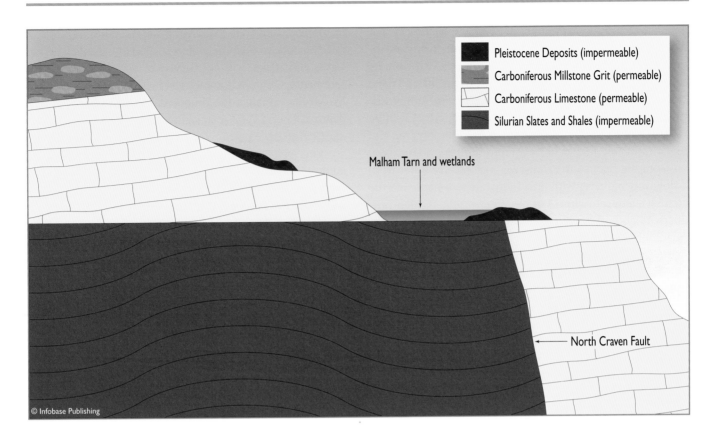

Pleistocene Deposits (impermeable)

Carboniferous Millstone Grit (permeable)

Carboniferous Limestone (permeable)

Silurian Slates and Shales (impermeable)

Malham Tarn and wetlands

North Craven Fault

© Infobase Publishing

Geological profile of Malham Tarn, a soligenous wetland in northern Britain. Water percolates through the porous limestone but accumulates above the impermeable shale rocks. Pleistocene glacial deposits of clay dam the lake and prevent the water from escaping into the layers of limestone below the fault line.

LAKE SEDIMENTS AND THEIR ORIGINS

Lakes and other open-water wetlands accumulate their water largely from their surrounding catchments, but water

Coastal wetlands, fed at least in part by seawater, are less influenced by the geochemistry of their catchments. Seawater is a rich source of many of the elements needed for living organisms, such as magnesium, calcium, and potassium, so any lack of these in the drainage water from the land is easily compensated for. Coastal wetlands, however, can only develop in regions of low tidal scour and wave action. They are often associated with estuaries, particularly if these are sheltered from waves by promontories or by shingle bars. Here, the geomorphology of the coastline is more important than the geology of the catchment in determining whether and what kind of wetland will develop.

is not the sole commodity they acquire from their catchments. The waters that flow in contain dissolved chemicals and suspended materials that are derived from the erosion of the surrounding soils. Once the water enters the wetland, the speed of its flow is generally reduced, allowing the suspended materials to sink to the bottom, where they accumulate as sediment. Materials derived from outside the wetland ecosystem are said to be *allochthonous,* and these contribute much to the buildup of many aquatic sediments.

The water body itself is occupied by a range of plants and animals that are permanently resident within the site. As plants age, they release dead parts of their structure—leaves, roots, stems, and so on—into the environment to be decomposed by fungi and bacteria. In waterlogged wetlands the rate of decomposition is impeded by a lack of oxygen, and portions of dead vegetable matter, together with animal detritus, contribute to the accumulating sediment. Materials derived from immediately local sources are said to be *autochthonous.* Between them, allochthonous and autochthonous materials comprise the sediments of wetland ecosystems.

The sediments of open-water wetlands generally contain both organic and inorganic materials. The precise proportion of these materials depends on a range of factors, such as the topography and vegetation of the catchment (whether, for example, it is steep and unvegetated or flat and forested); the energy of the streams that enter the ecosystem (fast-flowing or slow-flowing); and the vegetation and productivity of the lake and its surrounding wetland. A catchment that is actively eroded by rapidly flowing streams will contribute large quantities of inorganic sediments eroded from its soils. Such a situation often occurs early in the life of a river, close to its source and, since the headwaters may still be low in chemical components, the wetland productivity will be relatively poor, so the lake sediment has little organic and much inorganic matter in its composition. A lowland lake, far from the river source, however, may receive water from a slow-moving river with little inorganic sediment, and the relative chemical richness of the water results in high productivity both of the plankton and the larger aquatic plants (macrophytes), leading to a higher proportion of organic matter in the sediments.

The inorganic material entering a wetland consists of mineral particles of various sizes, and the proportions of the different sized particles (see sidebar "Particle Size," page 21) are referred to as the *texture* of the material. Small particles (silts and clays) are slow to sediment, so they arrive suspended even in low-energy rivers and are promptly deposited into the central parts of large wetlands. Sand particles are larger and arrive only when carried by high-energy streams; they are usually deposited close to the site of entry. In steep-sided catchments in cold conditions, even larger particles of gravel and rock debris may be carried into lakes, especially when trapped in winter ice. The texture of the sediment can provide an indication of the conditions of erosion in the catchment, and this fact provides a means of working out the history of wetlands by observing changes in the nature of the stratified sediments.

Cores of lake sediments sometimes reveal laminations—fine layers that usually alternate light and dark. Often these have been formed annually as the spring productivity of organic material creates a dark layer, and the winter erosion of soils and inorganic materials creates a pale layer. When these were first described in mountain lakes from glacial regions, they were termed *varves*. Counting annual laminations in sediment cores is an efficient method of dating different levels.

Plant fragments compose a major part of the organic fraction of the sediments of many open-water wetlands. Leaves, fruits, and even the trunks of trees can be washed into lakes and incorporated into their sediments. In the case of small wetland basins surrounded by forest, coarse vegetable debris of this type may form the major proportion of the sediment, but in larger lakes these coarse materials tend to be less apparent. Not all of the larger plant fragments are allochthonous in origin. Some are derived from local aquatic plants, such as water lilies, reeds, and sedges that grow around the margins of the wetland.

Although decomposition of these materials is slower in waterlogged conditions because of the lower microbial activity, the main limiting factor is the diffusion of oxygen in the water. In well-oxygenated waters, decomposition is relatively active and the softer portions of plants, such as leaves and stems, will not survive to be incorporated into sediments. Only the tougher residues, including wood and hard fruits, remain undecomposed. Whole leaves, twigs, and delicate flowers survive in sediments only under extremely *anoxic* (lacking oxygen) conditions.

One plant component that does survive well in wetland sediments is the pollen grain. The outer coats of these microscopic structures are very resistant to decay, so they accumulate in large numbers in most types of deposit. They may be allochthonous or autochthonous in origin, and their analysis provides a means of investigating both the local history of a wetland as well as the vegetation changes over an entire catchment or beyond.

Microscopic organisms, including plankton, also contribute to the organic matter in lake sediments. Most algae are rather delicate and survive in sediments only rarely. The star-shaped colonies of the green alga *Pediastrum* are among the most distinctive of the algal components of some lake sediments. Filamentous green algae, such as *Spirogyra* and *Cladophora*, do not usually become incorporated into sediments, but their spores do, and these distinctive elliptical bodies are often present embedded in an inorganic matrix. Among the green algae, perhaps the most distinctive group is that of the stoneworts (Charophyta). Many of the species in this group accumulate calcium carbonate and secrete a lime deposit in their cell walls. For this reason, they are only common in lime-rich wetlands (although some stoneworts, such as *Nitella*, do occur in more acid sites). Their vegetative structures, especially the distinctive and easily preserved reproductive organs, are an important component of lake sediments in some wetlands, contributing to the development of a white lime deposit called marl.

Diatoms are another group of organisms that are frequently found fossilized in the sediments of lakes. They were once regarded as algae but are now classified as protists, and like stoneworts they also produce inorganic cases (called *frustules*) that survive after the death of the cells. Diatoms build their frustules from silica rather than from calcium carbonate. They occur in a range of different conditions and the sculpturing of their frustules permits their precise identification, so that the past ecology of the wetland can be established by analyzing the diatom content of different layers (see "Microscopic Stratigraphy: Diatoms," pages 39–40).

Animal remains in sediments are generally less abundant than those of plants, but the wing cases of beetles, the mouthparts of midges, and the bodies of mites are often present. Larger animal remains such as the bones of vertebrates survive only in calcareous waters, for in acid waters the lime from the bones is likely to be dissolved. The sediments of past wetlands, however, have been the source of many bones of extinct mammals, ranging from giant elk and mammoths from the end of the last ice age to those of the wild cow, the aurochs, that lived in Europe well into historical times.

PEAT DEPOSITS: ORIGINS AND CONTENTS

Peat is a largely organic sediment that accumulates under waterlogged conditions, although not necessarily beneath water. The water table in a peat deposit often lies beneath the surface of the peat. The water in peat may be replenished by groundwater flow together with the runoff water from a catchment, or it may be supplied solely by rainfall. The balance between these two sources of water, however, will profoundly affect the nature and constitution of the peat because a rich supply of drainage water from a catchment is likely to bring eroded mineral sediments into the deposit from surrounding soils, as is the case for lake sediments. If, on the other hand, the water supply is purely from rainfall, then the mineral content will be very low, consisting merely of dust and suspended materials in the air that are washed out by the rain. This is unlikely to be a high proportion (often only 1 percent by weight), but it can be significant in areas such as Iceland, where volcanic activity puts much ash into the atmosphere. Here, peats fed solely by rainwater can still include 20 percent inorganic matter.

The proportion of minerals found in a sediment can therefore tell us much about the way in which the material was formed. Sediments, in the course of succession (see "Succession: Wetland Changes in Time," pages 14–15), build up and are raised above the water table, and thus the proportion of inorganic material within them diminishes. The simplest way of measuring this proportion is by determining the loss on ignition. This involves taking a sample of the sediment, drying it thoroughly, weighing it, then heating it in a furnace of more than 1,000°F (550°C) until there is no further loss in weight. This combusts all the organic matter and leaves behind only the inorganic content of the material. It is a simple, commonly used method, but it does have the disadvantage that lime (calcium carbonate) may also break down at such high temperatures to release carbon dioxide, which results in a loss of some inorganic material. If a sediment is rich in lime, therefore, the loss on ignition method is inappropriate, and organic material must be destroyed by a gentler process that involves oxidation with an agent such as hydrogen peroxide.

The variation between sediments with a largely mineral content and those that are almost purely organic is a gradual one. Pure bog moss, *Sphagnum* peat, for example, is often 99 percent organic with only 1 percent mineral matter. If we are to define *peat* as a scientific term, however, we need to choose a cutoff point in this spectrum. Soil scientists usually take 65 percent organic and 35 percent inorganic material as the point at which the term *peat* can strictly be applied to a sediment, but commercial harvesters of peat still use the term to describe sediments with up to 55 percent inorganic matter.

The presence of inorganic bands in a peat deposit can be useful in the interpretation of the history of a site, as in the case of erosion episodes following severe storms, fires, floods, or the artificial clearance of catchment forest. The fire history of a catchment can be reconstructed using bands of charcoal in sediments. Bands of inorganic material can also arrive at a site as a result of volcanic eruptions that deposit layers of ash over the landscape. These ash deposits, known as *tephra*, can be distinguished from waterborne inorganic materials. The structure of tephra is very like glass particles, for the silica within tephra is subjected to high temperatures in the eruption and melts in the process. These small particles can be carried many thousands of miles by the wind; eruptions of volcanoes in Iceland, for example, can be detected in the sediments of bogs in Scotland and Ireland. Each eruption has its own chemical peculiarities, or signature, so chemical analysis of tephra allows the identification of precise eruptions from the past. In turn, this means that tephra layers can often be precisely dated.

Carbon particles in the form of charcoal and soot are also frequent components of peats. These may reflect local fires within the vegetation of the mire surface, or they may be derived from more distant fires as the charcoal is borne to the site by wind.

The organic constituents of peats are largely botanical. Whole plants that once grew on the spot where the peat was formed may be incorporated intact into the peat, particularly such robust mosses as *Sphagnum*, the coarse growth of cotton grasses (*Eriophorum*), or members of the Restionaceae. These remains may be identifiable to the taxonomic level of species. Among flowering plant remains, roots predominate because these penetrate into deeper layers of the peat, where decomposition is slower (see "Decomposition and Peat Growth," pages 54–56), and hence are more likely to survive intact. Woody portions of stems, or the coats of fruits and seeds, also survive better than the soft parts of plants. This is because they are less palatable to fungi and bacteria, so they persist in the deposit.

Most of these large plant fragments have a local origin—they are derived from the resident plants of the mire surface. However, some of the smaller parts of plants, such as the small wind-borne fruits of birch trees, can be carried many hundreds of yards and may originate from vegetation surrounding the mire or even farther away. Technically, these remains are all *fossils* in the sense that they are the dead, persistent relics of organisms from the past (even though "the past" may be only a matter of hundreds or thousands of years). The plant parts mentioned above can all be recognized without recourse to microscopy, so they are called *megafossils,* or *macrofossils.* This distinguishes them from the smaller microfossils, which can only be observed and identified with the aid of a powerful microscope. Among the plant *microfossils,* the most abundant are the pollen grains, which may have a local origin or may have traveled considerable distances before sedimenting from the atmosphere onto the peat surface.

Animal fossils in peats include microfossils, such as rhizopods (creatures related to *Amoeba* but with tough shells that survive in peats), mites, and parts of small crustacean and insect exoskeletons. Animal microfossils may also be found, the most frequent of which are the wing cases of beetles, often conspicuous because of their iridescent colors, which are very apparent when peat is broken open. Some experts are able to identify beetles on the basis of their wing cases and can reconstruct the beetle communities of the past.

Bones of vertebrate animals are rare in peats because these deposits are generally acidic in their reaction and calcium carbonate is dissolved. Some fen peats contain bones, but in true bogs the bodies of vertebrates are more often represented by hair and hide rather than bones. In the case of human bodies, recovered in considerable numbers (running into hundreds) from the bogs of western Europe, for example, hair, skin (even fingerprints), fingernails (rich in tough, keratin protein), and leather or linen clothing survive well. The particularly gruesome appearance of the faces of these bodies is often the result of the bone structure being dissolved from within so that the facial expressions are distorted. The fact that many of these bog bodies experienced violent deaths, either as murder victims or by official execution (such as by strangulation) or perhaps even as sacrificial victims to a peatland deity, may well also have contributed to their generally unhappy expressions.

■ WETLAND STRATIGRAPHY

Succession consists of a change over time of vegetation and its associated animals that is both directional and predictable (see "Succession: Wetland Changes in Time," pages 14–15). One remarkable and, for ecologists, extremely convenient characteristic of wetlands is that they leave a record of their own development in their sediments. No other ecosystem has, within its own structure, a complete documentation of its history.

The sediments that underlie most wetlands have accumulated during the course of the development of that site, and these sediments are generally stratified, that is, layered, in a sequence that relates to time. The deep sediments are generally older than those near the surface. The fossil plants and animals contained in the sediments record the vegetation and animal communities of the past, and it is possible to reconstruct the sequence of changes in the ecology of a site as a result of the analysis, identification, and interpretation of these fossil assemblages.

The accounts of the vegetation changes and developmental processes leading to the formation of different wetland types described in chapter 4 are largely based on this type of information. Whereas, in most terrestrial ecosystems, the course of successional development over time must often be inferred from spatial patterns, such as *zonations,* of vegetation, in wetlands we have sound evidence of changes derived from the sediments. The zoning of vegetation types around a water body, for example, may provide clues to the possible sequence of vegetation types over time, and ecologists can actually confirm this by examining the sediment sequence underlying the supposed later stages in succession.

There are, however, problems to be faced when attempting to use sediments as a guide to successional development of wetlands. The first problem concerns the actual sequence of sediments. Do they follow a strict time sequence? In shallow lakes, for example, sediments may settle from suspension in the water body, only to be resuspended by the action of wind or water currents. The final resting place of the sediment may be far removed from its original location, often in the deeper parts of the lake. This process is called *sediment focusing,* and it can confuse the stratigraphic sequence of a site.

The in-wash of allochthonous materials into lakes, marshes, and fens may interrupt or contaminate the development of a stratigraphic profile. Soils, or even peats, from surrounding ecosystems can be eroded by storms and then

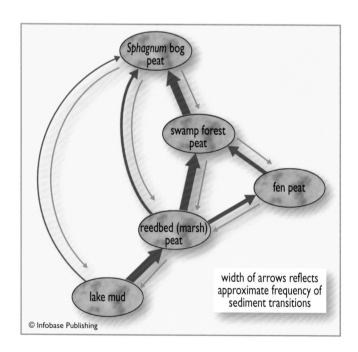

Sedimentary sequences in British wetlands. By studying the layers of sediments in wetlands, it is possible to trace the ways in which they develop.

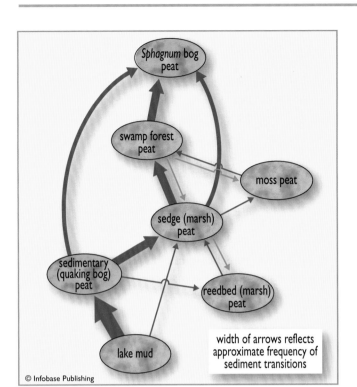

Sedimentary sequences in North American wetlands. Floating carpets of marsh and bog are more common in North American wetland development than in Europe.

redeposited to form a distinct layer in the profile of the wetland receiving the water flow. When such in-washed material contains fossils, these may show no relationship to the general sequence of the community development at the site. Even the general input of allochthonous material from a wetland catchment can be confusing, and paleoecologists (ecologists who study the historical development of ecosystems and landscapes) must try to separate local events at the site of sedimentation from the events of the surrounding catchment and wider region.

Just as a sediment sequence at one site may be interrupted by added material from elsewhere, so other sequences may have gaps resulting from erosion episodes. Parts of the time sequence may be missing from the record. When this occurs at the top of a sediment profile, the profile is said to be truncated. Erosion episodes in sediments are often evident in the form of sharp transitions from one type of material to another.

Stratigraphic studies of mires, despite these problems of stratigraphic interpretation, do supply a wealth of knowledge concerning general development in wetlands. This is especially true in the temperate regions, where the number of surveys far exceeds those in the tropics. Each individual wetland site has its own peculiar characteristics, depending on biogeographical location, climate, geology, soils, and so on, but there do seem to be some sequences of vegetation

succession in wetlands that are more common than others. The diagrams summarize surveys of stratigraphical studies carried out in Britain and in North America, which have provided ecologists with an opportunity to propose general wetland successional sequences.

The following is a number of important general conclusions that can be reached from these surveys:

1. There is confirmation of the tendency for open-water wetlands to become infilled by sediments and for these sediments to show a general sequence from limnic (lake) sediments through telmatic (reed swamp and marsh muds) to terrestric (fen and forested swamp deposits). This process is termed *terrestrialization*.

2. Successions often proceed beyond this swamp-forest stage and become wetter once more as mosses, especially bog mosses, assume dominance. This process is called *paludification*.

3. The idea that wetland succession (sometimes called *hydroseral succession*) can end in dryland forest is erroneous. Nonetheless, this imaginary sequence is often illustrated in ecology textbooks. The idea probably originates from observations of zonation patterns of vegetation around ponds which have transitions to dryland forest at the edge of their basins. Studies of sediments reveal that these zonation patterns cannot always be used as a guide to successional processes.

4. Reversals in successional developments can occur, taking the succession in an apparently opposite direction.

5. Successional shortcuts can also occur, seemingly skipping one or more of the "normal" stages in a succession and proceeding directly to a later or final stage.

6. A major difference between British and North American successions is the abundance of sedimentary peats (accumulated beneath the floating carpets of quaking, acidic vegetation) in North America. In Britain, the sequence from open water to reed bed is more common, and the terrestrialization process is consequently more gradual.

These schematic diagrams are extremely generalized and apply only to the particular regions from which they were taken. Other general sequences still need to be investigated in other parts of the world, but the use of stratified deposits containing fossils within the wetlands offers the opportunity for such broad successional studies.

Besides allowing the development of general models, stratigraphic analyses can also give paleoecologists an opportunity to examine detailed aspects of successional sequences. In work on the blanket peats of Wales, for example, detailed analysis of the basal few inches of the peat, which often overlie mineral soil, can provide many clues that help to reconstruct the events leading up to the initiation of peat. Wood charcoal can be found, along with the remains of certain mosses normally associated with slightly more nutrient-flushed conditions than those typical of the blanket peat themselves. A picture emerges of tree loss and paludification, accompanied by burning and nutrient enrichment of the original upland scrub woodlands, prior to the development of the *Sphagnum*-dominated bogs. This thin layer of basal peat may contain the vital clue to understanding this type of succession.

However, in all studies of this type, where stratigraphy is used as a tool for revealing succession, care must be taken concerning the scale of the analysis compared with the scale of the interpretation. A single coring from a wetland will provide many stratified fossils, and it may be tempting to reconstruct the site's history on that basis, but many more such cores are needed, preferably arranged in a linear grid covering an entire basin if a full and reliable understanding of the development of a site is to be achieved.

■ STRATIGRAPHY AND CLIMATE

Wetlands, because of their sensitivity to water supply, respond rapidly to alterations in climate. An increase in the water entering a wetland from catchment drainage will result in changes to the composition of the incoming suspended material and may also result in an overall increase in the water level of the basin. A change in overall water level, however, can only occur if there is no overflow from which the excess water is immediately discharged. If, for example, a lake discharges its water into an exit stream, then an increased water supply will lead to more water moving out of the lake via the stream, but it will not cause an increase in lake water level. If the stream becomes blocked by the development of peat-forming swamps, however, water level in the lake may rise, but not simply in response to changing climate.

Water-level changes in wetlands responding to climate change are best recorded in closed basin sites where there is no discharge stream; water is lost simply by evaporation. In such cases, an increase in water supply is recorded by more rapid sedimentation of eroded material, and a lower water supply may even lead to the drying out of the lake and the production of an evaporite—a deposit in which chemicals such as gypsum or common salt crystallize in a crusty layer as the water evaporates. This is particularly common in some of the saline wetlands of the continental interiors (see "Inland Saline Wetlands," pages 116–118), where such layers in the stratigraphy record past arid phases. In these sites, former high water levels may also leave behind elevated shorelines carved out of the surrounding landscape. Studies

of past lake levels in regions that are currently dry (such as Saharan Africa, the Arabian Peninsula, and the southwest region of North America) have proved useful in the reconstruction of past climatic changes going back hundreds of thousands of years.

The type of sediment that is brought into a wetland varies with a number of factors, one of which is climate. More rainfall may mean a more vigorous erosion of the catchment soils, bringing higher loads of sediment into the basin, and this is recorded in the stratigraphy by bands of coarser material—sands and gravels may be deposited over the silts, clays, and muds of the less erosive periods. Extreme cold can have a similar impact because frost can cause soils to become unstable and move down slopes into streams and lakes. This climatic effect is illustrated by the so-called Younger Dryas event (see sidebar) at the close of the last ice age.

The Younger Dryas Event

An example of wetland stratigraphy providing the first evidence of an important climate change in the past is afforded by the discovery about 50 years ago of an unusual sequence in Denmark, in deposits dating from about 14,000 years ago. At that time, the last of the great ice advances was over and the climate was rapidly warming. This was evident in the sediment record because clays, eroded from surrounding glacial deposits and carried into shallow lakes by streams, began to give way to organic muds as the local water plants increased their productivity with the rising temperature. Then, very abruptly, the dark organic muds were replaced once again with clays containing angular fragments of rock, suggesting that masses of soil had become destabilized and had slumped into the lake. This reversal of the expected sediment sequence has subsequently been recorded in many parts of Europe (from Spain to Scandinavia) and also along the eastern seaboard of North America. It clearly represents a sudden cooling that was geographically widespread about 12,700 years ago, centered on the North Atlantic. The event lasted about 1,200 years, and the postglacial warming then began again. Its global impact has now been confirmed by many other sources of evidence, but the existence of this unusual climatic event, which has become known as the Younger Dryas (from the presence of an arctic-alpine plant called *Dryas octopetala* found fossilized in the sediments) was first detected as a result of the study of wetland stratigraphy.

Stratigraphic evidence of climate change, however, is not always as simple to understand as this example of the Younger Dryas. Soil erosion and coarse layers of sediments can result from warm episodes in climate history. In wetlands fed by glacial streams, for example, periods of warmth can bring greater volumes of water and their attendant load of sediments. The peat stratigraphy of the Austrian mire shown in the diagram on page xx demonstrates how such sites can provide a record of warm episodes in the past.

All of the situations considered so far relate to rheotrophic mires—wetlands that are supplied by groundwater from their catchments as well as rainwater from the sky. In these conditions, many factors can modify or even obscure the impact of climate on stratigraphy. In the ombrotrophic, rain-fed mires, on the other hand, one should be able to discern climate change with greater resolution and greater confidence. The raised bogs, for example, are hydrologically independent of their catchment areas and are dependent entirely on precipitation (or, more strictly, on the ratio between precipitation and evaporation) for their water supply. If rainfall increases, or if temperature (and therefore evaporation) falls, more water becomes available at the bog surface and the vegetation responds accordingly. A wetter bog surface, for example, should result in a greater frequency of bog pools, together with the development and growth of different species of plants, most particularly different species of *Sphagnum,* which have very specific water requirements. The stratigraphy will record any such changes because these plants leave fossil traces of their former presence behind them when they die.

Climate change may leave even more evident marks in the stratigraphy of a bog. When examining an exposed face of peat, it is often possible to detect distinct layers with lighter or darker colors and with very different consistencies. This is a result of different degrees of decomposition, well-decomposed layers being generally dark and compacted, while poorly decomposed layers are light, less compacted, and have clearly visible plant fragments within them. Such banding in peat profiles, especially when the bands extend over the whole mire surface rather than simply being associated with local pool and hummock conditions, can provide very sensitive evidence of past climates. A wet bog surface means that even the acrotelm (see "Hydrology of Peat," pages 24–27) is saturated for much of the time, so decomposition within this most active layer is reduced. Dead plant litter then accumulates rapidly, soon becoming incorporated into the catotelm, where decomposition is reduced even further, and a pale, fresh-looking peat develops. A dry period, on the other hand, means that plant remains spend a long time in the aerated layers of the acrotelm and are heavily degraded by the time they reach the safety of the catotelm.

Some of the earliest work in reconstructing climate history was conducted in Scandinavia late in the 19th century and was based on the study of peat stratigraphy. By analyzing many peat profiles, researchers were able to propose a sequence of climatic periods for northern Europe since the end of the last glaciation—a pre-boreal and boreal period that were warm and relatively dry, followed by an Atlantic period that was warm and wet, then a sub-boreal period that was drier and cooler, and a sub-Atlantic one that was both cooler and wetter. More recently, climatologists, using both peat stratigraphy and other sources of evidence, have modified and elaborated these basic findings. The last 3,000 years of climatic history, in particular, have been the focus of much attention, and peat stratigraphic studies have revealed a number of particularly wet and cool periods recorded in the bog profiles.

■ STRATIGRAPHY AND WETLAND CATCHMENTS

Although the stratigraphic profiles of wetlands record changes in the general progress of vegetation succession and changes in climate, they also record a whole range of other processes that occur in their vicinity. Human land use, fires, and sea level changes all have their impact on wetland stratigraphy.

Climate is not the only factor that affects the amount of water draining from a catchment into a wetland basin. The removal of trees, for example, can increase the flow of a stream by up to 40 percent (see "Hydrology of Wetland Catchments," pages 23–24). This factor will in turn affect the amount of water entering a wetland and the nature of the sediments that are deposited, for faster flowing streams will carry larger loads of heavier sediments into the wetland. Tree removal from a catchment may take place as a result of human activity. Throughout the history of human agricultural activity, people have modified habitats to make them more suitable for the growth of their chosen crops or the grazing of their domesticated animals, and this has often entailed the removal of forests. In Europe, this process has been affecting natural temperate forests for 8,000 years or more, and many wetlands have been affected as a consequence. Indeed, certain wetlands, such as blanket bogs that spread over hillsides and valley mires, that are restricted to hollows, often owe their existence to catchment clearance by people (see "Blanket Bogs," pages 107–111).

The existence of a band of coarse sediment in the stratigraphy of a wetland profile, therefore, may imply climatic change, or it may indicate the destruction of the catchment vegetation. Vegetation destruction itself can come about through natural as well as human-induced causes. Wind can destroy areas of woodland and lead to subsequent erosion. Disease may cause extensive tree death, although this usually affects only one species of tree at a time, so the likelihood of whole catchment deforestation as a result is fairly remote.

Fire may be natural or human in its origin and is often involved in deforestation processes. If fire is implicated, then it can usually be detected in the wetland stratigraphy by the presence of layers of charcoal that show up as black bands in an exposed section or in a core. Charcoal particles are usually too small to be identified, so it is not normally possible to determine what types of tree or shrub have been burned. The size of the particles, however, can give an idea of whether the fire was local or whether it took place at some distance from the wetland itself. Large particles are unlikely to travel far by wind action, whereas small, dustlike particles of charcoal may travel great distances. By extracting the charcoal from a sediment and determining the proportions of the different-sized particles, it may be possible to determine the approximate distance of the fire. Studies of a number of neighboring sites can indicate the actual extent of a fire.

Deciding whether a fire was caused by lightning or by humans is much more difficult. If people are to blame, then the fire is often followed by a period of different land use, which may be apparent from the stratigraphy, particularly from the microfossils contained in the sediments (see "Microscopic Stratigraphy: Pollen," pages 40–43). Forest clearance by people is likely to be followed by the development of grasslands for grazing animals or by the plowing of land for arable crops. In either case, the pollen grains of grasses or of crop plants and weed species should be present in the sediments along with the charcoal, but the absence of such evidence does not preclude human activity. In the case of preagricultural human societies, for example, burning areas to drive game animals, or even to encourage fresh vegetation growth that will attract such animals as deer, was frequently practiced, but such a case is very difficult to distinguish in the sediment records from natural fire.

Often associated with charcoal in sediments are indications that wetlands have been enriched in their chemistry as a result of fire in the catchment. Just as mineral fragments are washed into wetlands following fire, so are chemical elements that stimulate the productivity of the wetland vegetation. Calcium, potassium, and phosphorus, all elements in demand by wetland plants, are often more abundant in postfire detritus than in that from other sources. These elements stimulate plant growth, including the growth of algal occupants of shallow lakes and ponds, which leave a record of their abundance in the form of very recognizable spores that become incorporated into the sediments. The filamentous green algae, such as *Spirogyra* and *Debarya,* are particularly characteristic of sediments deposited immediately after a catchment fire, and they respond to the flush of nutrients into the wetland—a process called *eutrophication.*

The stratigraphy of some coastal wetlands may be influenced by changes in sea level. Salt marshes, in particular, are affected in this way and may suffer erosion during high sea-level episodes, followed by regrowth when the marsh adjusts itself, by succession and renewed sedimentation, to the new conditions. More complicated stratigraphic sequences are found in the profiles of coastal peat bogs which have been periodically submerged by the sea during their history. These bogs usually occur in sheltered estuarine locations. A rising sea level floods the surface of the peat, leaving a band of silt to record the event. If the marine incursion is relatively brief, peat may continue developing over the top of these sediments, but if the sea-level rise is sustained, then the peats may remain buried beneath sea sand, submerged under the ocean.

The illustration shows the profile of an estuarine raised bog in western Europe. The mire has developed over an estuarine salt marsh as a result of sedimentation. Following the development of the raised bog peats, there has been a profound alteration in the height of sea level relative to land, with the result that salt water has flooded the bog and led to salt marsh formation again. In time, however, the bog vegetation has again taken over and raised the dome of peat yet higher.

Many areas of western Europe have submerged peat beds of this kind, both along the coast and farther out from the shore in the shallow waters of the North Sea between Britain and continental Europe. When this region was flooded by the sea some 4,000 to 7,000 years ago, forested swamps occupied much of the area, and fossil logs are still frequently washed up around the shores of the North Sea. Fishing boats sail over these sunken forests, in which the upright ancient tree stumps can still be seen. Many legends, including the Atlantis myth, may well have their origins in these submerged wetlands. The scientific study of the

Profile of an estuarine raised bog. A rise in sea level during the course of the bog's development led to marine flooding, leaving a band of marine clay within the peat deposits.

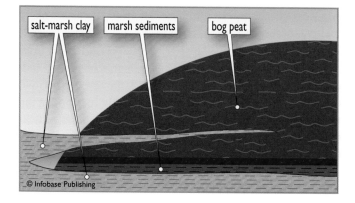

undersea wetland sediments, including the establishment of their dates of growth and submergence, is important for the reconstruction of the history of rising sea levels experienced in this region since the last ice age. Not only does this work supply an indication of how fast the ice caps melted as temperatures rose, but it also provides a record of the way in which the Earth's crust recovered as it was relieved of the weight of ice that had depressed it. The actual level of the sea relative to the land is the outcome of these two processes and is critical to the development of wetlands in coastal regions.

■ STRATIGRAPHY AND ARCHAEOLOGY

Of all the fossil materials found in wetland sites, the remains of human beings and their artifacts must rank among the most interesting. Just as we can trace the history and paleoecology of many plants and animals in the sediments of wetlands, so we can learn something of our own history and interaction within this distinctive habitat. The historical association of people with wetlands is discussed in chapter 6, but here the nature of the evidence, stratified in the peats and clays of the wetlands, will be considered.

Archaeologists interested in prehistoric societies and cultures are dependent on recovering either the remains of humans or their tools in order to reconstruct something of their way of life. In soils, where decomposition is generally efficient and where mixing of layers often takes place, the survival of such evidence depends on the existence of tough materials that have withstood the activities of chemical and microbial degradation. Bones may survive in alkaline soils. Stone tools and pottery are among the most resistant and therefore persistent materials, and metal artifacts may also survive if they are not chemically corroded. In the waterlogged deposits of the wetlands, however, other materials, including clothing and wooden implements, may also survive. The quality of preservation and the lack of disturbance in wetland sediments provide the archaeologist with great opportunities, and this has led to the development of the distinct science of wetland archaeology.

Bones are found only where the acidity of the environment has not dissolved away the lime they contain. The sediments of lakes, swamps, and fens may contain bones, but they are poorly preserved in acid bog peats. The bones and antlers of the great Irish elk (*Megaceros giganteus*), for example, are relatively common in the shallow lake sediments of Ireland, dated by radiocarbon back to the close of the last ice age, around 13,000–10,000 years ago. At least one of these specimens has been recovered in which a flint arrowhead is present, embedded into the leg bone of the elk. The stratified, datable evidence of these remains provides a firm basis for establishing the occurrence of hunting peoples in the area and also tells us something of the nature of their prey.

The more delicately preserved materials from bog peats include human corpses, which provide the most direct evidence possible of our ancestors. The discovery of human corpses in peat deposits is itself not new. Perhaps the first human corpse to be discovered in a bog was on June 24, 1450, in Germany. Peat cutters discovered the head of a man sticking out of the peat in a bog near Boensdoerp and sought the advice of their parish priest, only to be told that the victim had probably been lured to his death by bog elves and so could not be reburied in consecrated ground. Records of the recovery of other bodies in Germany, Denmark, and Ireland are fairly frequent through the 17th, 18th, and 19th centuries, and documentation indicates how finely preserved they were. The hair color was often discernible (usually dark reddish); the clothing could be identified (sheepskin caps, leather jackets, deerskin tunics); and the cause of death could occasionally be determined and was often violent (strangulation, cut throat, fractured skull). In more recent excavations, even the contents of the stomach have been examined and have revealed the nature of the victim's final meal (wheat, barley, oats, and seed of the goosefoot family, Chenopodiaceae).

The stratified position of a body in peat means that the vegetation and the environment of the time of death can be reconstructed from pollen in the deposits (see "Microscopic Stratigraphy: Pollen," pages 40–43), and the organic material intact in the body allows it to be dated very precisely using radiocarbon methods. Very many of the North European bog bodies date from the Iron Age and probably belonged to executed criminals or were possibly sacrifices to a wetland deity. In Ireland some corpses appear to have been staked to the underlying peat either before or after their death. Although much archaeological detail is available concerning the European bog bodies, the rituals surrounding their deaths must remain an area of speculation.

Wetland cemeteries containing bodies of those who have died naturally have also been discovered. In Florida, for example, at Windover, near Cape Canaveral, a 7,000-year-old burial pool has been excavated in which the bodies of more than 160 people have been discovered, each staked to the muddy bottom of the pool with tree branches.

The preservation of wood in peats has permitted the survival of many wooden tools and manufactured articles that would otherwise have been lost to the process of decay. Among the most remarkable discoveries have been the wooden roads and trackways of Europe and Asia. In 1818, a peat digger in the Netherlands discovered a buried roadway in a peat bog that consisted of a series of sturdy wooden logs laid side by side and stretching for a total of 7.2 miles (12 km). The wood has subsequently been radiocarbon-dated

Radiocarbon Dating

Geologists need to place a precise date on the horizons found in wetland stratigraphic profiles so that they can determine the speed of succession at its different stages. If there are fossils or archaeological artifacts embedded within a section, then these also need to be assigned to a particular point in history. If the sediments are from a lake that shows regular annual banding due to the presence of seasonal changes in sedimentation, features called *varves,* then these bands can be counted to calculate the date at any particular depth, but often this is not the case. Radiocarbon dating is an alternative method that is much more flexible in the types of sediment where it can be applied. It is based upon the analysis of carbon, which is an abundant component of all organic matter and originates from atmospheric carbon dioxide that has been fixed by plant photosynthesis. Carbon occurs in nature in three forms, or isotopes, ^{12}C, which is by far the most abundant type, accounting for 99 percent of the Earth's carbon, ^{13}C, which accounts for most of the remaining 1 percent, and ^{14}C, which is very much scarcer. ^{14}C is produced in the upper atmosphere when cosmic rays from the sun collide with nitrogen atoms. The ^{14}C atom is unstable and radioactive, gradually decaying to form stable nitrogen ^{14}N once again. The decay of ^{14}C occurs at a very steady rate, which is best described in terms of its half-life. This is calculated to be 5730 years (plus or minus 40 years), which means that in this period of time half of any given quantity of ^{14}C will have decomposed. So, an ancient particle of organic matter will contain less of the radioactive isotope than a modern sample, and the ratio of the two indicates the age of the ancient sample. The relatively short half-life of radioactive carbon, together with the scarcity of this isotope, means that samples more than about 40,000 years old have so little ^{14}C left that it is impossible to measure it accurately. The radiocarbon dating method can thus be used only for geologically recent materials, but it is very suitable for studying wetlands, most of which have developed during the last 10,000 years or so.

There is one defect in the radiocarbon dating method, and that is the assumption that the atmosphere has always held the same proportions of ^{14}C and ^{12}C. By dating samples of annual rings in ancient trees, whose age can be accurately determined by counting back from the present, it becomes obvious that this assumption about a steady carbon equilibrium in the atmosphere is not strictly correct. There have been times in the past when more cosmic radiation has resulted in rapid ^{14}C production and other times when it has been slower. The problem has largely been overcome, however, by accumulating a long series of radiocarbon dates for wood samples from annual rings, leading to the construction of a calibration curve for the radiocarbon method. A raw radiocarbon date is now recorded in uncorrected radiocarbon years, but once corrected it is expressed in regular, or solar years.

---

to 500 B.C.E. (the Iron Age). This was a time at which many sources of evidence (including peat stratigraphic studies—see "Stratigraphy and Climate," pages 34–35) point to a climatic change to colder, wetter conditions that must have necessitated the strengthening of any track across an increasingly wet peatland.

Similar trackways in southern England have been found to date from an even earlier period (the Neolithic, about 4,000 B.C.E.), and an examination of the branches used in these trackways has identified the tree from which they were cut (hazel, *Corylus avellana*) and the woodland management system used in their production. Hazel trees can be cut back almost to ground level, whereupon masses of new shoots grow from the stool. This sustainable form of wood production from hazel is termed *coppicing* and is still used in parts of Britain 6,000 years after its first record.

The presence of trackways, and even the wooden wheels of carts preserved in wetland deposits, show that these habitats were traversed regularly by travelers in prehistoric times. Some of the wetter sites may well have been valuable to people as a means of boat transportation, for wooden canoes have often been found in peat deposits. Even houses have been found, forming extensive wetland villages (see "Wetland Bronze and Iron Age Settlements," pages 182–184).

The attractions of wetland sites for archaeologists are twofold. First is the degree of preservation possible, especially for the soft tissues (such as the gut) of human remains, and for wood, which was such an important yet such a perishable material. Wooden tools, bowls, musical instruments, works of art, and talismans have all been found, well-preserved, in wetland sites. The second reason for the importance of wetland sites is the strict stratification of the sediments, which allows the precise reconstruction of the sequences of human cultures. These factors have combined to make wetlands a major archaeological resource.

MICROSCOPIC STRATIGRAPHY: DIATOMS

Macrofossils in the layered sediments of wetlands tell their own stories. These larger, visibly recognizable fossils mainly reflect the local changes in plant (and, to some extent, animal) life of the developing wetland. Microfossils, on the other hand, can provide information about both local and more regional events. Protist and algal microfossils generally record the conditions within the water body in which they lived, whereas pollen grains reflect a much wider environment. The stories told by all these microfossils are of importance if the full history and development of a wetland is to be understood.

Among the protists and the unicellular algae, many have relatively delicate cell walls surrounding their vegetative bodies and consequently succumb to decomposition soon after death. Only the spores, which are surrounded by a resistant coat, tend to survive as fossils. One group of protists is an exception to this, however, namely the diatoms, or Bacillariophyta. These are unusual organisms in that their unicellular bodies are surrounded by a shell made of silica (the same tough material that forms the basis of sand grains). Found in abundance in both marine and freshwater locations, they absorb dissolved silicon from their watery surroundings and construct their stiff coats from it. It is these coats that survive after the death of the cell within and sink to the bottom to be incorporated in the stratified records of the sediments.

The diatom coat (or frustule) is quite complicated in its structure. It actually consists of two sections, called valves, one of which is slightly larger than the other and fits over it like the lid of a box. The actual shape of the box varies; some are elongated or oval (the Pennales), and others are circular (the Centrales). The surfaces of the two valves of the frustule are delicately sculptured, often with radiating series of lines, and the elongate forms have a groove running down their center along which the cell contents flow, allowing the cell a degree of movement. The shape and pattern of sculpturing on the frustules permits their precise identification, which is important if they are to serve as indicators of past conditions.

Reproduction in the diatoms occurs in two ways. They can reproduce vegetatively by a form of binary fission. The two valves separate and a new valve is laid down, always inside the older valve. This means that the smaller of the two valves gives rise to a second valve that is smaller than itself. In this way, the population of diatoms gradually becomes smaller in its average size. The diminishing size of the diatoms is eventually interrupted, however, by a sexual reproductive phase in which two diatoms fuse, divide, grow, and produce a new, larger frustule once more.

Diatoms comprise an important part of the plankton of most water bodies, from large lakes to small ponds. They can also grow in the wet films surrounding rocks, on the surfaces of aquatic plants, on the wet muds on lakesides, or on estuarine mudflats, and each species has its own particular habitat preference and its range of tolerance to physical conditions. Diatoms fix light energy by the process of photosynthesis, so they all need light, but some are able to live in more shaded (or deeper) conditions than others. They take in their chemical nutrients from the surrounding water, and they differ in their particular chemical needs. In the case of water acidity, for example, some are able to tolerate more acid waters than others, just as some terrestrial plants prefer acid or alkaline soils according to species.

The main chemical elements needed by plants and protists, including diatoms, consist of nitrogen, phosphorus, potassium, calcium, magnesium, and, in the case of diatoms in particular, silicon. The different diatom species have varying requirements for these elements. Some, for example, need high phosphorus or nitrogen levels in their waters, and others manage better under nutrient-poor conditions. The acidity of the water (measured as the pH—see sidebar on page 20) is important because it influences how all of the other elements behave and how efficiently the diatom can absorb them from its surroundings. Consequently, acidity is one of the most important overall factors in the environment that affects the species composition of a community of diatoms. They are also influenced by the salinity of the habitat, some species being confined to saline waters and others being found only in freshwater. Water temperature is also important in determining their range, so their fossils can provide an indication of past climatic conditions. That diatoms are generally well-preserved, are identifiable to species level, occur in high numbers (so are appropriate for statistical counting and analysis), and are sensitive to environmental conditions makes them ideal subjects for study when trying to reconstruct the history of a wetland site.

One of the problems faced by many wetlands in developed countries in the Northern Hemisphere is that of acidification resulting from atmospheric pollution, especially from oxides of sulfur and nitrogen released by the burning of fossil fuels. Many wetland sites in northern Europe and Canada, especially those situated on acid rocks that are incapable of neutralizing acid rain, have been decreasing in their pH (that is, increasing in their acidity) in recent times. The process of acidification and, by historical implication, its possible causes, can be reconstructed using the diatom stratigraphy in the wetland sediments.

An example of this approach comes from a Scottish loch in which the diatom content of the top 34 inches (85 cm) of lake sediment has been analyzed in detail by a diatom expert, Rick Battarbee of the University of London, England. The diatom stratigraphy is shown in the diagram. In the

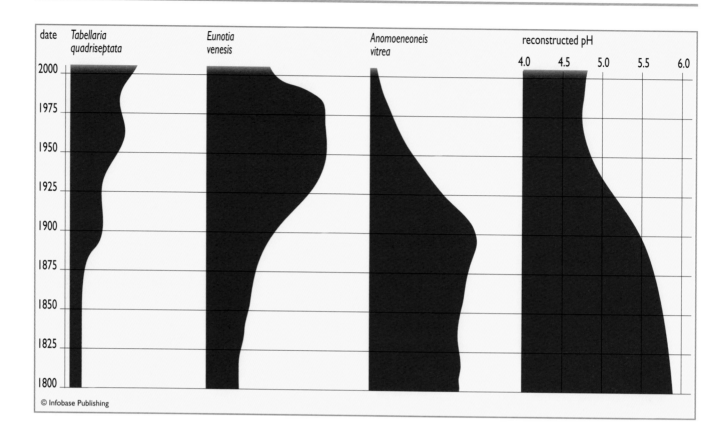

| date | Tabellaria quadriseptata | Eunotia venesis | Anomoeneoneis vitrea | reconstructed pH |

© Infobase Publishing

Diatom stratigraphy from the sediments of a lake in Scotland. Since about 1900 C.E. there has been a change in the composition of the diatom plankton community. Because the pH preferences of the different diatom species are known, it is possible to reconstruct the recent history of the lake in terms of its increasing acidity, shown on the pH scale.

sediments from below about 20 inches (50 cm) of depth (dating from the 16th century and earlier) the dominant diatoms in the fossil communities include mainly those that prefer higher pH (more neutral/alkaline) conditions. These include such types as *Fragilaria*, *Anomoenoneis*, and *Cyclotella*. This collection of diatom types suggests that the pH of the wetland at that time was above 5.5 (only weakly acid). Changes in the diatom composition begin after this time but become most apparent after 1900 when the types named above all declined sharply and the community became dominated by *Tabellaria* and *Eunotia*, diatom types that are more typical of acid waters. These changes suggest that the lake pH had fallen to about 4.8 (acid). The sudden acidification at this particular point in history provides strong circumstantial evidence for atmospheric pollution in the current century as the cause of the acidification of this particular wetland.

A similar investigation into wetland change has been conducted at Lake Washington, near Seattle, where the waters are becoming enriched with nutrients (eutrophication). The sediments of this site contain fossil diatom assemblages that indicate quite nutrient-poor conditions in this lake up to about 1850 (when the development of Seattle began) and then become moderately eutrophic until 1950, when diatoms that require high nutrient conditions (including *Asterionella* and *Fragilaria*) became much more abundant. The eutrophication of this site is clearly a very recent impact, probably resulting from increased human pressures involving the discharge of sewage effluents (untreated between 1910 and 1930) especially during the past half century.

Diatom microfossils, then, have great potential in tracing the development of wetland sites and in indicating the likely causes of current changes. Such work is of great importance if such sites are to be conserved and managed wisely.

◼ MICROSCOPIC STRATIGRAPHY: POLLEN

Pollen grains are the male reproductive cells of higher plants (including flowering plants and conifers), and they contain the genetic material that must be carried to the female of the species (see sidebar on page 41). The delicacy of this genetic material and the hazards of transport have resulted in the development of a particularly tough and resilient wall that surrounds the pollen grain, and it is this structure that survives the fossilization process and becomes stratified in lake sediments and in peat deposits.

Pollen Grains

Pollen grains are small, roughly spherical bodies ranging in size between about 0.008 and 0.04 inches (0.0002 to 0.001 cm). They are the male reproductive bodies of flowering plants (angiosperms) and the gymnosperms, a group of plants that includes conifers. Their main function is to carry the male genetic material from the anthers of the parent plant to the stigma of a receptive plant. Here, the pollen grain will germinate, producing a tube that penetrates the stigma tissue, travels down the style, and eventually reaches the ovary, where male nuclei from the pollen tube enter and one of them fuses with the female egg cell. The tissues within the pollen grain, therefore, are delicate and need to be protected from damage, particularly from desiccation, during their aerial journey. The structure used for this protection is a tough coat that almost completely envelops the reproductive cells, reducing the rate of evaporation of water and the risk of drying out. The chemistry of this coat is not fully understood, but it is believed to be a polymer of carotenoids (substances related to vitamin A) and carotenoid esters. It is a highly resistant material called sporopollenin that is even able to emerge unharmed after immersion in sulfuric acid, and this durable substance survives in a fossil form in lake sediments and peat. Bacteria can destroy sporopollenin in time, but only in the presence of oxygen; in the anoxic environment of wetland sediments, it can remain unchanged after thousands or even millions of years.

The coat of a pollen grain can be smooth or highly sculptured, depending in part on the pollination mechanism involved. Some pollen grains are dispersed by wind, and these usually have smooth surfaces, as in the case of the grasses, all of which are wind pollinated. Other pollen grains, such as thistles and sunflowers, are carried from one flower to another by insects, and these often have coats with complex sculpturing, having projections and spines on their surfaces. Botanists have put forward many ideas about how these projections might assist the pollen grains to adhere to the hairy bodies of their insect carriers. One possible mechanism is electrostatic forces that concentrate on the tips of spines and are attracted to opposite charges on the hairs of the insects. Whatever the mechanism, these elaborate sculptures on the pollen surface make the grains highly recognizable. Often it is possible to identify a pollen grain to a precise plant species, and most pollen grains can be identified to family level.

In addition to differences in sculpture, pollen grain coats also vary in the number and form of apertures over their surface. The presence of holes or cracks in the coat may seem surprising because the chief function of the pollen wall is the prevention of water loss, so apertures would appear to be disadvantageous. On arrival at its destination, however, the pollen grain needs to germinate by sending out a pollen tube, so an aperture is helpful in this respect. Also, since the pollen grain cannot control its orientation on the stigma, it is useful to have a few apertures so that one is likely to be in contact with the stigma surface. This still does not explain why some pollen grains may have dozens of holes in them. The likely reason for so many holes is that the pollen grain needs to communicate with the surface on which it lands. On arrival, the pollen grain exudes protein messengers through its pores, and these interact with proteins from the stigma, providing a "recognition" mechanism. Pollination is not always an accurate process, and it is quite possible that the pollen grain has landed on the stigma of the wrong species, in which case it needs to be rejected before the invasive tube begins its descent. So the pores provide communication channels so that any mistakes can be quickly rectified and the germination process inhibited. This rejection interaction with pollen is familiar to anyone who suffers from hay fever because pollen landing upon the mucous membranes of the sufferer produces its recognition proteins and the recipient responds with strong allergic rejection reactions.

The apertures and their arrangement, together with the surface sculpturing of pollen, are the features that make them highly recognizable to a scientist involved in pollen analysis, or *palynology*. The abundance of pollen in the atmosphere and the relatively close correspondence between pollen abundance and vegetation means that these tiny fossils, surviving in ancient sediments, preserve a record of the history of vegetation over wide areas.

Some pollen grains are produced in large numbers and are carried by the wind from the male to the female organ. These types will be widely dispersed and will be more commonly fossilized than those produced in small numbers and transported by insects. This differential production and dispersal of pollen has to be taken into account when fossil assemblages are interpreted, for the abundance of a pollen type is not a simple reflection of the abundance of the plant from which it originated. The distance of the pollen source must also be considered, for a frequent pollen type may mean that the plant is abundant at some distance or that it has a single individual source close to the site of preservation.

The transport of pollen to the source of sedimentation is also important. Whereas diatoms are all derived from the local body of water, pollen grains may arrive at a wetland by falling from the air above the site (or, more likely, being washed out of the air by rainfall) or may be carried into a wetland by input streams. Studies on the origin of pollen in water bodies suggest that up to 85 percent of the pollen entering a lake comes from streams and is therefore washed from the hydrological catchment. Only 15 percent comes directly from the air. This will vary, of course, from one wetland to another; a forested swamp may have a large proportion of its pollen input arriving directly from the plants growing on the site, whereas a saline wetland in a continental interior may have most of its pollen carried in by water. In a rain-fed, ombrotrophic bog, pollen largely arrives directly from the atmosphere, with a small input from the plants growing on the surface of the mire.

The collection of pollen grains found at any particular depth in a wetland deposit, therefore, contains elements of the local flora growing at the site, other elements that are derived from the surrounding catchment, and a further component that may have been carried hundreds of miles by atmospheric movements. Fortunately, this long-distance component is usually very small. From the stratified layers of fossil pollen in an aquatic or a mire site, a picture can be reconstructed of both the local changes in vegetation and more regional changes. The local changes may reflect successional developments, while the regional changes indicate the impact of climate change and land-use management by human populations on surrounding vegetation. Combining such records with those derived from macrofossils, diatoms, and other sources (such as sediment chemistry and geological stratigraphy) allows very detailed historical reconstructions to be carried out. Pollen grains are generally most useful in their record of more distant, regional changes rather than local ones.

The pollen stratigraphy of deep wetlands may record the changes in surrounding vegetation since the last glaciation or even earlier. It is possible, on the basis of pollen analysis, to follow the invasion of different trees as they adjusted their ranges to changing climate. In New England, for example, spruce was an early invader following the close of the last glacial episode, expanding its populations (and, therefore, its pollen input to wetlands) about 12,000 years ago. Beech was later, beginning to increase in abundance about 8,000 years ago and accelerating its population growth some 6,000 years ago. Hickory was still later in attaining importance,

dating from about 5,000 years ago. Similar information has been obtained from many wetlands all over the world, giving some very detailed pictures of the changing patterns of vegetation distribution in response to changing global climates. By dating the arrival of different trees at various wetland sites, as recorded by their fossil pollen, it is even possible to calculate how fast trees can migrate following climate change. Even large-seeded trees, such as oaks, are able to attain overall spread rates of about 400 yards (366 m) per year.

Early human impact can also be detected from the pollen record. In northern Europe, the arrival of Neolithic people with their agricultural methods (about 4,000–5,000 years ago) began a process of deforestation that has continued erratically right up to the present day. It is often the pollen of weed species, especially wind-pollinated weeds such as plantain, that provides the first clues to the arrival of agricultural peoples in a region. As forests were cleared, weeds invaded and flowered in the disturbed soils. Also, the overall abundance of tree pollen diminishes and is replaced by grasses as woodland is opened up and grazing animals are introduced to an area. The pollen of crop plants can sometimes be detected, including flax and hemp. Cultivated cereals are more difficult to identify on the basis of their pollen grains. Being members of the grass family, Poaceae, their pollen is similar to that of other grasses, having a smooth surface and bearing a single pore, but the pollen of cereals is generally larger than the wild grasses and has a bigger pore with a wider rim around it.

In North America, of course, the impact of intensive farming is much later in its first occurrence than in Europe and Asia, arriving with the European settlers and intensifying during the 19th century. The impact of farming cultures can again be observed in the pollen record (see the diagram on page 43) not only by changing abundance of trees but also by the presence of the weeds that accompanied clearance and cultivation. Ragweed, sorrel, and plantain, together with the grasses, can be used as clearly recognizable indicators of the arrival of intensive agricultural developments, often from about 1850 onward. Careful examination of deeper sediments, however, often reveals episodes of less severe impact on the forests, as shown by pollen of grasses and sometimes by the pollen of the crop plant, maize (*Zea mays*), which has a pollen grain that is much larger than most wild grasses. Often these minor clearance episodes are accompanied by an increase in charcoal in the wetland sediments.

The sediments that lie beneath wetlands, therefore, provide us with the opportunity of tracing not only the history of the wetland itself but also the changing environmental conditions of the surrounding areas. Such information is particularly important in understanding the development of landscapes and vegetation cover, especially for periods before historical records were kept. Even during historical

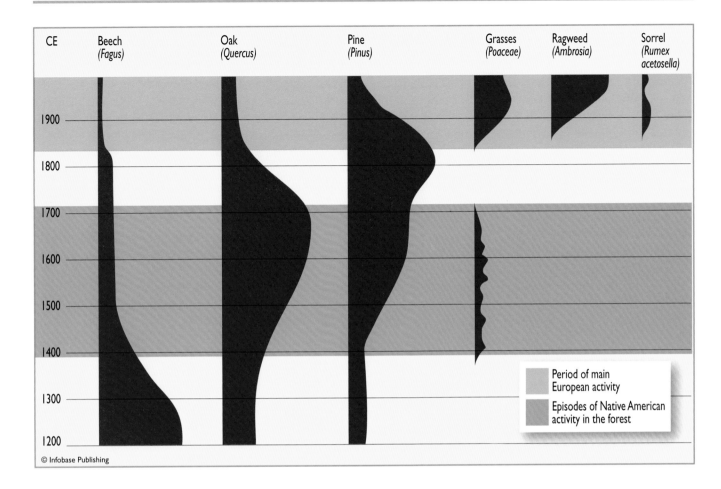

| CE | Beech (Fagus) | Oak (Quercus) | Pine (Pinus) | Grasses (Poaceae) | Ragweed (Ambrosia) | Sorrel (Rumex acetosella) |

1900
1800
1700
1600
1500
1400
1300
1200

| | Period of main European activity |
| | Episodes of Native American activity in the forest |

© Infobase Publishing

Typical pollen profile of the sediments from a bog in New England, North America. The breadth of the curve for each type of pollen grain indicates its relative abundance at different times in the past. The presence of grasses and weeds, such as ragweed, indicate periods of forest disturbance.

times, detailed information about the vegetation of an area or the pattern of landscape reclamation for agriculture is rare, so the archives of the wetland sediments provide an irreplaceable supply of evidence concerning our past use and misuse of the environment.

■ CONCLUSIONS

Wetlands are underlain by rocks of many different types, sedimentary, igneous, and metamorphic. Over very long periods of time, these rocks are linked together in a rock cycle in which sediments accumulate, are compressed into hard structures, and may then be raised up by tectonic forces. Once exposed to the atmosphere and the elements, the rocks are weathered, eroded, and leached, leading to their gradual destruction. As they are destroyed, particles and dissolved chemicals move through the activity of

water and are carried to wetlands where many particles are deposited in new sediments and where dissolved chemicals may be taken up by living organisms and circulate within the wetland ecosystem.

Wetlands form where water movement is slowed, its flow sometimes becoming almost stationary. This takes place where the landscape is gentle, where the water supply is adequate, and where the geology resists the soaking of water into the underlying rocks. Impervious rocks, such as slate, mudstone, and clay, often lie beneath wetland habitats. The pattern of water movement in a wetland catchment is important in determining the type of wetland that develops; both the geology and the vegetation of a watershed affect the quantity and movement of water into the wetland. Different kinds of sediment accumulate within wetlands depending on their location and the stage they have reached in succession. Sediments dominated by organic matter are termed peat, and the hydrology of this material is complex. The upper layer of peat, the acrotelm, is loosely compacted and has a high hydraulic conductivity, meaning that water flows freely though it. Below this is the catotelm, which is a mass of highly compacted and water-saturated peat, having a very low hydraulic conductivity. Water movement through the catotelm is much slower, so rainwater arriving from above is mainly diverted laterally over a bog surface rather than descending vertically through the peat mass, and this hydrological structure allows the development of large bodies of peat.

Wetland sediments contain evidence of the past history of a site. Particle sizes tell of the degree of energy and the speed of water moving through the habitat, and fossil materials in the sediment record the vegetation of the past. Fossils include large fragments of plants and animals, the macrofossils, and microscopic particles, such as diatoms and pollen grains, the microfossils. Some of the fossils are derived from the plant and animal communities that have occupied the wetland in the past, so the layers of sediment provide a full record of changing conditions during the course of succession. Other fossils, especially pollen grains, record vegetation change over much larger areas, so wetland sediments can provide evidence about the history of entire landscapes. Evidence relating to human history is present in the sediments, both directly in the form of archaeological remains and indirectly through the impact of settlements on wetland chemistry and catchment vegetation, which is reflected in the deposition of pollen and charcoal from catchment fires.

3

The Wetland Ecosystem

Every individual organism on the Earth interacts with its surroundings. A terrestrial organism, living on the surface of the land, is in contact with the atmosphere and may also be in contact with the soil, but it will also interact with other members of its own and different species. An individual organism, especially if it is a relatively advanced animal, will have social interactions with other members of its species, and a collection of individuals of the same species interacting in this way is called a *population*. An organism also interacts with other living creatures belonging to different species. Some of these may be its prey, while others may be its predators; some will be harmful parasites, while others compete with it for the same food resources. A collection of different species living together and interacting in all these different ways is called a *community*. Finally, a collection of living organisms coexisting in a community interacts with the nonliving world that forms a setting for all life. Plants are rooted in the soil, from which they absorb the minerals they need to grow, and they take up carbon dioxide gas from the atmosphere, which they convert into sugars in the process of photosynthesis. Animals take up most of the energy and the chemical elements they need in their food, which may consist of plant materials or other prey animals. Animals also drink water or absorb it through their surfaces and supplement their mineral intake in this way; some may even eat soil if they run short of certain elements. A community of different species of animals and plants living in the physical and chemical setting of the nonliving world is called an *ecosystem*.

WHAT IS AN ECOSYSTEM?

The concept of an ecosystem is an extremely useful one to ecologists and conservationists. It provides an approach to the study of the natural world that can be applied at a range of different scales. A single rotting log in a forest can be regarded as an ecosystem. Using this approach, one can study the ways in which the energy and mineral elements contained in the dead wood are decomposed by fungi and bacteria. These microbes are then eaten by invertebrate animals, which in turn may be fed on by carnivores, such as beetles, and these larger creatures attract visiting woodpeckers, which consume the beetles and thus harvest the energy that they contain. Each organism is obtaining energy from its food and uses some of this energy in such processes as growth and movement (see the sidebar on page 46). As the woodpecker flies away with its prey, it removes energy from the rotting log ecosystem and transports it elsewhere. In this example, the ecosystem concept is applied on a very small scale. It is possible, on the other hand, to regard the entire forest as an ecosystem, in which case the fallen logs are simply a part of a greater whole in which the photosynthesis of the trees is trapping the energy of sunlight, storing it in wood, and eventually providing an energy source for the microbes and animals inhabiting the decomposing materials, including the logs, of the forest floor. The woodpecker is now part of the same ecosystem, performing its own part in the organization of the whole.

Although it is possible to use the ecosystem idea at many different scales, all ecosystems have certain features in common. All ecosystems have a flow of energy through them. The source of energy for most ecosystems is sunlight, which is made available to living organisms by the photosynthesis of green plants. There are some bacteria that can photosynthesize, and there are some that can obtain energy from nonsolar sources, such as chemical reactions with inorganic materials, including the oxidation of iron, but these are generally of little significance in most terrestrial ecosystems when compared with the contribution of green plant photosynthesis. Some ecosystems, such as a mudflat in the estuary of a river, import energy from other ecosystems. Dead plant materials, rich in the energy derived ultimately from sunlight, are brought into this type of ecosystem, and these imported sources of energy and chemical elements supply the needs of the animals and microbes that feed on them. As energy flows through an ecosystem, one can distinguish certain groups of organisms that play

Energy

Energy is the capacity to do work. All animals, plants, and microbes need energy for their daily life because they need to grow, or move, or reproduce. When an animal forages for food, it uses some of its energy in the pursuit of yet more energy. When a plant absorbs elements from solution in the soil water, it expends energy in the process. Many of the biochemical processes that take place in individual cells demand the expenditure of energy.

Energy can be divided into two major types, kinetic and potential. Kinetic energy is the outcome of the motion and the mass of an object. Water running along a stream, a rock falling down a mountain slope, or the wind blowing in the trees are all exhibiting kinetic energy. Potential energy is stored in an object, ready to be released. A rock perched on the top of a cliff has the potential to release kinetic energy if it falls over the edge. Many chemical compounds contain potential energy in the bonds that hold the molecules together. A sucrose molecule, for example, will burn to release its component carbon atoms as carbon dioxide and releases kinetic energy in the form of heat as it does so.

Living organisms can capture the energy they need in either kinetic or potential form. A green plant, for example, absorbs electromagnetic radiation, which is a form of kinetic energy, when it intercepts sunlight. The green pigment chlorophyll is able to capture this energy and transfer it to chemical potential energy for use in the trapping of carbon dioxide molecules and reducing them to sugar. When a herbivore eats part of a plant, it removes energy-rich chemicals, including sugars and starch, which can then be transformed into animal tissues or respired to provide the kinetic energy needed for movement and other activities.

Energy occurs in many different forms and can be converted from one form into another. A burning piece of wood produces light and heat; falling water can turn a turbine and generate electricity. The movement of energy between its many forms obeys two fundamental laws, the laws of thermodynamics. The first law of thermodynamics is concerned with energy conservation. It states that energy cannot be created or destroyed, so each energy transfer can be described by an energy-budget equation that must balance. The second law of thermodynamics states that no transfer of energy from one form to another can be 100 percent efficient. A turbine can never capture all the kinetic energy released by falling water, and the electricity generator will never be able to convert all of the energy released by the rotation of the turbine to electricity. All energy transfers are accompanied by energy losses, often in the form of heat. When a person eats a potato, the energy contained in the stored starch is not completely transformed into human flesh. Wastage and losses may account for as much as 90 percent of the energy intake, and only a small proportion of the total energy is captured by the consumer. The second law of thermodynamics explains why the transfers of energy from Sun to plant, from plant to herbivore, and from herbivore to carnivore all involve losses and wastage.

different roles. The plants are *primary producers,* fixing solar energy into organic matter; they are said to be *autotrophic* in their nutrition, which literally means that they can feed themselves. Some of the energy they trap from the Sun is used in the energy-consuming activities of the plant, such as the uptake of chemical elements against a concentration gradient from the soil. Energy needed for such purposes is released from storage and is liberated by the process of *respiration.* The remaining energy is used in building new plant materials, leaves, stems, roots, flowers, and seeds. Herbivores are *primary consumers;* they are dependent on plants for energy, so they are said to be *heterotrophic,* meaning that they need to be fed by others. Predatory animals are also heterotrophic. They too depend ultimately on plants but indirectly because they feed on the herbivores or on other animals that eat herbivores. These are *secondary* and *tertiary consumers.* They occupy different positions in a hierarchy of feeding, sometimes referred to as a *food web* in the ecosystem. All of these organisms release energy as they require it by the process of respiration. The waste materials produced by living organisms and the dead parts or dead bodies of those individuals that escape consumption by predators and survive long enough to die a natural death are used as an energy source by the decomposer organisms. Animals that eat dead plant and animal materials and derive energy from them are called *detritivores,* but ultimately it is the microbial *decomposers* in the ecosystem, the bacteria and the fungi, that are responsible for respiring all the residual energy-rich materials in the process of decomposition. Nothing is wasted; nothing is lost. All the energy entering the ecosystem is finally used up and is dissipated as heat, released by respiration.

While energy flows through the ecosystem, chemical elements are cycled. Carbon atoms, for example, are taken up by plants in a gaseous form as carbon dioxide, and these atoms become incorporated into carbohydrates. They may be stored in this form or as fats, or that may be converted to proteins by the addition of the element nitrogen derived from the soil. Carbon compounds may also be converted into different kinds of molecules by the addition of other elements, such as phosphorus to make phospholipids, or sulfur to construct some types of amino acids, the building blocks of proteins. The materials built into plant bodies are consumed by animals, and a proportion passes through the body of the consumer to be voided as waste, while some becomes incorporated into the body of the animal. Respiration results in the release of carbon back into the atmosphere as carbon dioxide gas, while the nitrogen and phosphorus, together with other elements, are released by excretion or by death, and these enter the soil in the process of decomposition. Chemical elements are thus recycled and can be used over and over again. Energy passes once through the ecosystem and is finally dissipated, but chemical elements may cycle round and round the ecosystem indefinitely. The constantly turning wheel of element motion is called a *nutrient cycle*.

In addition to the cycle of nutrients, however, there is also usually an import and export of elements to an ecosystem. A stream entering an ecosystem will bring dissolved minerals from the ecosystems in the catchment area. Rainfall will also bring a supply of elements, its richness depending on how close the site is to the ocean. Animals may migrate into an ecosystem and bring elements from outside, such as the salmon that migrate up rivers to breed and die, bringing elements that they have collected in their ocean feeding grounds. Plant materials, including twigs and leaves, may be washed or blown into a pond, supplying a source of elements from another ecosystem. Just as these processes bring elements into an ecosystem, they can also take them out. Streams can leave an ecosystem, animals can move out, and plant material may be blown away. So the study of an ecosystem involves a consideration of its energy and nutrient imports and exports in order to understand the balance of supply and loss in both energy and mineral elements. Before an ecosystem can be fully understood it is necessary to calculate its energy and nutrient budgets.

HOW DO WETLAND ECOSYSTEMS WORK?

There are many different types of wetlands, each of which can be considered as an ecosystem with its own particular pattern of energy flow and nutrient cycle. There are, however, certain features that all wetlands have in common and that make them different from most other ecosystems. The most obvious feature is their wetness, which leads to unusual patterns of energy flow and storage. The main problem with an excess of water in an ecosystem is that oxygen may become difficult to obtain. All living organisms (with the exception of some very specialized fungi and bacteria) need a supply of oxygen so that they can respire. In the absence of oxygen, the energy contained in complex molecules cannot be completely released. When yeast (a type of fungus that can survive in the absence of oxygen) is supplied with sugar in an oxygen-free environment, it is able to obtain only part of the available energy by adopting a biochemical system called *anaerobic respiration*. Instead of oxidizing the sugar to carbon dioxide and water, it is only able to take the reaction as far as ethyl alcohol and consequently is not able to obtain all of the energy present in the sugar molecules. The energy present in organic compounds can only be tapped if oxygen is available so that the carbon from sugars and other sources can be fully converted into carbon dioxide and water.

Oxygen is readily available in air because it makes up about 21 percent of the atmosphere, but under water, oxygen becomes less easily available. Oxygen dissolves in water, and fast-flowing waters may be rich in dissolved oxygen, but when the water is stationary, oxygen movement is dependent on the process of diffusion. Diffusion is a kind of migration of molecules from areas of high density to areas of low density. Oxygen moves gradually from the surface layers, where the water is in contact with air, down into the deeper layers, where oxygen is being consumed by the decomposers in the mud at the bottom of the wetland. For a dissolved oxygen molecule, moving through water is a very slow process. In water, oxygen diffuses 10,000 times more slowly than it does in air, so as oxygen is consumed by the respiration of organisms in stagnant water, it is replaced only very slowly and is therefore often in short supply.

The most serious consequence of the slow diffusion of dissolved oxygen in water is that the decomposition of dead organic matter may be incomplete. In most ecosystems, all of the residual organic matter that falls to the floor and enters the soil is eventually decomposed and is lost, but in wetlands the slow decomposition can lead to an imbalance in the flow of energy, resulting in some energy-rich organic matter becoming permanently stored in the sediments of the ecosystem. It is this growing reservoir of organic carbon that has resulted in the formation of coal in the ancient wetlands and that leads to the buildup of mud and peat in many modern wetlands. Because of this imbalance, wetlands act as a "sink," or storehouse, for atmospheric carbon (see "Wetlands and the Carbon Cycle," pages 73–77). As a wetland develops and accumulates increasing quantities of sediments rich in organic matter, it builds up a reservoir of undecomposed material that is rich in energy.

The pattern of energy flow through a typical wetland ecosystem is shown in the diagram on page 48. The movement

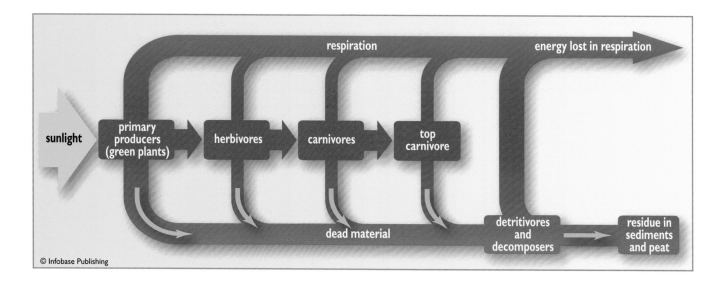

Energy flow in a wetland ecosystem. Energy enters the ecosystem as sunlight, which is used by plants to fix atmospheric carbon. Energy-rich carbon compounds are passed from one feeding level to another, and some of the energy is respired. Much of what remains enters the chain of decomposition and is respired by microbes, but a small residue accumulates as organic matter, forming peat.

of energy through producers, consumers, and decomposers follows the general pattern that is found in all ecosystems, but some of the energy fails to move through the entire system and remains stored in the organic matter of the sediments. This storage of materials and the energy they contain is what makes the wetland ecosystem different from all habitats.

■ PRIMARY PRODUCTIVITY OF WETLANDS

In wetlands, as in almost all of the other ecosystems of the Earth, the ultimate source of the energy that supports the mass and diversity of living things is the Sun. The trapping of the Sun's energy and its incorporation into organic compounds in the course of photosynthesis is the means by which living things take advantage of this resource. The amount of energy that the green plants of an ecosystem can tap is therefore influential in determining how much energy becomes available to the grazing animals, as well as to the detritivores and decomposers. When all of these members of the wetland community have taken what they require, the remains accumulate as the organic component of the wetland sediment, often in the form of peat.

The process of photosynthesis is carried out mainly in the canopy of the plants, that is, in the leaves, as well as the green stems and twigs held in well-lit positions. The precise amount of energy that can be trapped by an ecosystem is therefore dependent in part on the efficiency of the architecture of the canopy, as well as the leaf area that is exposed to sunlight. An extensive canopy with many leaf layers is more likely to be productive than one in which there are few layers of leaves. These leaves may be held vertically or horizontally and may be deciduous or evergreen (see sidebar "Why Be an Evergreen?" page 155).

The first stage in photosynthesis is the reception of light by molecules of chlorophyll in the chloroplasts of the cells. Energetically excited electrons return to their normal state by passing on some of their energy to molecules that are able to act as short-term storage reservoirs for that energy. Longer-term storage of energy, however, means building up more complicated and more stable sugar molecules, and these need carbon atoms for their construction. This carbon is acquired as carbon dioxide from the atmosphere, and this means that the plant must have open pores through which the gas can be absorbed. An inevitable consequence of open pores in the leaf is that water is lost in transpiration, but this is not normally a problem for wetland plants as water is one resource that is usually present in abundance.

The sugars initially built in this process are rich in energy and may be used by the plant as an energy resource for its own needs. When the plant absorbs elements from the surrounding water (taking elements from low concentration in the water and accumulating them in the plant body), it requires energy, as it does for the building of more complex molecules using organic and inorganic building blocks. So, some of the energy trapped in photosynthesis is used up in these processes, leaving the energy-rich materials that are evident to the observer, mainly plant biomass, to accumulate in the ecosystem. The observed increase in energy after the plant has taken its own requirements is called the *net*

Reed and cattail swamps are among the most productive ecosystems in the world. *(Jan Tyler)*

primary production. This is the energy input into the ecosystem that is available to consumer organisms.

Measuring the rate at which wetland vegetation grows is actually very difficult. Vegetation is naturally patchy, which means that sample areas for study have to be large and must be replicated in order to achieve consistent results. The most frequently used method of productivity measurement is harvesting plant material that has accumulated above ground over a particular period of time, but this method neglects root growth and also destroys the vegetation itself so that it cannot be measured again. Add to this the fact that insects and microbes are constantly taking their share of the productivity and that other parts of the plant may be dying and falling off, and it can be seen that values for the productivity of vegetation can only be an approximation. Some examples of the kind of figures that have been obtained by wetland ecologists are given in the following table.

A number of features emerge from these data. The Arctic and alpine wetlands are generally low in productivity, which is not surprising since they have a very limited growing season. Bogs have low productivity, even when they bear a forest cover. Reed beds and cattail marshes (see illustration), on the other hand, have a very high productivity, falling not far short of that of tropical rain forests. Tropical swamps have an even higher level of productivity.

The low productivity of bogs may in part be a consequence of their high latitude and restricted growing season, but this does not explain why they have lower productivity than marshes and reed beds at similar latitudes. It is more likely that the growth of plants on these sites is limited by the supply of nutrients, particularly phosphorus (see "Soil Chemistry in Wetlands," pages 65–67), which is often in short supply, being available to these ombrotrophic mires only through rainwater input. The fact that herbaceous marshes are often more productive than forested swamps in the same climatic zones is also difficult to explain. It is likely that tree productivity becomes limited because increasing proportions of the physical structure of a tree (such as trunk and branches) are nonphotosynthetic. Energy is

Primary Productivity Values for a Range of Wetlands in Pounds per Square Foot per Year in Dry Weight	
WETLAND	NET PRIMARY PRODUCTIVITY
Arctic sedge marshes	0.03–0.09 (0.1–0.3)
Alpine sedge meadows	0.03–0.12 (0.1–0.4)
Temperate raised bogs	0.09–0.21 (0.3–0.7)
Raised forested bogs	0.13 (0.5)
Temperate forested swamps	0.12–0.35 (0.4–1.2)
Canadian sedge marshes	0.13–0.29 (0.5–1.0)
Temperate reed beds	0.29–0.73 (1.0–2.5)
Florida saw-grass marsh	0.41 (1.4)
Cattail marshes	0.13–0.73 (0.5–2.5)
Tropical papyrus swamps	0.59–2.93 (2–10)

(figures in brackets are in kg m^{-2} y^{-1})

being invested in nonproductive tissues. In the cattails and reeds almost all aboveground parts of the plant are green and contribute to the process of energy trapping. Different types of plants thus vary in the ways they allocate new material to various parts of the plant body. This division of the products of photosynthesis between different parts of the plant is called *resource allocation* (see sidebar).

The most productive wetland ecosystems are not necessarily the most diverse in either their plant or animal components. Reed beds, in fact, are often virtually a plant monoculture. So, although productivity may seem to be an attractive aim for a conservationist, it is not always the most important goal, for productive ecosystems can sometimes be species-poor.

Productivity does, however, provide energy for other members of the ecosystem and can support higher levels of animal biomass. The factors that limit the productivity of a wetland ecosystem may include temperature, for if this is too low the enzymes in plant cells fail to function efficiently. This restricts the growing season of Arctic mires even within the period when light intensity is adequate for

growth. Alpine mires are similarly restricted by the cold, but they tend to have higher daytime temperatures than those in the Arctic because of their lower latitude and hence the higher angle of the Sun. Nutrient availability seems to be the other main limiting factor in wetland productivity; rheotrophic mires are generally more productive than ombrotrophic ones in the same climatic region. Where nutrients such as phosphorous have been added to ombrotrophic mires, their productivity has increased, but this also results in a change in the plant species composition as a result of which more nutrient-demanding species assume dominance.

■ FOOD WEBS OF WETLANDS

The energy trapped by wetlands in the process of photosynthesis is available to animals, but there are many different ways in which these animals can avail themselves of that productivity. Some consume the plants themselves. For example, butterfly caterpillars may eat leaf tissue, stem-

Resource Allocation

As a plant gains in weight, the newly acquired material must be moved to appropriate parts of the plant body. In the early stages of plant growth, the new products of photosynthesis may be directed toward the construction of new leaves, in which case all of the investment is used for the development of productive organs, in which more photosynthesis can occur. However, root development must keep pace with the growth of the shoot, so that the plant remains stable in the soil and a supply of water and mineral nutrients is maintained for continued transpiration and growth. As the shoot continues to grow, it needs extra support, so, instead of making photosynthetic organs, such as leaves, the plant must expend an increasing amount of its energy on support structures, including stems and branches. In the case of shrubs and trees, support tissues include wood, which holds the plant relatively rigid and assists in the development of a complex canopy, enabling the leaves to reach the light and continue their productive function. This division of new material among different parts of the plant is called *resource allocation*.

As plants grow, therefore, they must devote an increasing amount of their energy to nonproductive support tissues, especially roots and rigid stems. In its early stages of growth, a plant invests energy almost

at a compound rate of interest. This means that almost every portion of new material is reinvested in productive, photosynthetic tissue, so the growth is potentially logarithmic, but the older the plant becomes, the lower the proportion of new production (the "interest" gained) can be reinvested in this way. As more support tissues and roots are needed, the growth rate falls from a logarithmic one and becomes slower, eventually reaching a plateau as the plant reaches its maximum size.

Eventually the plant begins to reproduce, and the investment of new material is directed away from both productive and support tissues into the development of flowers and then fruits. In many annual plants, the proportion of energy invested in reproduction is very high because such species rely heavily on seed production for survival into the next growing season. Long-lived perennials devote proportionally less energy to seed production.

Grazing animals, which rely on plant products for their food, thus have a number of options for feeding. They may attack leaves directly or may feed on roots or timber. Alternatively, they may gain energy from nectar, pollen, or the seeds that are subsequently produced by the plant. Resource allocation by plants thus presents consumer organisms in the ecosystem with a range of feeding strategies.

Elk often graze at the edge of marshes, taking advantage of the high productivity of sedges.   *(Robert Karges II, U.S. Fish and Wildlife Service)*

boring insect larvae may consume the inner parts of stems, nematode worms may feed on roots, and beetles may eat the flowers and fruits. Each part of the plant may thus act as a support to certain types of animals. Larger, vertebrate grazers, including such creatures as hippopotamus, waterbuck, elk (see photograph above), hoatzin, and proboscis monkey, eat many different parts of the plant. Microscopic zooplankton in the water ingest the entire bodies of the photosynthetic unicellular phytoplankton that are responsible for much of the aquatic production. Other groups of animals may concentrate on the dead parts of plants that fall into the water or to the peat surface. Detritivores, such as the annelid worms, consume dead plant organic matter and also ingest the fungi and bacteria, which are in turn deriving energy from the decaying materials. Occasionally the food web has links that reverse the general trend, as

in the case of carnivorous plants (see photograph below), which supplement their normal photosynthetic processes by digesting and assimilating insects.

All of these plant-feeding animals have their own parasites and predators. Caterpillars are often the victims of parasitic wasps, which lay eggs in their living bodies. These hatch and the wasp larvae feed on nonessential organs until, when the caterpillar pupates, they kill the host and emerge. Herbivorous invertebrates may also suffer a more rapid demise if they are consumed by insectivores, such as sedge warblers, marsh wrens, frogs, or water shrews. These, in turn, are a food resource for higher predators such as herons, snakes, harriers, or terns (see photograph on page 52). Different species are thus linked in webs of feeding relationships, which can be very complex. Examples from different types of wetlands are shown in the diagrams.

Insectivorous plants, such as this greater sundew (*Drosera anglica*), reverse the normal flow of energy and nutrients in a bog ecosystem. By trapping and digesting insects, they become predators and part of the consumer food chain. *(Peter D. Moore)*

Royal terns are fish-eating birds of open-water wetlands that obtain their food by flying above the water and diving on their prey from a height.   *(South Florida Wetlands Management Department)*

It is convenient to think of each animal as occupying a particular feeding level (or trophic level) in this scheme, including such roles as herbivore, first-level predator, second-level predator, top predator, and so on. This is a simplification of the real state of affairs, as can be seen from the examples shown. A frog, for example, may consume a herbivore, a carnivorous insect such as a water beetle, or even a detritivore such as a worm. It cannot be simply assigned to one feeding level. This is why it is better to think of ecosystems as having interlacing "food webs" rather than linear "food chains." It is possible, on the other hand, to trace chains of energy movement through the food web, each chain having a series of links.

A single chain from the web rarely contains more than about five links. For example, in the Okavango diagram (page 53, bottom), the sequence: phytoplankton ➔ zooplankton ➔ insect larvae ➔ fish ➔ fish eagle, has five links, but this is a very long chain. The limit to chain length is determined by the amount of energy lost at each transfer. In the aquatic ecosystem, only about 10 percent of the energy in an organism is actually incorporated into the body of its predator—the rest is lost. This means that very little energy survives in an animal's body after a course of five transfers.

All three of the examples of wetland food webs shown here are highly simplified. Such categories as "crustaceans"

Simplified food web of a temperate shallow pool with slow-moving waters. Energy flows through the ecosystem, and some energy is lost at each point of transfer.

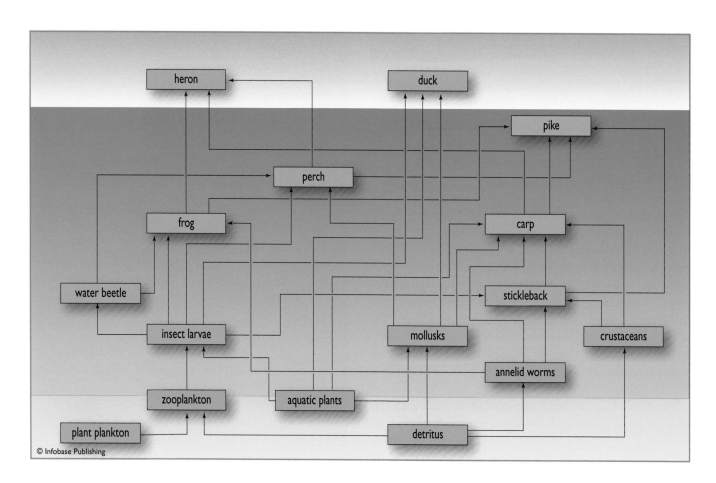

© Infobase Publishing

ity present in the tropical swamp as compared to temperate freshwater wetlands. The fact that energy is lost during transfers means that a high initial productivity in vegetation is likely to support more transfers as well as a greater bulk (biomass) of consumer organisms.

Similarly, the shallow pond wetland food web is more complex than the fast-flowing stream system, and this is also because of the differences in initial energy input. Still waters support more submerged aquatic plants and plankton, while fast-moving waters have a relatively low photosynthetic energy input, partly from detritus arriving from other ecosystems (leaves, etc.) and partly from algae attached to stone surfaces. An ecosystem with poor productivity, therefore, is likely to have less complex and lower biomass consumer food webs. This is true also of the temperate bogs, where plants such as bog mosses have low productivity and animal life is restricted both in biomass and diversity.

One consequence of the inefficiency of energy transfer in ecosystems is that the amount of energy available gradually decreases as it moves up from one trophic level to the next. There is a larger quantity of biomass of vegetation in a cattail marsh, for example, than there is of herbivores.

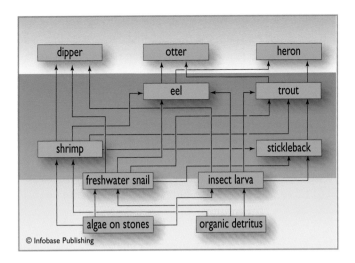

Simplified food web of fast-flowing wetland. Plankton is less important because it is constantly carried away by the flow. Only algae attached to stones avoid removal. Organic detritus, however, may be imported from other ecosystems.

and "insect larvae" are comprised of many species, so the complexity of real food webs is much greater than can be portrayed in a diagram of this kind. It is apparent, however, that the Okavango Swamp food web has more representatives of "high" trophic levels as well as more large-bodied, vertebrate herbivores than the temperate sites. This is due to the higher levels of primary (photosynthetic) productiv-

Simplified food web of a tropical swamp. This example is based on the Okavango Swamp in southern Africa. Tropical wetlands are rich in their biodiversity, and their food webs are consequently complex.

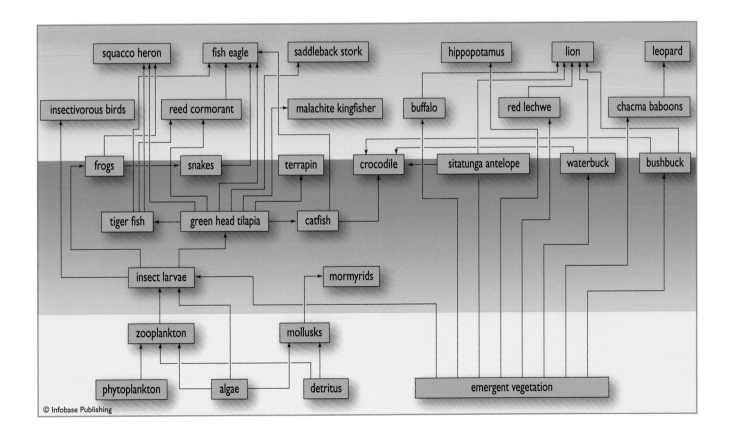

Similarly, if one could gather together all the herbivores in the marsh, from caterpillars to deer, their combined mass would be several times greater than that of their predators. The loss of about 90 percent of the energy at each stage of transfer means that the biomass present within each trophic level declines with each step up the food web. By the time these resources reach the top predator level (in a cattail marsh, this may be a great egret, as shown in the photograph below, or an alligator), there is only a very small quantity of living material present per unit area of the marsh. This arrangement is known as a *pyramid of biomass*. The energy pyramid is broad at the base, supported by the mass of vegetation, and becomes smaller at each trophic level until the top predators form the very small amount of biomass supported at the summit. This energy structure is common to most terrestrial ecosystems, including the wetlands.

In the case of wetland ecosystems, however, there are certain types that do not obey the normal rules. The open-water aquatic stages in the wetland succession often have an inverted pyramid of biomass. In a pool that lacks any large aquatic plants, the primary production is carried out by microscopic algae, the phytoplankton. These are consumed by microscopic animals, the zooplankton, which are then eaten by insect predators and fish. If it were possible to collect and weigh each of these trophic levels, the outcome would reveal that the fish biomass is greater than the zooplankton biomass and that the zooplankton outweighs the phytoplankton. So the pyramid of biomass is inverted. The reason why this state of affairs can be maintained and remains stable is the fact that one-celled phytoplankton reproduce at a very high rate. The cells are constantly dividing and replicating, replacing the cells that are being

The great egret is a fish-eating wetland bird with a very wide global distribution, being found in North and South America, Europe, Asia, and Australia.   *(South Florida Wetlands Management Department)*

consumed by the zooplankton. The small size and the fast reproduction make up for the fact that the total biomass of primary producers present at any given time is small. In the terrestrial stages of wetland ecosystem succession, however, the primary producer biomass is high and the standard pyramid of biomass model applies.

Another question that has long occupied the minds of ecologists is whether ecosystems with high diversity and high food-web complexity are more stable than simple ones. In other words, are they more robust and able to withstand disruption such as fire, drought, disease, or the loss of certain species? The intuitive response to such a problem is to suppose that complexity provides stability, for an animal with several choices of possible prey may switch from one to another if the first becomes scarce. Theoretical work with computer models, however, has proved confusing and indicates that complex ecosystems can easily become chaotic, losing their stability. To solve this dilemma, the obvious approach is to experiment with real ecosystems, observing the effect of removing species one by one. Such studies are now being conducted (mainly using very simple ecosystems) and have so far confirmed the original hypothesis, that complexity and diversity in an ecosystem render it more stable.

■ DECOMPOSITION AND PEAT GROWTH

All ecosystems follow a general pattern of energy relationships. The Sun's energy normally is fixed by green plants and stored in the chemical form of complex carbon compounds, built from the carbon dioxide (CO_2) absorbed from the atmosphere. These energetically rich compounds are then either consumed by grazers or die and enter the decomposer chain, where detritivorous animals, together with decomposing microorganisms (fungi and bacteria), tap these energy resources within the dead material and eventually transform them back to carbon dioxide that reenters the atmosphere to continue the cycle. In a mature ecosystem, the entire process is in equilibrium; the amount of CO_2 released to the atmosphere by respiration and decomposition is precisely the same as that originally absorbed, so the quantity of organic matter in the ecosystem (either in the form of living plants and animals, or as dead matter in the soil) remains constant over time. Ecosystems that are immature and still in the course of succession may still be growing in their organic matter content (for example, as living biomass), so the equilibrium has not yet been attained.

In the case of peatland ecosystems, the organic matter in the "soil" (that is, the peat itself) continues to accumulate even when the vegetation has achieved an equilibrium state (steady biomass). This makes peatlands unique in their energy relationships. In a sense, they are always in a state of disequilibrium, always accumulating more carbon than they are releasing, and always increasing in the organic matter content of the ecosystem. This fact, incidentally, makes peatlands a "carbon sink" and means that they are important absorbers of the additional carbon that humans put into the atmosphere as we combust fossil fuels (see "Wetlands and the Carbon Cycle," pages 73–77).

The difference between peatlands and "normal" terrestrial ecosystems lies in the decomposition process. In a deciduous woodland, for example, leaves accumulate on the ground each fall, but these have generally disappeared by the following spring or summer. In the peatland, a considerable proportion of each year's litter remains, not only into the following year but even into the following century or millennium. Exactly how much survives depends on a number of factors, among them the nature of the litter, the oxygen supply, and the amount of water present in the peat. All of these are factors that influence the activity of the decomposers, the fungi and bacteria responsible for the breakdown of the litter.

A peat profile, exposed by peat cutting, reveals that the surface layers—the top 8–15 inches (20–40 cm)—are more loose and friable, are less compacted, and have a structure that is richer in cavities (through which air and water can move) than is the case with the lower layers. The importance of this distinction between surface (acrotelm) and deeper (catotelm) layers in peat with respect to hydrology is discussed on pages 24–27, and the differences in structure are also of profound importance in the process of decomposition. These layers are shown in the accompanying diagram.

The changing levels of oxygen availability with depth can be illustrated by the following simple experiment. If a polished silver wire is inserted into a peat deposit, left for a few minutes, and then withdrawn, the upper part of the wire remains untarnished, but the lower part is blackened by a coat of silver sulfide. It is therefore obvious why the catotelm of a peat deposit has sometimes been referred to as the "sulfide zone." The cause of this staining effect is the lack of free oxygen down in the deeper layers of the deposit, and consequently sulfur is present in the form of sulfide (reduced) rather than sulfate (oxidized), and in this form it reacts with the silver wire.

The catotelm layers are permanently waterlogged and tightly compacted by the weight of peat above, so very little oxygen is able to penetrate below the acrotelm. This affects the chemistry of the catotelm (as the silver wire experiment shows), and it also affects its biology, for most fungi and many bacteria can exist only in the presence of oxygen, which is why the sealing of foodstuffs in an oxygen-free atmosphere—usually one of carbon dioxide or nitrogen—is an effective strategy for food preservation and longevity.

Greatly reduced microbial activity means, of course, that the rate of decomposition is also curtailed.

The British wetland ecologist R. S. Clymo conducted a simple experiment to test the idea that decomposition might take place at different rates according to depth in a peat deposit. He took samples of bog moss, weighed them, enclosed them in muslin bags (the holes in the fabric are small enough to keep the moss in but also permit the entry of microbes), and carefully buried samples at different depths in a peat profile. He recovered them three months later and was able to show that weight loss (decomposition) in the top 8 inches (20 cm) (the acrotelm) was about 5–10 percent, whereas losses in the lower, catotelm layers were very much smaller, generally less than 1 percent. So the acrotelm is indeed the location of the bulk of the decomposition in peat.

These observations are of importance in considering what parts of plants are likely to survive as fossils in peat. Roots penetrate into lower peat layers, so when they die they are less likely to experience decay than the shoots and leaves that are deposited on the surface of the peat. The constant accumulation of plant litter means that, in time, the peat surface rises and, with it, the permanent water table and the oxygen-free zone. So even superficial materials will eventually enter the sulfide zone, with its relative security from decay, but the longer a fragment resides in the upper, aerated zone,

The upper layers of a peat profile. Plant and animal detritus falls on the surface and much of it decays, but what remains accumulates as peat. The top layer is loosely compacted and is penetrated by air channels; this is called the acrotelm. The lower layer is dense, compacted, and devoid of air; this is the catotelm.

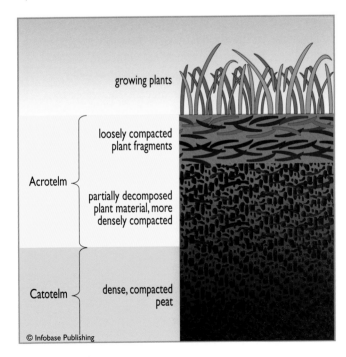

One other consequence of the pattern of decomposition at different depths in the peat is that an increasing thickness of peat leads to an increased total decomposition in the growing peat column. There comes a time when all additional input of peat is matched by decomposition, so the bog ceases its growth (see sidebar).

■ ELEMENT CYCLING IN WETLANDS

All organisms are made of chemical constituents derived from their environment, and the processes of growth and tissue replacement mean that these elements must be constantly available. Animals derive much of their chemical needs from food, whether from other prey animals or from plant tissues, but they also receive some directly from their nonliving environment, that is, from drinking water and sometimes even from soils and rocks. Most plants fill all their chemical needs from these inorganic sources, an exception being the carnivorous plants (see "Plants in Bogs: Carnivorous Plants," pages 151–154). The chemical constitution of the environment is, therefore, of great importance in determining what kinds of plants and animals can exist within a habitat.

Plants take up their chemical elements (apart from the carbon and oxygen used in photosynthesis and respiration, respectively) in solution from the water that is normally available in a wetland ecosystem. The sediments can also be a source of elements, but these nutrients must first be dissolved in water in order to be absorbed. The water of wetlands is thus ultimately the main source of chemicals for all of the living things in the wetland ecosystem. When dissolved in water, many elements are found as charged particles, called *ions,* the positive or negative nature of the charge being determined by the element involved. Thus, common salt, sodium chloride (NaCl), dissolves in water to give positively charged sodium ions (Na^+) and negatively charged chloride ions (Cl^-). Positively charged ions are termed *cations,* and negatively charged ions are called *anions.*

Among the most influential ions found in natural waters is the hydrogen ion (H^+). The concentration of hydrogen ions in water determines its acidity—the higher the concentration, the greater the acidity. Acidity is conventionally expressed on a negative logarithmic scale, called pH, where the higher the pH value, the lower the acidity (see "Acidity and pH," page 20).

Calcium ions (Ca^{2+}) are found in abundance only in rich fens, and their concentration is often inversely related to that of hydrogen ions. All plants and animals need calcium for the functioning of their cell membranes, but some need the element in large quantities, as in certain aquatic plants that

the more subject to decay it will be. More resistant items, such as tough fruits and seeds, survive their stay in the aerated zone better than the soft parts of leaves and flowers, so they are more commonly found in the peat as fossils.

become coated with lime and those animals that build bones or shells made of calcium carbonate. Snails, for example, live only in wetlands with an adequate calcium supply so that they can build shells, whereas slugs, having no shells, can exist even in the most acid, calcium-poor bogs.

Sodium ions (Na^+) are among the most abundant of ions in nature. In coastal wetlands they can dominate the chemistry of the site, as is also the case in inland saline sites (see "Inland Saline Wetlands," pages 116–118). Sodium is probably not required at all by plants, although most plants contain some simply because of its ubiquity. Some aquatic plants have quite large concentrations of sodium, although its biological significance within them is not understood. Animals, on the other hand, need sodium for nerve function and, where other sources of sodium are poor, they may graze on aquatic plants to supply their needs. This is probably why moose so often spend the early summer grazing in wetlands, as shown in the illustration.

Sulfate ions (SO_4^{2+}) are often associated with low pH (acid) conditions in waters, being derived from the solution of oxides of sulfur from the atmosphere to form sulfuric acid. Sulfur is an essential element for plants and animals because it is constituent of certain amino acids, the building blocks of proteins, but very high sulfate concentrations can be damaging to life, especially to the photosynthetic apparatus of plants. When conditions are anaerobic (that is, when there is a lack of oxygen dissolved in the water), the sulfate ion may become reduced to sulfide ions (S^{2-}). These ions may combine with iron to form ferrous sulfide (the black compound found in the detritus of stagnant ponds) or with hydrogen to form hydrogen sulfide, an obnoxious, smelly gas.

Magnesium (Mg^{2+}) is a common element in seawater and is therefore abundant in wetlands close to the sea. It is essential to plants because of its involvement in the construction of chlorophyll, and it is needed by animals for bone structure. Two other important elements for plants and animals are potassium (K^+) and phosphorus (which can occur in a variety of ionic forms), and the concentrations of these two ions are very variable in different wetland types. They are both derived from the weathering of rocks, so their abundance is often related to catchment geology (see "Geology of Wetland Catchments," pages 27–29), but they can also occur in abundance as a result of human land use and pollution. Phosphorus can be present and yet be very difficult for plants to obtain when the pH of the water is high (alkaline).

Nitrogen in the form of nitrate ions (NO_3^-) is also an element strongly influenced by human activities. Like potassium and phosphorus, it is vital for all plant and animal life and is therefore a major component of agricultural fertilizers. Unlike the other two, it can also be fixed from the atmosphere (where it is abundant) by the activity of both free-living bacteria and by microbes associated with certain plants (including several wetland species). It is an element that is often in short supply in salt marshes and in ombrotrophic bogs.

Iron may be present as Fe^{3+} when there is an abundance of oxygen in the soil but is reduced to Fe^{2+} when oxygen is scarce, as in permanently waterlogged, stagnant habitats.

Young moose grazing on aquatic plants at the edge of a marsh. Many aquatic plants, including the mare's tail (*Hippuris*) shown here, accumulate sodium, which is sometimes in short supply in inland habitats. *(Carolyn McKendry)*

This is because certain bacteria use Fe^{3+} as an oxidizing agent or an alternative to oxygen. It may prove toxic to some plants in this latter form and may exclude them from this extreme environment. Other elements may also prove toxic to sensitive species, particularly under acid conditions, such as aluminum (Al^{3+}) and zinc (Zn^{2+}). One of the reasons that acid rain and acidification of wetlands has proved so damaging to many plant and animal species may be related to their sensitivity to aluminum toxicity, as this element is more soluble under acid conditions.

In nature, the elements that compose living plants and animals are constantly being recycled. An atom of nitrogen taken up by a wetland plant, for example, may become incorporated into proteins and used within the plant as an enzyme for metabolic reactions or for the construction of membranes. When the plant, or part of the plant, dies, it will be broken down by fungi and bacteria and reused in the proteins of these microbes. If the plant is eaten by an animal, on the other hand, then the protein is digested and the nitrogen-containing amino acids will be absorbed and made into new proteins within the animal. If the nitrogen exceeds the animal's requirements, it will be excreted back into the environment and reenter the waters of the wetland to be taken up by plants once again.

All ecosystems have such elements cycling within them, but their precise patterns vary according to the nature of the physical environment and the types of plants and animals present. Even within different types of wetland, the patterns of nutrient cycling vary considerably, depending on the hydrology of the wetland and the precise element involved. A generalized scheme is shown in the diagram on the facing page.

The chemicals needed for living organisms are contained in the waters of the wetland, and from here they are available to the vegetation, but there are different ways in which these elements actually arrive in the mire water. The most obvious source is rainwater. The rain and snow that fall through the atmosphere are never pure but contain dissolved elements that have been taken up as the water droplets pass through the air. Gases, such as carbon dioxide, sulfur dioxide, and oxides of nitrogen, dissolve in the falling water, together with fine particles of solids, including salts carried from the sea and dust picked up from soils, fertilizer applications, and industrial pollutants. This collection is known as the *aerosol*, and precipitation brings this "cocktail" of substances directly to the wetland. This source of elements is common to all wetlands, both rheotrophic (groundwater-fed) and ombrotrophic (supplied only by rainfall).

The rheotrophic wetlands also receive additional supplies of elements in drainage water entering the mire through streams and surface flow, together with underground soaks from the deeper aquifers within the rocks of the catchment. This water includes rainwater collected from the wider region of the catchment and water that has passed through

(opposite page) Nutrient cycling in a wetland ecosystem. The main sources of nutrients are stream input and precipitation. In the case of ombrotrophic wetlands, precipitation is the sole source of incoming elements.

vegetation canopies and through soils on its way to the wetland and has consequently been enriched by those elements dissolved during this passage. In catchments where human activity is significant, the water may also receive an input of agricultural additions. Fertilizers and pesticides used in agriculture may be carried by groundwater flow from the fields into the wetland. The waste products of domestic animals may add to the nutrient load of the water, and changes in land use, such as when grassland is plowed for arable agriculture or when forest is cleared, may also result in the enhancement of the nutrient supply in drainage waters. There is also the direct impact of humans, both in the form of road and street drainage (including salts from treated roads and detergents from washed vehicles) and the input of human sewage, either in a treated or untreated form.

The supply of drainage water to a wetland can thus add significantly to the nutrient supply, especially where human activity is involved. Rheotrophic wetland systems are thus generally richer in nutrients than the rain-fed ombrotrophic systems. The management of a rheotrophic wetland involves the control of the entire catchment if the supply of nutrients is to be regulated effectively. Such catchment management is not as critical for the conservation of raised, ombrotrophic mires.

The concentration of elements within different mires will, therefore, vary considerably depending on the pattern of water movement and the source of drainage water. The supply of elements to any particular plant is partly dependent on this concentration, but it also varies with the speed at which water moves past it, replacing those elements that plants have taken from the water. A plant growing in a mire where the element concentration is low but the flow rate is fast may be as well supplied with nutrients as one growing in richer but more stagnant waters. The consequences of these two factors in creating patterns of vegetation within mires is best seen in the valley mires (see "Valley Mires," pages 95–97), but they can also be observed in the complex mosaics of blanket mires.

Plants expend energy in accumulating the elements that they particularly need, therefore they often conserve their reserves and protect them from loss. In a nutrient-poor mire, for example, plants often mobilize such elements as potassium and phosphorus as leaves age and die and move them to the growing, nutrient-demanding parts for reuse before shedding the old leaf. In nutrient-rich environments, such conservation is not so necessary, although phosphorus

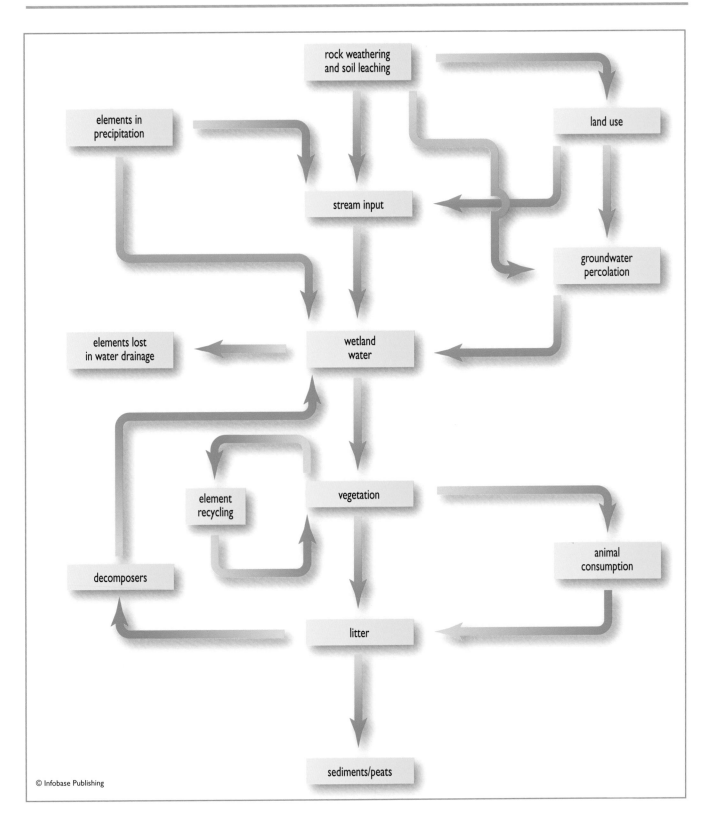

is a scarce element in relation to the plant's need and is often withdrawn and retained as leaves die.

The litter of dead plant and animal remains is the major resource for the decomposer organisms, fungi and bacteria. Many of the residual elements within the dead tissues are thus liberated back into the environment during decompo-sition, particularly into the surrounding waters where they are available to the plants once again. However, decomposi-tion in wetlands is slower than in most ecosystems because of waterlogging and lack of oxygen. Consequently, a propor-tion of the nutrient content of the dead tissues is locked away in the peat and remains in this growing element-reservoir until conditions change or until the very slow process of anaerobic decomposition in the lower peat layers (catotelm) releases this reserve back into circulation.

The fact that a proportion of the nutrient supply to wetlands becomes locked away in the sediments and peat may be of little consequence if the initial supply is rich, as in many rheotrophic systems. If, on the other hand, the wetland is ombrotrophic or has a very poor input of elements from drainage waters, this drain on the nutrient capital of the ecosystem can lead to further impoverishment for the vegetation and animal life living at the surface. Not only is the supply poor, but a proportion of those elements captured by the vegetation becomes unavailable to the next generation of growing plants. It is not surprising, therefore, that nutrient supply is a limiting factor for the growth of many organisms in such circumstances.

■ RAINFALL CHEMISTRY

Wetlands that receive chemical inputs from groundwater, either through stream flow or from springs and soaks around their edges, are generally little influenced by the chemistry of rainfall. Groundwater usually has a higher concentration of most elements than does rain and is available in greater quantities, so it generally provides a more reliable resource for nutrients. The only exceptions are those rheotrophic wetlands that lie upon very acidic and nutrient-poor rocks, where the rainfall may, as a consequence, be a significant source of some elements. The groundwater supply of minerals may also serve to neutralize any excess acidity in the rainfall; the presence of calcium carbonate in a wetland, for example, derived from the weathering and erosion of limestone in the catchment soils, reacts with excess hydrogen ions in the precipitation and neutralizes the effect of acidic rainfall. Only where the catchment rocks are themselves acid in reaction will a rheotrophic wetland be influenced by acid rain, as can be observed in some of the acid, nutrient-poor wetlands of Canada and Scandinavia.

Rainfall is naturally acidic because it dissolves some of the carbon dioxide gas in the atmosphere to form carbonic acid. This is a relatively weak acid, but its presence means that, even in the absence of human pollution, rainfall has a pH that is lower than 7.0 (that is, more acid than neutral). Atmospheric pollution resulting from human activities has led to increasing acidity of rainfall (see sidebar).

As explained previously, wetlands receiving groundwater are likely to be buffered against such extremes of acidity in rainfall, for they will contain greater concentrations of compounds such as lime that will react with the acids and neutralize them, but the raised and blanket bogs that receive their water only from precipitation (ombrotrophic mires)

Acid Rain

Human activity, particularly industrial activity, adds many chemicals to the atmosphere that result in the pH of rainfall becoming lower. Additional carbon dioxide is produced by all kinds of combustion, from firewood to fossil fuels, and this leads to more carbonic acid being contained in the precipitation. More important still is the production of sulfur dioxide as a result of the burning of fossil fuels. The sulfur contained in these fuels (especially in coal) is oxidized to sulfur dioxide gas, which dissolves in rainfall and reacts with oxygen and atmospheric ozone to produce sulfuric acid, a much stronger acid than carbonic acid. The nitrogen present in organic matter, including coal and oil, is also oxidized during combustion to form several different oxides of nitrogen, which again react with oxygen and ozone in the atmosphere and dissolve in water to produce nitric acid. So, the outcome of human industrial pollution of the atmosphere is a sort of cocktail of pollutant gases, all of which lower the pH of rainfall—they produce acid rain.

It is possible to trace the history of rain acidification by analyzing the layers of ice that have accumulated in recent times in the world's ice caps. The concentration of sulfates in the ice of the Greenland ice cap, for example, has increased by about four times between 1850 and the present day. This is just one of the components leading to the increasing acidity of rainfall during this period. Acidity levels as extreme as pH 1.6, a consequence of extreme pollution episodes, have been recorded in Scotland during individual storms in recent years, and this degree of acidity, sufficient to cause human suffering and death, may be expected to have an impact on the chemistry of wetlands.

will be much more sensitive to such changes in the chemistry of rainfall. It is possible that the world's ombrotrophic mires, particularly in regions where industrial air pollution is high, have been acidified by the changes in precipitation chemistry over the past 150 years. A simple method of determining whether bog acidification has resulted from acid rain is to analyze surface bog waters and compare them with the chemistry of the local rainfall. This type of survey has been conducted in the British Isles, where there is a steep gradient in the acidity of rainfall from the relatively low degree of acidity in the unpolluted, oceanic west portion of the

Rainfall Chemistry

LOCATION	SODIUM Na	POTASSIUM K	CALCIUM Ca	MAGNESIUM Mg
United States				
Long Island (coastal)	126 (142)	6 (7)	9 (10)	17 (19)
Hubbard Brook (inland)	2 (2)	1 (1)	3 (3)	1 (1)
United Kingdom				
Cornwall (coastal)	176 (198)	4 (4)	8 (9)	18 (20)
Kent (inland)	18 (20)	3 (3)	9 (11)	4 (4)

Table showing the quantities of certain elements delivered to an ecosystem by precipitation each year, contrasting coastal with inland sites. Data are expressed as pounds per acre (kg/ha).

islands, to the more industrialized and therefore polluted regions with acidic rain in the central and eastern districts. The research has shown a strong correlation between the pH of rainfall and that of the surface water of raised and blanket bogs. The waters of the western bogs were, as predicted, less acid (higher in pH) than central and eastern bogs. It does appear, therefore, that atmospheric pollution has an impact on the water quality of these ombrotrophic mires.

Just as the acidity of rainfall affects the water of raised and blanket bogs, so does the quantity of other ions present in rainfall. Precipitation in areas close to the sea has a very different chemistry from that of inland areas (see table). As might be expected, sodium and chloride ions are more abundant, as is magnesium, an element that is relatively common in seawater. The wind whips droplets of seawater into the atmosphere; they become incorporated into rainfall and are then carried into the waters of inland wetlands. Raised bogs on the west coast of Ireland, for example, have more than twice the concentration of sodium ions in their waters than those of the east of Ireland. Bogs on the west coast of Wales have five times the magnesium ion concentration of the bogs of eastern England, so the quality of precipitation clearly has an impact on the chemistry of bog waters.

It is more difficult, however, to determine whether these chemical differences, resulting from proximity to the sea, have any influence on the vegetation or animal life of the bogs. There is one plant in the British Isles, the bog rush (Schoenus nigricans), which is found only on those bogs that are located in the most oceanic conditions, and studies of its physiology and nutrient requirements suggest that its biogeographical distribution is limited by the chemistry of the precipitation.

Some of the elements derived from industrial atmospheric pollution may be toxic to the plants of ombrotrophic mires. Sulfur, in too great quantities, can affect the growth of many plants, especially mosses such as the bog mosses (Sphagnum) and some algae. Fungi are also sensitive to sulfur, so aerial pollution can influence the decomposer food chain. Aluminum also becomes a problem under acidic conditions. It is more soluble in acid soils and becomes toxic to many plants.

Other "pollutants" that enter the atmosphere (and hence the rainfall) as a result of human activity are actually elements needed by plants and animals for their growth. Nitrogen oxides, for example, dissolve to form nitric acid and react to give nitrates, which is the form in which most plants absorb their nitrogen from the soil. Nitrogen is a major component of agricultural fertilizers, and its presence in rainfall may, therefore, be expected to be advantageous to plant growth. In general, this is true, but some plants are better adapted to take advantage of higher nitrate levels than others. Most bog plants have evolved under conditions of low nitrates and are able to maintain their populations under this form of stress. If nitrate levels rise, other plant species with less specialized physiology may be selectively advantaged and may assume dominance to the detriment of the typical bog species, such as the bog mosses.

■ EUTROPHICATION

The enrichment of the mineral nutrient supply to a habitat is termed *eutrophication*. It is a process that often happens as a result of human activity, usually an accidental discharge, loss, or wastage of certain elements into the environment. An example of such eutrophication is the discharge of nitrogen compounds into the atmosphere as a result of burning fossil fuels. This is a form of aerial eutrophication that ultimately enriches wetlands as the compounds are washed out of the atmosphere by rainfall.

A more obvious form of eutrophication is the enrichment of waters draining from land surfaces, some of which may affect the nutrient balance of flow-fed, rheotrophic wetlands. Changes in wetland vegetation may result, especially if the nutrients received were formerly in short supply or even limiting to the growth of some of the plants present. Nitrogen, in the form of nitrate ions, and phosphorus, as phosphates, are the two most important elements in the eutrophication process, but other elements, such as potassium, calcium, and, in aquatic systems, even iron and silicon, can also affect the nutrient balance of a wetland ecosystem.

The results of eutrophication in a wetland can be very apparent and can occur rapidly following the input of nutrients. The arrival of a rich supply of nitrates and phosphates to a shallow-water wetland, for example, can result in very rapid growth of algal populations. Phytoplankton "blooms" may take place, and dense masses of filamentous green algae, including *Cladophora* and *Spirogyra* species, can rapidly cover the surface of a water body with a carpet that obscures the water below. Appropriately, such algal masses are often given the generic name *blanket weed*. If phosphates enter a wetland without accompanying nitrates (as, for example, when detergents are discharged into the system), it is often the blue-green bacteria (cyanobacteria) that benefit. This is because this group of photosynthetic organisms can fix their own nitrogen and so are not limited by supplies of this element.

Eventually, these blankets of algae begin to die and decompose, leading to a surge of activity by decomposer organisms, which consequently consume the oxygen dissolved in the water. Beneath the surface layers of algae, the waters and sediments become anaerobic, and those living organisms sensitive to lack of oxygen, including many invertebrates and fish, also die and decompose. Black, anaerobic muds, rich in ferrous sulfide, form, and bubbles of hydrogen sulfide gas are produced in the anoxic (oxygen-depleted) waters as the biodiversity of the ecosystem rapidly falls.

Eutrophication can result from a very wide range of causes, some of which are natural. Fire in the catchment, for example, liberates elements from the living organic matter (biomass) into the environment. Some of the elements involved, such as nitrogen and sulfur, are discharged mainly in the form of gases, but other elements, including phosphorus, potassium, and calcium, enter the soil and can be washed into wetlands. Natural storms and winds can also lead to the erosion of soils from catchments, and such events can often be detected in the sediment stratigraphy of wetlands (see "Sediment Chemistry and Historical Monitoring," pages 69–71).

Eutrophication is particularly associated, however, with human activity in the catchment. Clearing and burning forests for timber production or for agricultural development releases elements into soil in much the same way as natural catastrophe. In an experimental clearance of a forested catchment in the Appalachian Mountains, where the stream chemistry was monitored, it was observed that nitrate discharge levels were raised to between 40 and 60 times the normal level for the first two years after clearance. Potassium levels were raised by a factor of 16 and calcium was four times as high, following clearance. Land-use changes in a catchment can therefore have profound effects on the chemistry of water discharge into wetlands.

Farmers are often blamed for the eutrophication of waterways because of what some see as their excessive or inappropriate use of fertilizers. Nitrates, in particular, are poorly retained by soils because of their high solubility and the fact that the nitrate ion is negatively charged, as are clay particles in the soil. Clays (and organic matter, including peat) have negatively charged surfaces, and they attract and retain positively charged ions, such as potassium and calcium, but repel negatively charged ions such as nitrate. If nitrate fertilizers are used in excess of the crop's requirements at the time of application, they are flushed quickly from the soil. This is not in the economic interest of the farmer nor in the conservation interest of the wetland manager. It can also create problems for water supplies as high nitrate levels in drinking water cause human health problems, such as poor oxygen carriage in the bloodstreams of infants, and nitrates may also be implicated in some types of intestinal cancer. Farmers should restrict fertilizer applications to the time when the crop is growing most rapidly and then use only as much nitrate as can be taken up by the crop at that time.

Agricultural eutrophication may also occur as a result of changes in land use in a catchment. Just as the deforestation of a watershed can result in nutrient flushing into wetlands, so can the plowing of grasslands for the development of arable farming. Grassland establishment is often used as a means of allowing soils to recover their structure and nutrient content following arable cultivation—fields are said to be left *fallow*. During this treatment, organic matter builds up in the soil and forms a rich reservoir of organic nitrogen. When the field is plowed again, the organic matter rapidly decomposes as a consequence of increased fungal and bacterial activity, so nitrates are released into the soil in a surge of nutrient flushing. This type of land management is a major cause of eutrophication of wetlands in many parts of the world.

Wetlands may respond to eutrophication by subtle or sometimes quite marked shifts in the composition of vegetation. Biodiverse, calcareous fens, for example, may give way to marshes that are poorer in species and are dominated by a few nutrient-demanding, robust, and competitive plant species. There may even be a shift from herbaceous to tree domination of the vegetation. In the development of quaking bogs, the influx of nutrients into acid lakes during eutrophication can result initially in floating mats of vegetation over the surface of water and ultimately in the growth of floating bogs. The sensitivity of rhetrophic wetlands to changes in the land use of their catchments means that wetland conservationists must concern themselves with the ecology and land use of wide areas surrounding the site they are responsible for managing. The health and general development of wetlands can also provide an indication of the quality of water draining from a catchment, and changes in the wetland vegetation or fauna can provide an early warning system of eutrophication problems that might eventually affect human health.

■ VEGETATION AND WATER CLEANSING

The chemical elements dissolved in water are essential for all the living plants and animals of the wetland ecosystem, but an excess of some elements can lead to problems of eutrophication, a form of pollution that results from an overabundant supply of nutrients. Because the nutrients involved in eutrophication are growth stimulators (nitrogen, phosphorus, potassium, etc.), their loss into waterways and, subsequently, wetlands, is unfortunate both from the ecological and from the agricultural and economic point of view. These elements are needed by crop plants and are expensive to obtain for the production of fertilizers, so agriculturalists share with conservationists a desire to avoid such losses. The human health problems associated with high concentrations of certain elements (such as nitrogen) in drinking water add to the demand that water should be cleansed of such elements before it is released into streams and rivers.

There are, however, both technical and economic problems associated with the avoidance of these element losses, whether from sewage or agricultural drainage. Wastewater treatment systems are generally most concerned with the removal of organic detritus and pathogenic organisms from the water before release into rivers. Sedimentation and decomposition can be employed as a means of reducing organic matter content, but this results in further release of nitrates, phosphates, and other ions into the wastewater, and the removal of these ions by physical or chemical means is expensive. The reduction of nutrient loads in discharged water after human use is, therefore, a major problem facing environmentalists, especially in poor areas where expensive treatment systems cannot be employed and where the associated nutrient losses can least be afforded.

One possible answer is to exploit the very efficient system that wetland plants use to extract and concentrate these nutrient elements from water. All plants need such elements as nitrogen, phosphorus, potassium, calcium, sulfur, magnesium, iron, and others, but some plants are much more demanding than others. The plants of ombrotrophic mires, where the water is poor in such elements, generally have low demands but the plants of rheotrophic, nutrient-rich swamps often have very high demands for nutrients and combine this with a high capacity for the concentration of these elements within their tissues.

Taking up an element such as phosphorus from a low concentration in the surrounding water can significantly drain a plant's energy stores. Energy, derived ultimately from photosynthesis, has to be expended in capturing and transporting the element from the outside of the cell membrane of a root cell into the living tissues of the cell itself. The absorption of the element is so important for continued plant growth, however, that it is worth investing energy in maintaining its supply. In this way, the plant becomes richer in the element, and the surrounding water consequently becomes poorer. In the case of cattail (*Typha latifolia*) marshes, for example, the rate of absorption of phosphorus from water can attain values as high as one thousandth of an ounce per square yard per day (0.02 g m^{-2} day^{-1}). In a dense stand of cattail marsh, therefore, large quantities of this important source of eutrophication can be extracted every day, enhancing the growth of the plants and purifying the water.

Highly productive wetland plants, such as cattails and reeds (*Phragmites australis*), can, therefore, provide simple, inexpensive instruments for the extraction of unwanted elements from polluted water. Their rapid growth and wide ecological tolerance of different conditions (reeds will even tolerate a degree of salinity) means that they can be used for this purpose in a wide variety of conditions and climates. Fast-growing, nutrient-demanding, flood-tolerant woody species such as willows and poplars can also be used in this way, although alders are less appropriate because they have nitrogen-fixing bacteria associated with their roots and so may add to the eutrophication problem. In many parts of the temperate and tropical world, artificial wetland ecosystems are being used commercially for wastewater purification. They can also be used along roadsides in sensitive areas to catch accidental spillage of organic pollutants and prevent them from contaminating natural wetlands and waterways.

There are two additional problems associated with this application of wetland ecology in the service of environmental cleansing. First, there is the complication that could result if element recycling is permitted. All ecosystems, including wetlands, have decomposer systems operating within them that break down organic tissues and release not only their carbon but also their other components, including nitrogen, phosphorus, and potassium back into the environment. In wetlands, decomposition is slower than in most other ecosystems (see "Decomposition and Peat Growth," pages 54–56), but it still occurs, and this could reverse the nutrient capture process that is intended. The second problem, especially where nutrient elements are in short supply for agriculture, is how to recover elements in plant tissue so that they can be reused as fertilizer.

Both problems can be solved if the vegetation that has accumulated the elements is harvested and then burnt (supplying biomass energy as a by-product) or digested in biogas systems (where organic matter is decomposed to give methane gas), and the ash elements from burning can then be recovered. Drying and burning harvested vegetation involves some loss of elements in smoke (especially nitrogen and some phosphorus), but much remains in the ash, particularly potassium. This ash, or the slurry (a watery mixture of insoluble matter) from biogass digesters, can then be used directly as a fertilizer application to arable fields.

Natural wetlands can serve a very similar function, where human wastewater passes through them. In the tropical lakes of East Africa, for example, increased urbanization and agricultural intensification are causing eutrophication problems. The ultimate outcome is the lowering of oxygen levels and the loss of sensitive animal species, including fish from the lakes. Around the fringes of many of these lakes, however, extensive floating rafts of papyrus (*Cyperus papyrus*), shown in the photograph below, grow rapidly, breaking up into floating islands and colonizing new areas (see the diagram on page 65). These floating swamps have a very high demand for the nutrients entering the lakes from rivers and streams. In a study on their effect on water qual-

ity at Lake Naivasha in Kenya, ecologists found that water emerging from these swamps was distinctly lower in the concentration of many elements when compared with the water entering the ecosystem. The concentration of potassium declined by 9 percent, phosphorus by 57 percent, sulfur by 58 percent, iron by 86 percent, and manganese by 94 percent. Emergent swamp water is evidently purer when it leaves such a system than it was when it entered.

These floating rafts of papyrus, however, accumulate masses of organic detritus among their rhizomes and roots within which much of the extracted nutrient load is stored. In time, this falls as particulate matter and becomes incorporated in the lake sediments below. Some of this material may decompose in transit and release the nutrients back into the water, but much remains sealed in a fossil form and locked away in the sediments of the lake. The value of these natural swamps as a purification system should greatly increase political and economic pressures for their conservation, despite the problems they sometimes pose for navigation. The nutrient cycling of wetlands can thus be used in the service of humanity.

Floating papyrus marsh at the edge of Lake Nabugabo in Uganda, East Africa   *(Peter D. Moore)*

Profile of a floating papyrus marsh at the margin of an East African lake. Water rich in nutrients and agricultural runoff enters from a stream, and a large proportion of these nutrients is absorbed by the floating carpet of vegetation.

SOIL CHEMISTRY IN WETLANDS

The roots of most wetland plants penetrate the firm substrate, whether it happens to consist of lake sediment, peat, or saturated mineral soil. The conditions encountered by the roots are almost invariably wet and usually saturated, and the chemistry of soils under these conditions differs from that of drier soils. The upper layers of the profiles of some wetland soils may be well aerated for much of the time, but deeper layers invariably encounter poor air penetration. As a consequence, they become anaerobic.

Oxygen is the element that is most critical in determining the function of these lower layers. Oxygen diffuses much more slowly (about 10,000 times more slowly) when dissolved in water than is possible within a mixture of gases, such as the atmosphere. So, when soils are laden with water, oxygen diffuses very slowly and, since it is consumed by most living organisms, it is often in short supply. Lack of oxygen affects the respiration of both plant roots and the microbial populations of the soil and can lead to reduced rates of decomposition. Such a lack can also influence the chemical behavior of different elements in the soil, which in turn may influence their availability to plant roots and hence to the whole ecosystem.

The process of respiration actually involves the movement of electrons along a chain, the final acceptor normally

being molecular oxygen, O_2. When oxygen becomes scarce, microbes turn to other possible acceptors of electrons, and this influences the chemistry of the soil. One of the options available is the nitrate ion, NO_3^-. This can be used as an electron acceptor and, in the process, it becomes reduced to N_2O and eventually to nitrogen gas, N_2, which escapes from the soil. The process is called "denitrification," and the result is that this vital element is lost from the soil, to the detriment of plant growth. This is particularly likely to happen under high pH (alkaline) conditions, so agriculturalists dealing with very wet soils would be unwise to add lime as it encourages nitrate loss.

Iron can also act as an electron acceptor, so, under anaerobic conditions, the ferric, Fe^{3+}, ion becomes reduced to the ferrous, Fe^{2+}, ion, which is more soluble than the ferric ion. If there is movement of water through a soil, therefore, iron can be flushed out, and the effect on the general appearance of the soil profile is a bleaching. This is because the presence of oxides of iron in its oxidized state (Fe^{3+}) give a rusty red color to the soil, and their loss removes this coloration. When such iron-rich water emerges from a wetland soil, it may become dark brown as the dissolved ferrous iron becomes oxidized once more. Sometimes iron is redeposited lower in the soil profile, resulting in a series of layers, or horizons, developing. The result is a profile that soil scientists call a podzol, which is shown here diagrammatically and also in a photograph, where the podzol soil profile has become buried and fossilized beneath a depth of peat in a Welsh blanket bog.

Oxides of the reduced form of ion (Fe^{2+}) give a blue-gray color to their surroundings, so a waterlogged soil that has stagnant water within it often has this color. Wetland

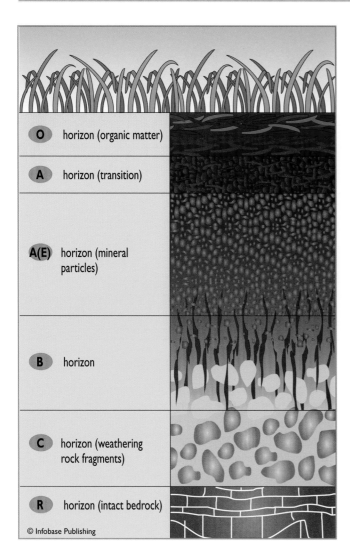

Profile of a podzol soil. Plant litter accumulating on the surface is a source of organic acids that percolate through the soil and carry away certain elements, such as iron and aluminum from the A horizon. These elements are subsequently deposited in the B horizon, where they may become concreted and consequently interrupt soil drainage. When this happens waterlogging follows, and peat begins to accumulate.

mineral soils that experience this saturation by groundwater are called *gley* soils, as shown in the illustration. One of the distinctive features of gley soils is the presence of root channels, the hollow tubes which the roots of wetland plants have formerly occupied. After the death of the root, an open channel remains and air, including oxygen, diffuses through these channels. The linings of the channels, therefore, have a relatively good oxygen supply, and the effect of this is the oxidation of iron in the soil to the ferric form, leading to an orange color surrounding the old root locations.

Another element that can act as an electron acceptor in waterlogged soils is manganese. This element may occur as Mn^{4+} or Mn^{3+} under oxidizing conditions but is reduced to Mn^{2+} following the acceptance of electrons. Also, sulfates may accept electrons and become reduced to sulfides, giving a black appearance to sediments when they combine with the reduced form of iron to produce FeS, ferrous sulfide. Hydrogen sulfide gas, which smells of rotten eggs, may also be produced under acid conditions.

So, the process of respiration, the constant shunting of electrons, continues despite the lack of oxygen as a receptor. As the process of oxidation in a soil becomes more difficult (measured on a scale termed *redox potential*—an electrical measure of the oxidation level of a soil), so different elements become utilized in the respiratory process. Oxygen is used first, then manganese, then iron, and finally sulfate. Sulfate reduction to sulfide, therefore, is an indication of severe oxygen lack associated with permanent waterlogging.

Podzol soil profile beneath a blanket bog in Wales
(Peter D. Moore)

Gley soil profile from eastern England   *(Peter D. Moore)*

The process of photosynthesis is, in several respects, the opposite of respiration, for electrons are consumed rather than generated by the organisms concerned. Some bacteria are photosynthetic and exist both in purple and green pigmented forms, and these can take advantage of the abundance of ferrous sulfide by using it as an electron source for photosynthesis. (Green plants use water as their electron source.) The photosynthetic bacteria can do this, of course, only in the presence of light, but in stagnant ponds where light penetrates, they form colonies on the submerged water/sediment interface and convert sulfides back to sulfates as they photosynthesize. The purple photosynthetic bacteria can produce spectacular coloration to wetlands, especially some of the saline wetlands and the temporary, evaporating pools (salt pans) of the deserts and coastal regions.

Some reduced compounds may be released from the wetland soil as a gas. Hydrogen sulfide, for example, is released in this way, particularly when a footstep compresses sealed pockets of the gas below the surface and results in the escape of bubbles. Methane (CH_4) may be produced from the incomplete oxidation of organic compounds; hydrogen is released by certain bacteria; phosgene (PH_3) may also be released by phosphorus reduction. All of these last three gases are inflammable, and phosgene in particular may spontaneously ignite, producing the "will-o'-the-wisp"—flames in the air above a wetland—that has provided such habitats with a certain mystique, exploited by some fiction writers (see "Bogs and That Sinking Feeling," pages 196–197).

More serious, from the point of view of the vegetation of wetlands, is the toxicity of some of the compounds and ions produced as a result of anaerobic conditions. Reduced forms of iron and manganese (Fe^{2+} and Mn^{2+}), for example, can prove toxic to some plants. Some of the more successful plants of permanently wet locations have developed tolerance mechanisms to these ions. The high cation-holding capacity of the organic matter in peaty wetland soils can also give rise to problems because toxic heavy metals can become bound into the soil. Copper, for example, may be retained in peaty soils, which can adversely affect the growth of some plants (such as the bog mosses) and could lead to general copper toxicity in the ecosystem. This fact must be taken into account if a wetland is drained and subsequently used for agricultural purposes.

■ NUTRIENT CYCLING IN SALINE WETLANDS

Wetlands vary widely in their degree of salinity, that is, in the salt concentration of their waters. Coastal wetlands that have direct contact with the sea have a salt concentration in their waters that corresponds approximately to that of seawater, depending on the inflow of water from land drainage. Similarly, certain inland wetlands are rich in salts if they have no outflow and are subject to the evaporation of (relatively pure) water from their surfaces, leaving the salts behind. The salt content of seawater has an approximate range of 3.3–3.7 percent. In some localities, a substantial inflow of freshwater combined with limited access to open seawater leads to relatively low salt concentration. This may be as low as 0.7 percent in the Baltic Sea, where melting snows in spring bring a rich supply of freshwater and the contact with the Atlantic Ocean, the source of seawater, is the relatively narrow channel between Denmark and Sweden. A high level of evaporation combined with a poor inflow of freshwater, on the other hand, can lead to exceptionally high salt concentrations, as in the case of the Red Sea, where salt concentration reaches 4.1 percent. This constricted sea is surrounded by the deserts of Egypt and the Arabian Peninsula.

Common salt, sodium chloride, is the main chemical involved in the salinity of seawater. Between them, the sodium and chloride ions represent about 85 percent of the dissolved ions in seawater. Other important constituents include sulphate (7.7 percent), magnesium (3.7 percent), calcium (1.1 percent), potassium (1.1 percent), together with smaller proportions of many other ions, including bicarbonate, bromide, borate, strontium, and fluoride. Of the 93 natural chemical elements, 73 are found in seawater, so the medium is a rich source of chemicals.

Its chemical richness might lead one to suppose that seawater would be an ideal source of mineral supply for plants and animals, but this is not generally the case. Apart from the algae and some one-celled organisms, most photosynthetic plants have evolved from land-based ancestors. A number of plants, including flowering plants, have invaded marine habitats and are now found in saline wetlands, but these are relatively few in number. Seawater presents higher plants with two main problems. First, there is the difficulty in obtaining and retaining sufficient water in their cells when bathed in a salt solution, and second, there is the unfortunate fact that the elements most abundantly found in seawater are not those needed in large quantities by plants. Sodium, in particular, presents difficulties because it constitutes about 30 percent of the dissolved ions in seawater yet is not needed at all by plants. This is not true of animals, because they need sodium for the function of their nervous systems, but even they find sodium in seawater present in much greater concentrations than is needed.

The tendency for plants and animals to dehydrate in saline environments results from the movement (diffusion) of all materials from areas of high concentration to areas of low concentration. Thus, if a cell contains a higher concentration of water (that is, a lower concentration of salts) than the outside medium, then water will tend to move outward from the cell. The greater the difference in concentration between the outside and the inside of the cell, the greater the force pulling on the cell's water content. Although water molecules are present in abundance, they are not easily available in seawater—a situation sometimes referred to as physiological drought. If a plant is to absorb water under such conditions, it must exert an even greater force to pull water molecules inward and extract them from the seawater. In the case of a plant growing with its roots in salt water and

its leaves above the surface, the evaporation of water from its surface makes it imperative that the water is replaced from the base, and this can best be achieved by building up even higher concentrations of chemicals inside its cells than are found in the surrounding water. In this way, the concentration gradient of the water is reversed—water is scarcer, in relative terms, inside the plant than outside. When this state is achieved, water molecules move out of the seawater and into the plant. This manipulation of diffusion gradients is the process of *osmosis,* and this enables both plants and animals in saline situations to maintain and control their water intake.

One further problem remains in the control of water, namely how to build up a high concentration of solutes (dissolved materials) in the cell without interfering with the general cell health and function. If, for example, sodium chloride were used to build up cell concentrations of solutes, the cell membranes and the enzymes that control cell activities would fail to function. Sodium chloride in high concentration is toxic to the cell: It is a compound that can be accumulated in the vacuoles (enclosed reservoirs sealed off by membranes within the cell) but not in the living cell tissue, the cytoplasm. An alternative, nontoxic material has to be used, and the amino acid proline seems to be one that has proved widely acceptable to both animals and plants. It does no harm to the cell, even in quite large concentrations, yet it provides the cell with the osmotic pull that it needs in order to absorb water from the outside.

The force required to maintain water balance varies with the degree of physiological drought experienced, being greatest in salt lake sites where high salt concentrations and high temperatures combine to put pressure on the organism. In Utah, for example, some plants have been recorded that are capable of exerting forces 10 times that of seawater in order to draw in water from their environment.

Even when the water problem is solved, however, the difficulty of element imbalance remains. The organism needs to absorb certain elements selectively from the environment to provide for its needs. One example of this is the selection of potassium in preference to sodium. In seawater, there are approximately 50 atoms of sodium to every atom of potassium, but both plants and animals require potassium in greater quantities than sodium. The answer to this problem lies in making the membranes of the cells selective in what they permit to enter the cell. They may concentrate certain elements by pumping them across the membranes in preference to other elements. This process consumes energy, but it is effective: Both seaweeds and the higher plants of coastal wetlands are able to maintain a potassium-to-sodium ratio in their cells of at least 1:1 by this means.

An alternative, or additional, method for maintaining this balance is to excrete excess sodium. This is achieved by means of the kidneys in higher animals and by specialized salt-excretory glands in the case of many higher plants. Cordgrasses (*Spartina* species), for example, produce concentrated droplets of salty water on their leaves from these glands, and the drying effect of the Sun may result in evaporation and the production of salt crystals on the leaf surfaces. The saline wetlands thus present a number of problems for plants and animals, and, as a consequence, the cycling of nutrients in these ecosystems takes on a distinctive pattern.

SEDIMENT CHEMISTRY AND HISTORICAL MONITORING

Wetlands are remarkable among ecosystems in that they leave in their sediments a record of their past history. The sediments (including peats) contain the remains of the plants that formed them and the materials that were washed into the site during their development. From this ordered sequence of organic and inorganic remains, it is possible to reconstruct a detailed account of the development of a wetland site. The chemistry of the sediments adds to this record, for the chemicals eroded from catchment soils, washed from the atmosphere by rain, and accumulated within the bodies of wetland plants and animals are also deposited in a stratified sequence and bear record to the past chemical history of the site.

The study of chemical stratigraphy shares all the problems discussed when dealing with fossil stratigraphy (see "Wetland Stratigraphy," pages 32–34). Sediments may be eroded from the profile and lost, leaving gaps in the record, or they may slump from one part of a basin to another, creating a confused sequence. Disturbance by burrowing invertebrate animals may also result in sediment mixing. Despite all these problems, chemicals are often found to be layered in a meaningful sequence that provides additional information about the ecology of wetland development. The chemicals preserved in peats and mineral sediments may be incorporated into the constituent mineral particles and the organic matter, in which case only destructive chemical analysis will release them, or they may be loosely attached to the surfaces of clays and organic fragments. The fine clay fraction of a sediment, together with the organic remains of plant tissue, generally carries a negative charge on the surface that attracts and bonds weakly with positively charged cations. So, when these particles become incorporated into the sediment, the chemicals remain loosely attached to them. Analysis of these superficially bound ions involves the leaching of samples of sediment using solutions of such compounds as ammonium acetate, which displaces the weakly held ions. These ions can then be identified and their concentration measured using conventional chemical techniques, such as atomic absorption spectrophotometry. The weakly bound ions are gener-

ally more interesting than those incorporated into the fabric of sediments because they more closely reflect the chemistry of the wetland waters contemporaneous with their deposition.

In shallow-water wetlands, the particles falling through the water and arriving at the sediment surface are subject to further decomposition and chemical alteration as a result of invertebrate and microbial activities. They may also be resuspended when the water body is disturbed by wind and storms and then settle once more. So it can take several years for a particle finally to become established in its permanent resting place in the sediment profile. In peat, such resuspension does not occur, but there is a greater likelihood of ions being carried down the profile by water movements, particularly in the superficial layers of the acrotelm where water moves more freely than in lower, catotelm layers (see "Hydrology of Peat," pages 24–27). The cation-binding properties of peat are considerable, so it is likely that those ions still free in the peat water are taken up by organic surfaces and bound to them by their opposing electric charges. In this form, they are retained in position within the developing peat profile.

Studies of aquatic sediment chemistry are particularly helpful in reconstructing the history of a catchment. Elements such as calcium, potassium, and magnesium are easily leached from disturbed, unstable soils, so their abundance in the stratigraphy of a wetland often indicates open vegetation, either a consequence of climate (as in the later stages of the last glaciation) or human clearance of a catchment. In peat profiles, it is sometimes possible to trace the change from rheotrophic to ombrotrophic conditions, especially in more oceanic sites, by analyzing chemical stratigraphy. At this point, there is a distinct fall in the abundance of most elements, as the peat surface becomes elevated and totally dependent on rainfall for its water and mineral ion input. The greater influence of precipitation as an ionic source, however, means that the proportion of those elements that are particularly associated with seawater, such as sodium, chloride, and magnesium, is greater in the ombrotrophic peat layers. The ratio of magnesium to calcium, in particular, can be seen to rise as the mire becomes precipitation-dependent.

Rheotrophic wetland chemical stratigraphy is particularly sensitive to change in catchment land use. The erosion of soils is often accompanied by peaks in a wide range of elements in the stratigraphic profile. In a study of lake sediments from northern Britain, analysis revealed high levels of the element iodine were noted at certain points in the stratigraphy, and researchers concluded that this was being brought from the Atlantic Ocean and reflected periods of higher rainfall. More detailed and extensive research showed that episodes of high iodine input were also periods of catchment soil erosion, so it is likely that this element was also being eroded into the basin from surrounding soils, along with the calcium, potassium, magnesium, and sodium that are typically associated with soil erosion.

Some of the most extreme examples of catchment disturbance that have been reflected in wetland sediment chemistry are those associated with industrial development and mining activities. The release of heavy metals into the atmosphere and, more important from the point of view of rheotrophic wetlands, into streams and rivers has often been recorded in sediment profiles. In Lake Washington, near Seattle, for example, the top eight inches (20 cm) of sediment has a very different chemical composition from that of the lower layers. These upper eight inches date from the end of the 19th century, when the lead level in the sediment began to rise rapidly, to the present. Up to that point, the lead level of sediments was consistently less than $40\mu g\ g^{-1}$. This means that for every pound of sediment there were 40 thousandths of a pound of lead, and its relatively high level reflects the lead content of the local soils eroding into the lake. Lead concentrations in the sediment then rose steadily to about $300\mu g\ g^{-1}$ by the 1970s as a consequence of industrial development, including the smelting of metals, in and around Seattle. The sediments thus provide a detailed record of the history of smelting in the region. Similar studies have been conducted in many different parts of the world, including Lake Constance (Bodensee) in southern Germany, where

the rise in concentrations of lead, zinc, and cadmium over the past century can be seen in the accompanying diagram.

In aquatic sediments, the elements stratified in the profiles arrived at the site either by groundwater or by aerial transport. The chemical stratigraphy of ombrotrophic peatlands, on the other hand, records only the materials brought to the site by atmospheric dust and rainfall. Nevertheless, many chemical studies of raised and blanket bog profiles have also shown evidence of increasing heavy metal pollution in relatively recent times. Research on the Pennine Mountain chain, which forms the spine of northern Britain, has shown that the lead content of peat from high altitude blanket bogs began to increase as long ago as 1400 C.E., with the development of medieval mining industries in the region. Dust containing a high lead content was evidently injected into the air in the course of this mining. In this particular region, the lead concentrations peaked in about 1700–1900 C.E. and then decreased as the lead mining industry collapsed during the 20th century.

Changing concentrations of metals at various depths in the sediments of Lake Constance (Bodensee) in southern Germany. Lead, zinc, and cadmium all show increases in abundance in recent times.

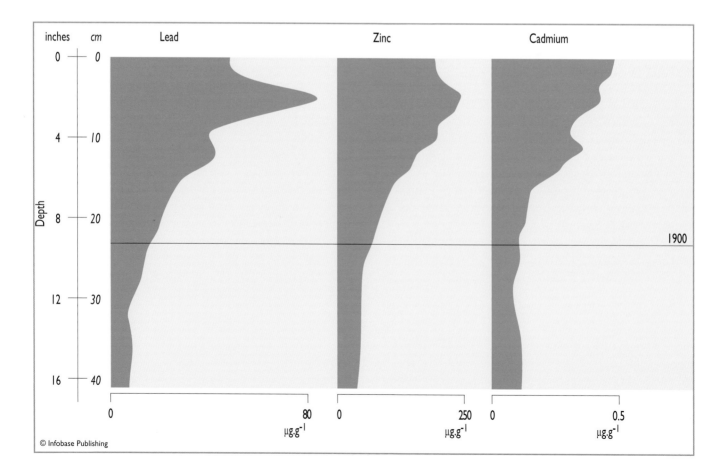

Interpreting the chemical stratigraphy of peat deposits, however, can be difficult because some elements do not remain at the level where they were first incorporated into the peat profile—they are vertically mobile. Elements such as phosphorus and potassium, for example, are in short supply, so the plant recycles them from dying tissues and takes them back into the growing parts, often at the upper tip of the plant. Such plants as *Sphagnum* mosses thus take these scarce and valuable elements up the profile with them as the peat deposit develops. Other elements, such as aluminum and iron, change their solubility as the oxygen levels of the peat deposit diminish. Often there is a change in the behavior of these elements at the boundary between the acrotelm and the catotelm. Some of the heavy metals, such as lead, cadmium, and zinc, may remain approximately at the stratigraphic location where they first enter the peat, so they can provide some indication of past environmental conditions.

WETLANDS AND THE ATMOSPHERE

The atmosphere is a gaseous envelope wrapped around the Earth. The gases of which it is composed become thinner and more diffuse with altitude above the Earth's surface, so there is no strict cutoff between the atmosphere and the space that lies beyond. Most of the atmospheric gas, however, lies within the volume limited by a height of about 60 miles (100 km) of the Earth's surface. Above that lie the upper stratosphere, the mesosphere, and the thermosphere, which are the zones where gas is scarce and within which space shuttles, satellites, and the aurora borealis operate. The lower part of the atmosphere is called the *troposphere,* and this extends to a height of only about nine miles (5 km). This layer contains all the mountains of the world, and it represents the approximate limit of the *biosphere,* the zone within which living things can maintain themselves. Most humans go beyond this level only when they travel in jet aircraft, which fly in the lower part of the *stratosphere.* It is the troposphere, therefore, that interacts most strongly with the Earth's living organisms and its habitats, including wetlands.

Dry air contains approximately 79 percent nitrogen, 20 percent oxygen, less than one percent of scarce, inert gases, including argon and neon, plus a trace of carbon dioxide. In addition, there is a variable amount of water vapor, depending on local conditions of humidity and temperature. Atmospheric wetness and the balance of precipitation and evaporation from the land surface have an important influence on the global distribution of wetlands and largely determine what type of wetland can develop in any geographic region (see "Climate and the Global Distribution of Wetlands," pages 4–9).

The major component of the atmosphere, nitrogen, is present as a stable gas with a molecular constitution of N_2. Although all plants and animals need this element in considerable quantities (it is a major component of proteins), most are unable to tap the atmospheric resource directly simply because the gaseous N_2 molecule has such a low chemical reactivity. There are, however, certain microbes which can capture these molecules and break the firm bond between the two nitrogen atoms using the enzyme nitrogenase. These microbes are then able to reduce the nitrogen atoms to ammonium ions NH_4^+, which subsequently can be incorporated into amino acids and then proteins. This ability to "fix" nitrogen is very limited in the biological world. Some of the microbes that are capable of the reaction are free-living, such as the cyanobacteria (sometimes erroneously referred to as blue-green "algae"), while others live symbiotically in the roots of various plants, where they cause lumps, or nodules, to form around their colonies. In wetlands, the cyanobacteria are an important group of organisms in fixing nitrogen that can then be passed on to other organisms after their death. Also important are some lichens—those that contain cyanobacteria as the "algal" component of the lichen symbiosis. In addition, there are some plants in wetlands that possess root nodules with nitrogen-fixing bacteria present. These include the alders (*Alnus* species) and the sweetgale (*Myrica gale*).

Oxygen is needed for respiration by almost all plants and animals. The most important exceptions in wetlands are those bacteria that can operate in the absence of oxygen, deep in the waterlogged sediments and that are responsible for the generation of such gases as methane and hydrogen sulfide as they oxidize organic matter incompletely (see "Wetlands and Methane," pages 77–79). The availability of oxygen to the plants and animals that need it in wetlands is not restricted by its abundance in the atmosphere, but rather by the ability of the organism to tap this atmospheric resource. Animals and plant roots living in the deeper sediments of wetlands may find the oxygen supply poor simply because the air does not penetrate effectively into the substrate, and oxygen in a dissolved form does not diffuse quickly through the water.

The inert gases in the atmosphere are of no interest to living organisms, but the trace of carbon dioxide is vital for the maintenance of life because this molecule is the raw material of photosynthesis and the very base of almost all food chains. Although it is present in the atmosphere at less than 0.04 percent by volume, it is this minute quantity of carbon dioxide that is demanded by all green plants for the construction of sugars and other organic molecules. This is true for almost all ecosystems (an exception is the deep-sea thermal vent environment), not just for wetlands.

Global Carbon Cycle

In recent years, there has been an increasing interest in the global cycle of carbon, mainly because of concern about the balance of carbon dioxide in the atmosphere and the need to know what factors control it. As a result of many studies throughout the world, it is now possible to place approximate figures on the quantities of carbon that reside in different "reservoirs" over the surface of the Earth (including the atmosphere, the oceans, the living organisms, the soil, etc.). A summary of these figures is shown in the accompanying diagram. Also shown is the approximate quantity of carbon that moves from one reservoir to another in the course of a year. This, of course, is even more difficult to estimate than overall quantities, as it must be calculated from many small-scale measurements in different parts of the world and then scaled up to global proportions. Consequently, there is likely to be a great deal of error in these preliminary figures, but they provide some indication of where the world's carbon resides and where it is likely to go.

The figures used for this type of exercise represent immense quantities of material. The units involved, therefore, are almost unimaginable in their size; one gigaton (Gt) is a billion (109) tonnes. A metric ton is roughly equivalent to a normal ton, but these figures refer to the weight of carbon alone, without reference to all of the other elements found in living tissues. The amount of living biomass on wetland surfaces has not been calculated in detail but is incorporated into the global biomass figure. It is possible, however, to work out how much carbon is locked up in peat deposits, and this comes to between 180 and 530 Gt, depending on whose data are used. An approximate and conservative figure of 200 Gt carbon is used in this diagrammatic version of the carbon cycle. This is nevertheless a very large quantity of carbon. It represents more than 25 percent of the carbon in the atmosphere, for example, or is equivalent to the carbon held in about a third of the world's biomass. If the higher estimate is correct, then the peatlands could contain about two-thirds as much carbon as the atmosphere and almost as much as the entire living biomass of the planet. This means that the peatlands play a very important role in the global carbon cycle and that their future maintenance, management, and possible loss could have a major impact on the balance of the various carbon reservoirs.

(opposite page) The global cycling of carbon, showing the main reservoirs where carbon is stored and including the global peat reservoir of the wetlands. The rates of movement (flux) of carbon between wetlands are also shown, indicating how rapidly or slowly carbon moves from one reservoir to another. The figures in boxes are in gigatons (Gt), i.e., billion (10^9) tonnes, and those beside arrows are in Gt per year. A metric ton is approximately the same as an imperial ton.

Most wetland habitats, however, behave in a different manner from "normal" terrestrial ecosystems in that they do not respire away all of the material that they accumulate; some of the organic matter produced is preserved as peat. This means that peatlands in particular are net absorbers of atmospheric carbon dioxide; they are net "sinks" for the gas. Peatlands, therefore, must be taken into account when the global carbon budget is being studied (see sidebar).

Knowing how much carbon resides within the peatlands of the world is important for a number of reasons. Atmospheric levels of carbon dioxide are rising, so environmentalists need to know exactly where the present reserves of carbon are and how big they are. These reserves are shown in the illustration. If people were to destroy the world's vegetation, then a considerable additional load of carbon would be shifted to the atmosphere (about 500 Gt). The same principle applies to the wetland sediments. If these were oxidized (by draining and natural decomposi-tion or by peat extraction and burning), then the impact on the atmosphere would be almost half as great as that caused by the destruction of all the world's vegetation. This presents an important argument in favor of the conservation of wetlands and their peat deposits. The horticultural use of peats in the developed world is as great an ecological error as the destruction of the tropical rain forests (see "Peat for Horticulture," pages 213–214). Wetlands also represent a carbon sink; that is, a system that absorbs and stores carbon from the atmosphere. In this respect they are doubly valuable, providing not only a reservoir for the carbon but also a means of removing it from circulation. The loss of any carbon sink on the surface of the planet is particularly undesirable while people are still using more ancient wetland deposits (the fossil fuels) as a major energy resource. Every sink available is needed to soak up the additional carbon dioxide that is released to the atmosphere whenever people combust fossil fuels.

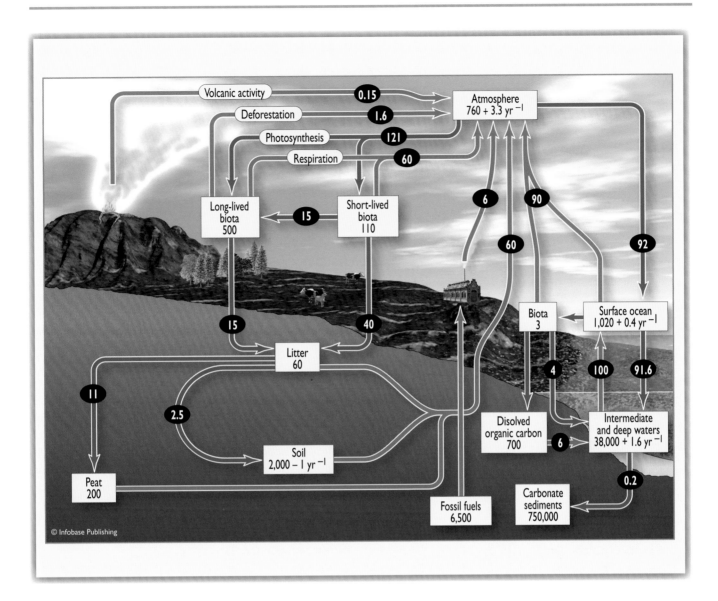

Volcanic activity
0.15
Deforestation
1.6
Photosynthesis
121
Respiration
60

Atmosphere
760 + 3.3 yr⁻¹

Long-lived
biota
500

Short-lived
biota
110

15

6
90

60

92

Biota
3

Surface ocean
1,020 + 0.4 yr⁻¹

15
40

4
100
91.6

Litter
60

11

2.5

Disolved
organic carbon
700

6

Intermediate
and deep waters
38,000 + 1.6 yr⁻¹

Soil
2,000 − 1 yr⁻¹

0.2

Peat
200

Fossil fuels
6,500

Carbonate
sediments
750,000

© Infobase Publishing

WETLANDS AND THE CARBON CYCLE

The peat deposits of the world thus represent a substantial reservoir of the Earth's carbon. This reserve is added to every year as new peat and other organic sediments accumulate, a buildup resulting from the failure of microbes to decompose all of the plant and animal detritus available in wetland habitats. Effectively, carbon is moving from the atmospheric reservoir into the peat reservoir, but there is also a movement in the opposite direction as peat deposits are drained as a result of human activity and are used for agriculture or forestry. The newly aerated peats become available to more intensive microbial decomposition and become oxidized into carbon dioxide. Some peats are also burned as an energy resource or used in horticulture as a soil conditioner. Both processes ultimately result in the transfer of carbon back from the peat into the atmosphere as the organic matter is oxidized. So peatlands are not a static reservoir for carbon but a dynamic one.

Measuring changes in the atmospheric content of carbon dioxide might seem an easy task. Admittedly, the gas is present at low levels, but techniques are available (infrared gas analysis) which are very sensitive, accurate, simple, and cheap to operate. The problem is that the concentration of atmospheric carbon dioxide varies with many factors. Close to the ground, the CO_2 concentrations tend to be high, for the respiration of animals, microbes, and plant roots in the soil constantly adds the gas to the atmosphere. At night the atmospheric concentration rises because the plants cease to take up CO_2 in photosynthesis yet continue to produce it during their respiration. The growth season of plants can affect concentrations, especially at higher latitudes where there is more seasonal variation in growth. In summer, concentrations of CO_2 fall and in winter they rise. Also, the

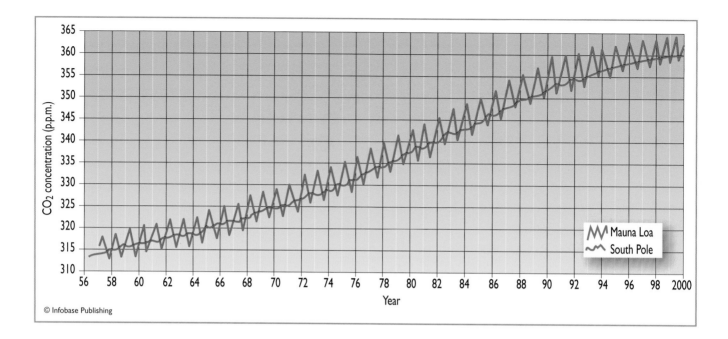

The changing concentration of carbon dioxide in the atmosphere in the Northern Hemisphere at Mauna Loa, Hawaii (blue line), and in the Southern Hemisphere at the South Pole (red line). There is a wider amplitude of annual fluctuations in the Northern Hemisphere because this hemisphere contains a larger plant biomass, which is responsible for the seasonal uptake and release of carbon dioxide.

concentration varies over different landscapes—industrial, rural, forested, oceans, etc. So, finding out how atmospheric CO_2 is varying is not as easy as it sounds. Long-term measurements at an isolated recording station in Hawaii and an even more remote location on the Antarctic icecap have proved the best solution for the determination of general trends, as shown in the accompanying diagram.

The "wiggles" in the curve are annual in nature, and are caused by seasonal uptake and release of carbon by vegetation in the Northern Hemisphere. The Southern Hemisphere has a far less wiggly curve because there is much less land and hence vegetation in the Southern Hemisphere to cause these seasonal variations. Despite this background Northern Hemisphere "noise," the general upward trend in carbon dioxide is clearly evident. It is possible to follow the trend further back in time by analyzing bubbles of atmospheric gas trapped in the great ice sheets of the world. Using this technique for the analysis of ice cores obtained from Greenland, researchers have shown that the current rise in atmospheric carbon dioxide concentration began in the 18th century—before which it was relatively steady—as shown in the diagram at the top of page 75. It is now generally accepted that the increase is due to human activity, partly in clearing the world's forest, but mainly in burning ancient deposits of coal, oil, and gas, thus bringing back into circulation carbon that has been stored up in geological reservoirs for millions of years. The diagram at the bottom of page 75 shows the estimated emission of carbon dioxide over the past 150 years from these two sources. Much of the atmospheric carbon input is therefore derived from the photosynthetic fixation activities of ancient wetlands (see "Coal-Forming Mires," pages 174–175). The current burning of fossil fuels will cause a return to the atmospheric conditions and the climate that existed prior to the carbon-fixing operations of those wetlands.

The importance of these findings lies in the fact that even small changes in atmospheric CO_2 concentration (a rise from about 0.031 percent to 0.036 percent has been observed over 50 years) could have considerable impact on the Earth's overall energy balance and climate. Carbon dioxide (along with several other gases, such as water vapor, methane, oxides of nitrogen, ozone, chlorofluorocarbons, and carbon monoxide) absorbs infrared radiation. This means that although the light from the Sun passes through the atmosphere easily, when it strikes the Earth's surface and is radiated anew as heat, the energy is more efficiently absorbed by the atmosphere on the way back out. This "greenhouse effect" is predicted to result in a general rise in world temperature, and a rise of almost 1°F (0.5°C) has already been observed over the last century. The relative contribution of different gases to the greenhouse effect is shown in the pie diagram. The implications of such a change are considerable, both for agriculture and for natural ecosystems, including wetlands.

Considerable research effort is currently being expended on understanding the sources and sinks of atmospheric carbon and how fast the different fluxes (element movements) operate. The diagram at the bottom of page 76 provides some

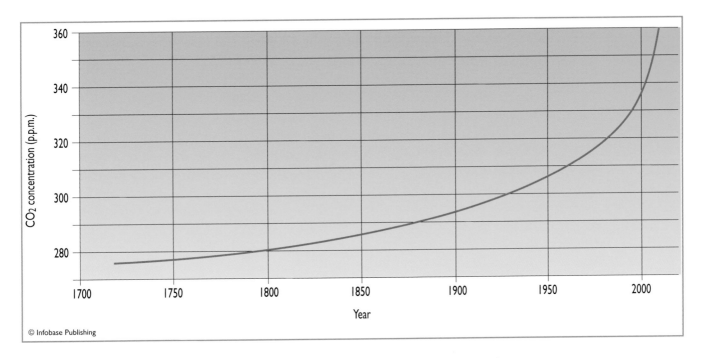

The change in atmospheric carbon dioxide over the past 300 years. Note the acceleration of the increase in this gas in the past 50 years.

the net carbon inputs to the atmosphere come from human activity. The figures relating to fuel combustion and forest destruction are reasonably reliable, so the overall input figure of in excess of 7 Gt per year is an acceptable one. It can

rough estimates of the rates and sources of carbon movement into the atmosphere, as well as some likely sinks. Volcanic activity supplies a small amount, but the bulk of

Estimated emission of carbon dioxide over the past 150 years separated into fossil fuel combustion and forest clearance.

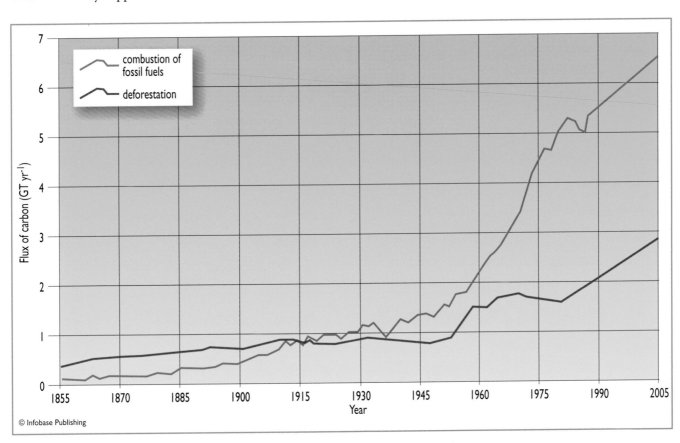

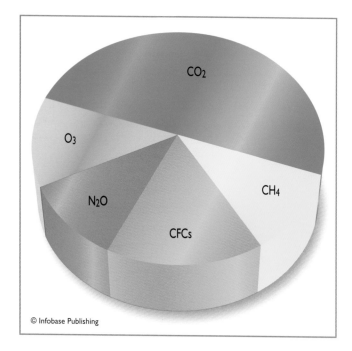

The estimated contribution of different gases to the greenhouse effect. Water vapor has been omitted because of its great variability both in time and in space.

around the world. There remains an unknown sink, sometimes called the "missing sink," which could be a result of expanding global vegetation biomass, of greater oceanic uptake than has been calculated, of chemical reactions with the world's soils leading to CO_2 absorption, or to an increase in global soil organic matter (including peat accumulation). Precise figures are very difficult to obtain because the measurement of any of these variables presents great problems on a global scale, but it is possible that the wetlands of the

also be calculated that almost half of this (3.2 Gt per year) accumulates in the atmosphere and can be seen in the concentration curve. The remainder may dissolve in the oceans or may be taken up by the new forests that are being planted

(below) An estimated carbon budget for the atmosphere in a single year. Figures are in gigatons (Gt) per year. A metric ton is approximately equal to an imperial ton. The "missing sink" consists of the carbon that is currently unaccounted for, but it may actually be the additional uptake of carbon by vegetation. The total biomass of vegetation may be increasing as a direct result of the fertilization effect of the raised atmospheric level of carbon dioxide.

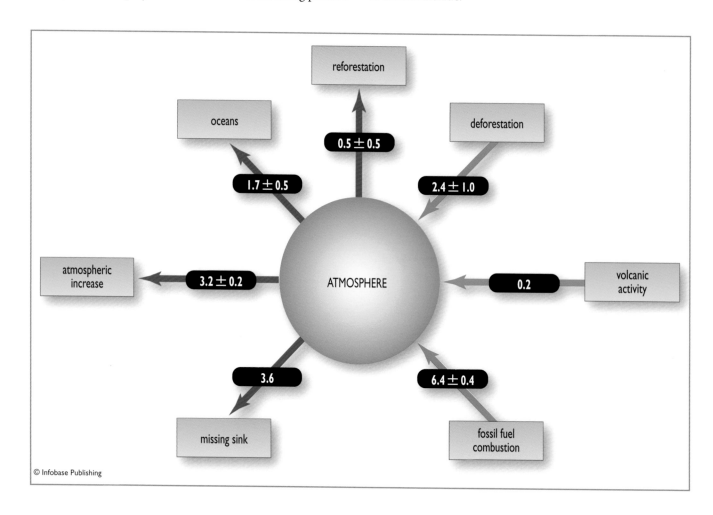

world are at least a part of the missing global sink of carbon that is currently protecting the atmosphere from an even greater accumulation of excess of carbon dioxide than is currently being observed.

It is difficult to estimate the total amount of carbon that accumulates as fossil organic matter in the wetlands of the world. Organic accumulation is most efficiently achieved by peatlands, and the bulk of the peatlands lie in the temperate regions of the Northern Hemisphere. There are about 1.4 million square miles (3.5 million km^2) of these northern peatlands and they have an average rate of peat formation of about 0.07 ounces per square foot (21 g m^{-2}) each year. This approximation is based upon a wide range of individual measurements from different parts of the world, but sample plots are usually small, while the area of peatland is vast, so scaling up results in considerable inaccuracy. The total amount of carbon stored by the peatlands each year, when calculated on this basis, comes to 0.07 Gt. This is a very small net storage rate, equivalent to only about 1 percent of the carbon released into the atmosphere by human activities. Nevertheless, wetlands are currently a sink rather than a source of carbon and for that reason they play a positive role in the global carbon budget. Human destruction of wetlands, and any climate change that might lead to drying of wetlands, could easily convert them from a sink to a source of atmospheric carbon and thus add to the greenhouse effect.

■ WETLANDS AND METHANE

The balance between wetlands being a net carbon sink and a net carbon source is a delicate one. Human mismanagement or a change in climate could influence the swing of the pendulum. There is, however, a further component in the equation that must be considered, namely the production of methane by wetland ecosystems.

Methane, CH_4, is produced during the anaerobic respiration of some bacteria, which derive energy from the breakdown of larger organic molecules but do not oxidize them to carbon dioxide. Like CO_2, methane is a greenhouse gas, absorbing strongly in the infrared region of the spectrum. In fact, it is a much stronger absorber than carbon dioxide: Methane acts about 30 times more efficiently than that compound in trapping infrared radiation. This means that methane is a much more powerful greenhouse gas. Fortunately, however, it is much scarcer in the atmosphere. Whereas CO_2 is found at a concentration of about 355 ppm (parts per million, equivalent to 0.0355 percent), methane is present at only about 1.7 ppm. Whereas CO_2 is responsible for about 50 percent of the greenhouse effect, methane is the cause of about 14 percent, so it is still a significant contributor to this problem despite its relative scarcity.

The methane in the atmosphere comes from a variety of sources. Leaking natural gas pipelines (22 percent), landfill sites and sewage disposal (11 percent), and biomass burning (8 percent) are all sources that can be directly related to human action. Methane is also produced by termites (4 percent) or, more specifically, by the bacteria contained in their digestive systems. Ruminant (cud-chewing) animals such as cows also contain methanogenic bacteria in their rumens (first stomach chamber), where compounds such as cellulose are broken down, and they are responsible for a significant methane release into the atmosphere (16 percent). It is natural wetlands (24 percent) and the artificially managed "wetlands" of rice paddies (12 percent) that between them contribute the biggest fraction of methane to the atmosphere. The diagram below shows that methane emissions are greatest in the Tropics, as a result in part of rice cultivation and termite activity, and also in the high latitudes because of the abundance of peatlands. As in the case of carbon dioxide, the atmospheric content of methane is increasing rapidly, and ice-core studies show that the increase has largely taken place over the past 100 years, before which methane levels were steady (see the diagram on page 78). Again, the human hand is clearly suspect in this major change in atmospheric chemistry.

Methane emissions to the atmosphere at different latitudes north and south of the equator. The effect of rice paddies is concentrated in the Tropics, while the natural wetland emissions form two peaks, one in the Tropics and a second in the northern boreal zone.

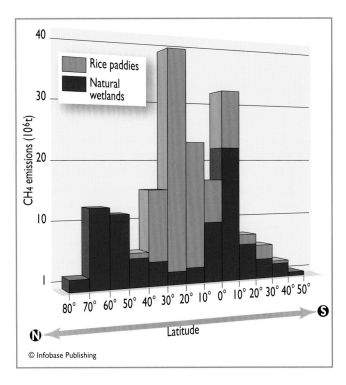

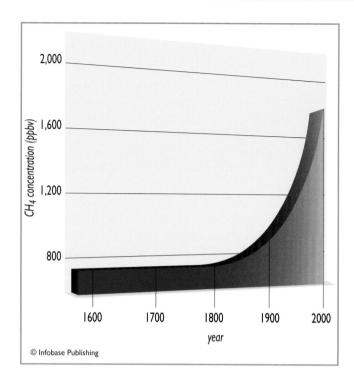

© Infobase Publishing

The concentration of methane in the atmosphere over the last 400 years based on the analysis of bubbles of gas in ice cores and also on historical records. Note the greatly increased concentration in recent years.

In wetlands, methane is produced within the deeper, anaerobic parts of the soil profile. The gas may reach the surface as bubbles through pores in the soil, lake sediment, or peat, or it may be transmitted through the hollow parts of plants. Thus, the hollow, tubular stems of reeds and the leaf petioles of water lilies, which carry oxygen down into the roots, can also transport methane from the depths of the soil into the atmosphere. It is possible to detect the diffusion of methane out of the cut stem of a reed in a wetland and to gauge the rate of efflux of the gas. Cut rice stems, following the rice harvest, similarly act as direct outlets for methane from the anaerobic layers of the saturated soils.

The map of global distribution of wetlands (page 4) shows their tendency to be most abundant in the high northern latitudes (50–70°N) and also around the equatorial regions. The diagram of latitudinal methane production shows similar peaks associated with these natural wetlands. In addition, however, there is also a tropical and subtropical band of methane production, particularly in the Northern Hemisphere, which is associated with rice cultivation in the hot but irrigated parts of the world.

The combined impact of rising CO_2 (predicted to reach 550 ppm by 2050) and CH_4 in the atmosphere, together with other greenhouse gases (see figure on page 76), is resulting in a gradual increase in global temperature. This will have considerable geographical implications for wetlands as well as many other habitats, one of which may be an increase in CO_2 and CH_4 production from the mires themselves. If this is so, then the atmosphere enters an an upward spiral, where raised temperature causes the release of greenhouse gases from wetlands that will accelerate the global warming yet further.

Raised temperature increases the speed of chemical reactions and, within a given range of tolerance, this applies to biochemical reactions in microbial communities also. It is to be expected that more decomposition will take place, and more carbon dioxide and methane will be produced, in a "greenhouse" world. The greater rates of methane production in tropical wetlands are apparent in the bar graph on page 77, and global warming would likely raise the output of methane from the temperate and arctic mires.

There may also be a more complex interaction between the two gases in the modified atmosphere, however. Plant photosynthesis, if temperature, light, and water supply are adequate, is known to be limited by the level of CO_2 in the atmosphere. A rise in CO_2, therefore, would accelerate photosynthesis in wetland plants, as has been demonstrated in many experiments. In one such experiment, using an ombrotrophic blanket bog site in Wales (a rain-fed peatland spreading over hillsides and plateaus), methane output was also monitored. Following the artificial raising of atmospheric carbon dioxide to 550 ppm in enclosed conditions, experimenters found that methane emission from the peat rose by a factor of three. Increased productivity on the part of the vegetation (mainly *Eriophorum vaginatum*, the cotton grass) led to increased root production and organic exudation from the roots, which stimulated methanogenic bacteria in the peat into higher levels of activity. This threefold increase in methane emission was achieved under controlled conditions so that temperature and water table were not allowed to alter. In a greenhouse world with higher temperature and lower water table, the methane output could have been elevated higher still.

There is, however, yet one further complication to the picture that could operate in the other direction. If plant photosynthesis (and hence production) increases as a result of higher carbon dioxide levels and higher temperatures, then this could result in the bulk of living plant tissues in the ecosystem becoming greater. In other words, the biomass itself becomes a more efficient carbon sink. There is now some evidence to suggest that this is happening. It is difficult to measure accurately any relatively small biomass change over wide geographical areas, but one novel approach has been to look at the "wiggles" in the rising CO_2 curve (page

74). The fall in the curve each spring caused by the "draw-down" of carbon as vegetation begins its growing season can be used as a measure of how much productivity is going on in the hemisphere. If the drawdown is regularly greater each spring, then it suggests that there is more plant biomass at work. A recent analysis of the drawdown data has shown that the amplitude of annual fluctuation has increased by 20 percent in a decade. There is more plant biomass in the hemisphere, and the increase seems to be most marked in the region between 45° and 70°N, where most of the wetlands currently lie together with the northern deciduous and evergreen forests. Forest and mire vegetation seems to be growing faster.

The production of methane by wetlands, therefore, complicates the role of these ecosystems in the carbon cycle and in the development of greenhouse conditions, but the evident growth response of vegetation to rising carbon dioxide may well compensate, at least in part, for increased greenhouse gas production by mires as global temperatures rise.

To summarize, wetlands, particularly peatlands, are acting as a carbon sink by transferring carbon from the atmosphere to organic material in the sediments. They may also act as a sink by increasing the biomass of their living vegetation (especially true of boreal forested mires). In both of these respects, we can regard them as beneficial to the health of the planet. The other side of the coin is that some of the bacteria (methanogenic bacteria) contained within the anaerobic peats are conducting a slow process of decomposition in which they generate the powerful greenhouse gas methane. Warmer climates may increase this effect and thus create more serious greenhouse conditions. One thing is certain, however: The continued drainage and exploitation of wetlands by humans is more likely to make matters worse than to improve them.

■ CONCLUSIONS

The concept of the ecosystem involves the study of all living organisms, plants, animals, and microbes in the setting of their physical and chemical environment. Ecosystem research requires the determination of the rates of energy flow through the ecosystem and the cycling of nutrient elements around its component parts, together with energy and nutrient budgets describing the processes of input, output, and storage. Wetlands can be viewed as ecosystems, and they differ from most terrestrial ecosystems by storing a reserve of energy in the form of undecomposed organic detritus, or peat. This results from the inefficient decay process in wetland environments because of a lack of oxygen supply to microbial decomposers.

Energy fixation by green plants (primary productivity) in wetland ecosystems varies greatly with the type of wetland involved. Reed, cattail, and papyrus marshes have very high productivities, resembling those of tropical rain forests, while temperate bogs have a low rate of primary productivity. The movement of this energy through food webs is also very variable, depending on the wetland type, but is most complex and diverse in tropical marshes and swamps.

The accumulation of peat in some wetlands follows a distinct pattern. Air permeates the upper layers of organic detritus (the acrotelm) because the water table falls in warm and dry conditions, allowing microbial decay to proceed rapidly. In the lower layers of compacted, waterlogged organic matter (the catotelm), however, decomposition is very slow. While plant litter formation exceeds the total decay in the peat column, then the peat mass continues growing. When the height of the peat is such that the total microbial activity is sufficient to match litter formation, however, the peatland ceases its upward growth. It will remain in that condition until climatic change to wetter conditions may encourage further growth.

Wetland chemistry depends on the geology of the catchment and the chemical content of the precipitation. Elements may become trapped in the growing peat and form a growing reservoir, but some scarce elements needed for plant growth, such as nitrogen, phosphorus, and potassium, are often mobilized prior to the death of leaves and reused by the plant. Excess nutrient may enter a wetland from the surrounding catchment, especially if agriculture or forestry is taking place. These lead to eutrophication, in which algae are stimulated to rapid growth, leading to the decay of excess organic material and the depletion of oxygen in the water. The high demand of some rapidly growing wetland plants, however, such as reeds and papyrus, means that these species can be used to extract nutrient elements from contaminated water. Saline wetlands present particular problems for the plants and animals that live in them because water is more difficult to extract from a solution of salts.

The accumulating sediments in wetlands leave a record of their past history, and the changing chemistry of wetland ecosystems is documented in these stratified deposits. Altered patterns of catchment use by people, climatic changes that result in changing oceanicity and precipitation chemistry, and industrial pollution with its associated air pollution and contaminated runoff can all be assessed and monitored by the analysis of wetland sediments.

Wetlands, because they accumulate energy and chemical elements in their sediments also have an effect on the global environment. They build up organic material and in doing so take carbon out of the atmosphere. They thus act as a carbon sink in the global carbon budget. Their contribution to the removal of atmospheric carbon is relatively small (about 0.07 Gt per year), but the considerable mass of carbon locked up in the wetlands could result in their becoming a very significant source of atmospheric carbon if decomposition increased globally. Changing climate and human destruction of wetlands could make this happen. The production of methane, a very influential greenhouse gas, could also increase if the wetland ecosystems of the world begin to release their stored carbon.

The ecosystem concept is thus a very useful one in wetland studies, helping ecologists to understand how wetlands function and how they may be conserved or even managed for the future good of humanity.

Different Kinds of Wetlands

Wetlands are found in many different parts of the world and therefore develop in a range of different climatic conditions. The geology of the areas in which wetlands are found is extremely variable, so the chemistry of the water that enters wetlands can be very diverse. Wetlands also vary in the depth of water that covers them or the position of the water table in the soil. Finally, wetlands constantly change over the course of time, a process termed *succession* (see "Succession: Wetland Changes in Time," pages 14–15). When all of these variables are considered, it is not surprising that there are many different kinds of wetlands in the world.

It is not an easy task to classify wetlands, partly because of their great variety and partly because the different types often grade into one another. A number of different characteristics have to be considered when wetlands are being classified. One of the most important of these is the source of water for the wetland, whether derived from the drainage of a catchment (rheotrophic) or entirely from rainfall (ombrotrophic). These concepts are explained in the section "Hydrology of Wetland Catchments" on pages 23–24). Some wetland ecosystems are also distinctive because of the saltiness, or salinity, of their waters, especially but not exclusively those in coastal regions.

The position of the soil or sediment surface in relation to the water level or the water table is also a valuable basis for wetland classification. This is one of the factors that changes during the course of wetland succession, leading to one type of wetland being replaced by another. The developmental history of wetlands is recorded in their underlying sediments (see "Wetland Stratigraphy," pages 32–34), and this can also be used to classify wetlands into different types. Finally, and closely related to these other features, the vegetation of the wetland provides a useful basis for classification, especially the structure, or architecture, of the vegetation. Floating aquatic plants are characteristic of shallow-water wetlands; emergent herbaceous aquatics are typical of marshes; and swamps are dominated by trees and shrubs with high canopies and relatively rigid, persistent structures.

Because so many criteria can be used in classifying wetlands, it is not surprising that ecologists have suggested a number of different schemes, one of the simplest and most logical of which forms the basis of the account presented here.

■ SHALLOW FRESHWATER WETLANDS

Perhaps the most widespread of freshwater wetlands found throughout the world are those in low-lying areas that receive the drainage water from surrounding basins and catchments. They are widespread simply because of their hydrology—the way in which they receive their water. A wetland that is fed entirely by rainwater, such as the raised bogs of the temperate latitudes, will be restricted in its distribution to areas where rainfall is high and evaporation low, but those wetlands that are fed mainly by drainage water are less dependent on climatic regimes. They can even be found in the relatively hot and dry parts of the world as long as they are fed by a reliable source of water.

Shallow-water bodies of this type that are found in cooler and high-rainfall conditions, however, are more likely to be permanent than those in hotter, drier regions. Where rainfall is low, the existence of a freshwater body is dependent on water being fed into the basin from a wide catchment. So the size and the stability of a water body in a dry region depends on the topography of the surrounding land, as well as the degree to which water soaks into the geological strata. The maintenance of fresh, rather than saline, water demands that the water body be drained by an outlet stream so that the entire system is constantly flushed by moving water. In more arid regions, this may not occur because the water entering a wetland may evaporate rather than pass through, and this leaves the salts behind to create a saline wetland (see "Inland Saline Wetlands," pages 116–118).

Shallow aquatic wetlands, then, can be found throughout the world. Around the edges of Arctic lakes (as in the

Alaskan tundra), forming once extensive areas in temperate regions, but now often drained and reclaimed (fringing a few remaining parts of the shores of the Great Lakes, as at Point Pelee National Park, Lake Erie, Ontario, Canada), and around the margins of tropical lakes (as at Lake Nabugabo on the western edge of Lake Victoria, Uganda), these flow-fed wetlands are widely dispersed in all but the most arid regions of the world. Deep freshwater aquatic habitats merge into what can be truly termed *wetland* as the waters become shallow (less than six feet [2 m]) when aquatic vegetation, rooted in the basal muds, is able to grow and flourish. Often this vegetation takes the form of water lilies (see photograph below) with thick, fleshy stems anchored in the sediment and with leaves on long stalks that float flat on the water surface. Many species are able to grow in water depths of 13 feet (4 m) or more, but they generally grow better in relatively shallow water, less than six feet (2 m) deep. Under favorable conditions, the floating leaves may form a complete cover over the water surface or even become so crowded that they erupt into an irregular mass of upturned foliage.

The platform developed by the floating leaves may be strong enough to support the weight of foraging birds, such as jacanas and gallinules. The larger floating-leaved aquatics, for example the Amazonian water lily *Victoria regia,* are even robust enough to support the weight of a small human. The flowers of these plants also float on the water surface, buoyed up by air tubes in the stalks, and these flowers provide additional resources for animals living in the aerial environment above in the form of nectar and pollen. At the same time, the varied underwater habitat provided by the patchwork of leaves is exploited by aquatic animals from microscopic crustaceans to fish and alligators.

The rooted aquatics may also be accompanied by unattached plants that can have a marked effect on the architecture of this type of wetland. Submerged forms, such as the Canadian pondweed (*Elodea canadensis*), live in patches where the light penetrates the carpet of floating leaves, and floating species occupy any vacant surface water. The floating aquatics include very small plants, such as the duckweeds (*Lemna* species), with circular floating stems only a few millimeters across, but also larger colonial types, such as the water-fern (*Salvinia molesta*) and the water hyacinth (*Eichornia crassipes*), shown in the photograph on page 83, both of which are major pest species of waterways in tropical wetlands.

Wetlands of this type, then, are able to develop wherever there is shallow freshwater, whatever the climate. So, unlike the peatland types, it may be found throughout the world, although the precise species of plants and animals involved in its composition will vary from one part of the world to another. Water hyacinth, for example, is restricted to the tropics and subtropics, as are alligators. Only the American and Chinese alligators manage to survive in the temperate regions. In the cooler, high-latitude areas, these wetlands may freeze during the winter, and both plants, such as water lilies, and animals, such as frogs, may survive these unfavorable periods in the muddy basal sediments where the temperature is likely to remain above freezing.

One effect of the growth of aquatic vegetation in shallow water is to slow the rate at which the water is able to flow. The drag created by the many underwater stalks of leaves and

Water lilies are rooted in submerged mud, but their leaves and flowers are held just above the water surface. They are pollinated by flying insects. *(U.S. Fish and Wildlife Service)*

Floating communities of aquatic plants, such as water lilies and water hyacinth, can cover the surface of open pools. *(South Florida Wetlands Management Department)*

the early observations on the impact of pesticides on the natural world came from studies of predatory birds, such as grebes, living in shallow-water wetlands. Even the amount of water flowing into these wetlands can be affected by human activity, for the clearance of forest from a watershed leads to increased runoff water and this often carries eroded soil that increases the siltation and infilling of the basin. Water may be removed from the catchment for agricultural irrigation or for fulfilling the industrial and domestic needs of people, reducing the input of water to wetlands and causing considerable changes in their vegetation.

The conservation of these shallow-water wetlands, therefore, demands a careful control of the entire watershed that feeds them. The wetland itself cannot be managed in isolation from its surrounding ecosystems, and this inevitably means certain conflicts of interests and often social and political problems in their conservation.

flowers and the tangle of submerged vegetation results in a slowing of otherwise fast-flowing currents. Whereas a fast current can carry sediment in suspension (derived from the erosion of surrounding land surfaces and soils by the drainage water), a slow-moving current has far less capacity for sediment bearing. As a result, part of the sediment carried by the moving waters becomes deposited around the obstacles (vegetation) and the process of filling in subsequently proceeds faster. Added to this input of inorganic particles derived from the wetland catchment is the litter—the dead and decaying organic detritus—produced by the aquatic vegetation itself. The exact amount of organic matter, the residue of the plant productivity after herbivores have eaten their fill, that actually accumulates in the sediment depends upon how fast decomposition proceeds. If the waters are well supplied with dissolved oxygen, then the microbes (the aquatic fungi and bacteria) will be able to make good use of the dead organic matter, which is then rapidly decayed. Relatively little may survive to become incorporated into sediments. If, on the other hand, the water is low in oxygen, perhaps having already passed through stagnant swamps rich in organic detritus, then decay may be impeded and some of the organic matter will become fossilized in the sediments of the water body.

Shallow, open-water habitats have long been important for human beings as locations for fishing. Prehistoric peoples hunted with harpoons in these waters and even built villages within them, perched upon stilts in order to cope with periodic floods (see "Wetlands in Prehistory," pages 180–182). These wetlands have also been sensitive to human activity in their surroundings because the water draining into them can carry whatever humans deposit on the catchments, whether fertilizers, pesticides, or other pollutants. Some of

■ TEMPERATE REED BEDS AND MARSHES

The shallow, open-water wetlands are often surrounded by a habitat dominated by a very different type of vegetation. Tall, narrow-leaved herbaceous plants, reeds, cattails, and sedges, their bases still within the water, form extensive and uniform swaying, flexible beds. The names given to this type of habitat vary in different parts of the world; *marsh* is often used in North America, whereas *swamp* is used in Europe (a term restricted in America to the forested wetlands). The marsh habitat develops in shallow-water environments by continued sedimentation below the water and expansion of vegetation from the lake edges.

There are two types of development that occur within the shallow-water wetlands, both involving the invasion of plants, termed *emergent aquatics*. These are herbaceous plants that have their roots and sometimes their stems (rhizomes) under water and have leaves and/or shoots emerging vertically above the water. The presence of emergent aquatics endows the ecosystem with a completely different architecture, for it now has a significant structure both above and below the water level; it has taken on a totally new spatial dimension that offers a range of opportunities for animals that live above the surface. The emergents can be one of two major forms, and this will affect what goes on below the surface. They may be rooted in the sediment, or they may float in rafts on the surface. These types of plants and their zonation around open water wetlands are shown in the diagram on page 84.

As the water becomes shallower as a result of sediment accumulation, some emergent plant species will be capable of taking root in the muds and silts and extending their shoots

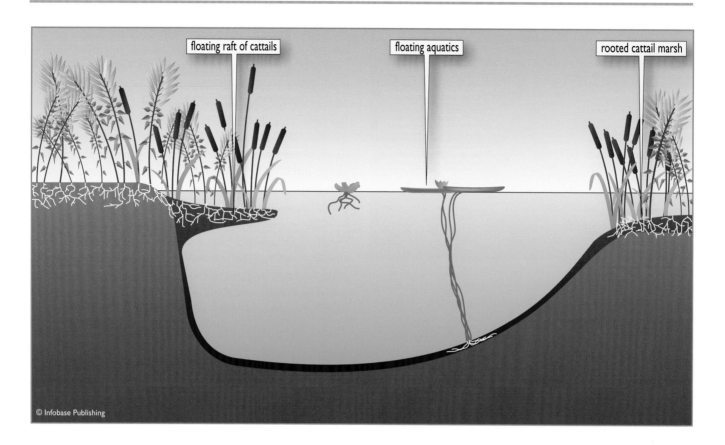

floating raft of cattails

floating aquatics

rooted cattail marsh

© Infobase Publishing

Cross section of a temperate marsh. Cattails and other emergent aquatic plants often form a floating mat extending over shallow water.

or leaves vertically through the water layers and into the air above. Among such plants, perhaps the most widespread throughout the world is the common reed (*Phragmites australis*). This plant can grow to a height of 10–13 feet (3–4 m) above the water level, as in the delta of the Danube River in southern Europe and along the Nile in Egypt. It is able to cope very adequately in water five feet (1.5 m) in depth but thrives in shallower water. At its most productive, it is capable of forming dense stands with up to 15 shoots per square foot (150 shoots per square meter) creating a bed of reeds that may be almost impenetrable by humans.

Other plants may play a similar role in other localities, such as the cattails (*Typha* species) and sawgrass (*Cladium jamaicensis*), so typical of the Florida Everglades. For birds, the dense architecture that these plants supply to the ecosystem opens up many possibilities for building nests among the reeds or hiding floating structures beneath their closed canopies, safe from the view of predatory raptors flying overhead. In North America the marsh wren (*Cistothorus palustris*) is a good example of such a bird, building its domed nest close to the water level to avoid detection. In Europe, the sedge warbler (*Acrocephalus schoenobaenus*) builds a deep, cup-shaped nest attached to several reeds, which is able to sway with the wind without spilling its contents.

Various herons also exploit the reed bed habitat, the most remarkable being the bittern (*Botaurus lentiginosus*), which, when disturbed, raises its bill in line with the vertical emergent plants and is thus well camouflaged. Its eyes are so aligned that it is able to see around the base of its beak and so remain vigilant even in such a remarkable pose. Herons, including the great blue heron (*Ardea herodias*), which is pictured on page 85, are also characteristic of this habitat.

Reed beds efficiently filter sediments from the waters that flow through them. One estimate from an English study suggests that a density of 10 to 15 shoots per square foot (90–150 shoots per square meter) is capable of trapping four and a half pounds (2 kg) of sediment each year per square foot (0.1 sq m). Sedimentation at this rate, of course, ensures that the water becomes rapidly shallower—a process that can be termed *terrestrialization,* or developing into drier land. The rate of sediment buildup will vary considerably depending on a number of factors, such as how close the site is to an input stream, how fast the water is moving, how much sediment is suspended in the water, how dense is the reed bed, and so on, but the presence of emergent aquatics undoubtedly accelerates the process of terrestrialization.

As the water becomes shallower, other plant species may begin to grow among the emergent reeds, enriching the diversity of the habitat from what may initially have been a virtual monoculture of reeds. It also allows the emergents to extend their growth out into the open water, invading the beds of floating aquatics and enlarging the reed bed. Under favorable conditions (good sediment supply and high productivity), the reed front may advance by 3.3 feet (1 m) every year. This encroachment, of course, takes place at the expense of the truly aquatic habitat and may not always be welcome to those wetland users who need to navigate the waters or to exploit them for fish.

An alternative development that also leads to the encroachment of sediment-rooted emergent aquatics into the open water is the growth of floating mats of tangled rhizomes (modified stems) forming extensive rafts of vegetation. Again, this may take place in almost any geographical area where freshwater occurs. The precise species of plant involved depends on such factors as climate and water chemistry. In the arctic and cool temperate regions, especially in acid waters, the bottle sedge (*Carex rostrata*) may assume this role, along with such plants as the bogbean (*Menyanthes trifoliata*) and some rushes (e.g., bulbous rush, *Juncus bulbosus*). These are often instrumental in the development of quaking bogs (see "Quaking Bogs," pages 102–103). Under more alkaline conditions, cattails (*Typha* species) often develop floating rafts, and these may form a substrate that is quite firm and strong enough to walk on.

Floating rafts of vegetation provide a microhabitat in which a range of fish can shelter. The layer of decomposing vegetation above produces a constant rain of organic debris, including the bodies of invertebrates, and this forms the feeding basis for many foraging fish. The floating cover can also supply a place where predators, such as pike, can lurk and from which they can dart to attack their prey, whether other fish or swimming birds.

Although the diversity of plants in marshes and swamps is low, these highly productive ecosystems nevertheless support a wide variety of fish and bird life. They have been exploited by humans for thousands of years, and there are still some peoples who rely on this habitat for their living. In eastern Britain, the extensive lakes and reed beds known as the Norfolk Broads have long been used to supply reeds and sedges for the production of thatch—bundles of dried reeds that are packed together to form roofing material. The development of more efficient and cheaper roofing materials, however, has led to the steep decline of this industry and, as a result, many reed beds have been neglected and, left unmanaged, they become invaded by wetland trees. In southern Iraq, the Ma'dan or Marsh Arabs have occupied extensive areas of marsh for more than 5,000 years and have developed a distinct culture based on the use of reed beds as building materials and fish (mainly barbel and carp) as a source of food. The basal shoots of cattails are also edible and are eaten in Iraq and in Egypt, but both the habitat and the way of life of the Iraqi Marsh Arabs is under threat as the area faces drainage and development.

The conservation of marshland involves the prevention of further succession leading toward swamp forest. This can be achieved by harvesting some of the organic production by mowing or even by periodic burning. Maintaining a high water table, which requires the management of the entire watershed, is imperative, and the quality of the water entering the ecosystem is of great importance to aquatic animal life. Reed beds, however, are efficient absorbers of fertilizers, such

The great blue heron feeds on fish and amphibians at the edge of open-water wetlands and marshes. *(South Florida Wetlands Management Department)*

as nitrates and phosphates, so they can be used to purify waters. Some sewage and water treatment facilities exploit this capacity of reed beds commercially.

■ TROPICAL MARSHES

Although the relatively dry subtropics are not well supplied with wetlands, the tropical latitudes closer to the equator are generally rich in wetlands. An abundance of rainfall ensures that low-lying, water-collecting parts of the landscape tend to become occupied by wetland ecosystems. As in the case of the temperate shallow-water ecosystems, their tropical counterparts become invaded by tall-growing emergent plants, which often form dense floating carpets as their shoots become intertwined.

By far the most impressive communities of floating emergents in the tropical regions, especially in Africa, are formed by papyrus (*Cyperus papyrus*). Growing to a height of 13–16 feet (4–5 m), papyrus communities tend to contain few other plant species. They form very thick rafts of peaty material beneath which there can be a considerable flow of water from the land surface into the water body. As well as

extending out into lakes, the papyrus rafts may break up into small floating islands that drift away from the shore and may then expand and link up again. This is particularly apparent on the western shores of Lake Victoria in Uganda, where such islands are termed *sudds* (see illustration on page 65). These floating rafts are independent of water depth and are sensitive only to the disruptive effects of wave action, so they are often found in up to 30 feet (10 m) of water. A cross section of a tropical marsh is shown in the diagram below.

In southern Sudan there is a vast area (40,000 square miles [100,000 sq km]) of papyrus-dominated marshland called the Sudd. For many years, this tract of wetland impeded attempts to navigate up the Nile River (see "Wetland Exploration—Africa and the Nile," pages 185–187). The breaking up of the papyrus rafts meant that channels were constantly changing so that no maps of the system had any lasting value. Associated with papyrus in the Sudd

Cross section of a tropical marsh. Reeds, cattails, and papyrus grow tall, creating a habitat with complex architecture, interrupted by open pools with floating aquatics.

© Infobase Publishing

is the tall, floating aquatic grass *Vossia cuspidata* that may well have been the "bulrush" in which Moses was hidden in the Bible story.

Papyrus marshes are able to survive even in conditions of very poor nutrients as they are able to accumulate chemicals from low concentration, but if the waters are too acidic, papyrus may be less competitive than some other wetland plants, such as the bog moss, *Sphagnum.* This is a collection of moss species that is much more typical of the temperate bogs than of the tropical swamps, but there are some localities, such as Lake Nabugabo in western Uganda, where unusually acidic conditions have permitted *Sphagnum* to proliferate and, in places, dominate areas of the floating vegetation of the fringing marsh.

In the water, beneath the floating rafts of papyrus, there is a constant rain of organic detritus that accumulates in the sediment below. As in the case of the rooted emergent vegetation, this buildup consists both of the products of local productivity on the part of the raft plants and also the input of inorganic material eroded from the surrounding land surfaces. Tropical swamps are among the most productive ecosystems on Earth, so there is a rapid production of large quantities of litter, but, just as in the case of temperate reed beds, the decomposition rate is also usually fast because of good oxygenation in the water, so the amount of organic matter accumulating in the basal sediment of the lake is relatively low. Nevertheless, siltation and terrestrialization do eventually occur, and the vegetation that once floated becomes rooted in the basal muds. Indeed, in shallow water and flooded valleys, papyrus may never form a floating raft but will invade waters in the same way as an emergent reed with its roots directly embedded in the mud.

The hippopotamus (*Hippopotamus amphibius*) is one of the large mammals often seen in the vicinity of the East African papyrus marshes. It actually spends most of the day in the shallow aquatic habitat along the edges of the swamps. A true papyrus-dwelling mammal, however, is the sitatunga (*Tragelaphus spekei*), a stocky, heavy antelope with unusually broad hooves that enable it to walk across the floating mats of vegetation. It feeds on papyrus and other aquatic plants and is entirely at home in the water and foraging within the dense swamp. Even the young are raised on platforms of downtrodden papyrus hidden within the swamp; their main enemies are crocodiles.

Floating "meadows" of aquatic plants are also found in the Amazon basin. Here there are few true lakes, but the flooding of the river creates extensive backwaters in which reed-like grasses, such as *Paspalum repens,* form floating swamp carpets. These carpets begin life rooted to the ground, but as the floodwaters rise, they are torn from the ground and float away. Many other floating aquatics, such as water ferns and water hyacinth, are carried along in these islands of marsh growth that can extend for several miles. Shallow

lakes in South America, such as those of Colombia, are often invaded by *Cyperus giganticus,* a plant closely related to the African papyrus, which also leads to lakes being filled in through enhanced sedimentation.

The marshes and swamps fringing tropical lakes are vitally important to local human populations, often providing the most efficient means of transport and also supplying food in the form of fish. The floating carpets of marsh are important breeding areas for fish, which are extremely diverse in these tropical lakes. In Lake Victoria, for example, there are about 170 species of cichlid fish, almost all of which are found nowhere else on Earth. Unfortunately, in an attempt to improve the fisheries, a large and voracious predator, the Nile perch (*Lates niloticus*), was introduced into the lake in 1960 and has subsequently decimated the native fish populations. This error has at least served to emphasize how sensitive the wetland ecosystem is to imposed changes.

Once drained, soils of the papyrus marshlands provide rich agricultural land, especially for crops that can cope with occasional flooding, such as rice. This poses a serious threat to the continued existence of herbaceous marshland in East Africa in particular.

Water itself is an important resource in those parts of the tropics where a dry season prevails, such as in the Sudan. The marshes of the Sudd are important to many tribes in this respect, but there is a scheme to build a canal that would bypass the Sudd marshes, thus rendering navigation easier and avoiding the dissipation of much of the Nile water into the wetlands, where it will evaporate. Although there are evident economic advantages in such a plan, it would deprive this unique wetland of much of its water supply and, in doing so, would also have adverse effects on the local peoples.

Just as the reed swamps of the temperate zone have been used for the purification of water by the removal of plant nutrients, so the papyrus swamps of East Africa have proved useful in mopping up the waste fertilizers that drain into lakes from agricultural land and the waste products and sewage from human settlements. At Lake Naivasha, near Nairobi, Kenya, floating papyrus rafts have been found to absorb more than half of some elements entering the lake, including phosphorus, sulfur, iron, and manganese. If these elements are allowed to enter the lake systems, they cause eutrophication.

■ FENS

Terrestrialization of herbaceous swamps and marshes may lead directly to the invasion of tree species or may result initially in the development of a less wet ecosystem still dominated by herbaceous vegetation. This is termed *fen* and is best defined by the depth of its water table. Whereas swamps and marshes are constantly inundated with high water levels, which are always above the ground or sediment surface, a fen has a summer (or dry season) water table that is below the ground surface. At the same time, it remains a rheotrophic mire, fed by the flow of water through the ecosystem. This means that fens are generally well supplied with chemical nutrients from the leaching and erosion of their catchments, but this is dependent on both the richness of the local rocks in plant nutrients and the rate of flow of water through the system (see "Hydrology of Wetland Catchments" pages 23–24). It is thus possible to have both rich and poor fens according to the rate of supply of chemicals to the mire.

Fens are less extreme environments than reed beds and marshes. They are not perpetually waterlogged, so air can penetrate at least the upper parts of their soils for part of the year, leading to the survival of many more species of plants that are less well adapted to extremes of waterlogging. Fens therefore tend to be diverse ecosystems, harboring many species of plants and invertebrate animals. The total biomass of fens is usually lower than in the swamps and marshes because the plants in fens are shorter. Even where plants such as reeds, cattails, and even papyrus survive in fens, they grow less tall and do not dominate the vegetation to the same extent. Invasion and subsequent dominance by tree species is always a possibility in fens, but herbaceous vegetation may persist over long periods due to factors that prevent this from happening. Fluctuating water tables leading to occasional submergence or intensive grazing by large herbivores or the occasional incidence of fire can all restrain tree invasion.

Since they are fed by groundwater, fens are relatively independent of climate, requiring only a sufficient movement of water into a topographically suitable location, such as a hollow, a valley, or a poorly drained slope, to provide waterlogging through much of the year. Peat builds up in the soil but is accompanied by the mineral materials derived from catchment erosion. Fens can be found, therefore, throughout the world, wherever the topography is suitable.

In temperate regions, fens are widespread in basins, floodplains, and low-lying coastal areas. As is the case with most wetland types, they are more common in oceanic regions where precipitation is higher and hence where runoff water is more common and more widely available throughout the year. In such situations they are often closely associated with the larger, ombrotrophic (rain-fed) mires, such as raised bogs. They fringe the edge of these structures, following the course of drainage water around their perimeters (see "Temperate Raised Bogs," pages 103–105). They may also be associated with blanket mire complexes (see "Blanket Bogs," pages 107–111), where fens (usually poor fens) are associated with slope sites that collect runoff water from surrounding, peat-covered catchments. In blanket mires, fen systems are partic-

ularly evident in oceanic sites where the rainwater itself is richer in nutrients, creating a greater opportunity for the development of chemically enriched flushes in the generally acidic bog vegetation.

Coastal fens develop from estuarine reed swamp ecosystems and consequently experience a saline influence from the proximity of the sea. Similarly, the small fens that develop in association with sand dune habitats (sometimes called dune slacks) can be very rich in species, especially when well supplied with calcium from local rocks or from the shells of mollusks in the sand. These dune fens may receive seepage of seawater from beneath the sand and hence have brackish or quite saline waters entering them, leading to the development of vegetation that has both fen and salt marsh elements. Dune fens are particularly evident around the Baltic Sea in Scandinavia and in the coastal dunes of Estonia, Lithuania, and northern Poland.

Temperate fens may be extensive in low-lying regions that receive floodwaters from major river systems. The eastern part of central England, for example, is known as the Fenlands, and most of the area lies close to or even below sea level. The history of human settlement, drainage, and reclamation here is long and complex (see "Wetland Bronze and Iron Age Settlements," pages 182–184), but the original habitat was likely to have been very extensive reed beds and fens with drier areas forested with wetland tree species, and in all probability with some regions of raised bog established. Almost all of this has now vanished as a result of agricultural development, but the original Fenlands must have been a remarkable habitat, covering hundreds of square miles and being virtually impenetrable except along narrow waterways.

The western part of Russia is rich in raised bogs, which are linked and surrounded by fen ecosystems, but as one moves eastward into temperate Asia, fens become increasingly important in the vegetation. Emerging from the forest and into the steppe regions, fens dominated by reeds, cattails, sedges, and mosses are the main wetland types associated with basins and floodplains. Many of these fens have developed in basins of 10–13 feet (3–4 m) depth. The mountains of east Siberia have a very continental climate with hot summers, very cold winters, and relatively low precipitation. Fens develop in the valleys where water collects, but they are often of the poor fen variety, dominated by tussock sedges and cotton grasses, and resembling a southern extension of the tundra. Often these fens become colonized by trees.

In North America there is a similar concentration of patterned fens (also called *aapa mires* or "string bogs") in the northern regions (see "Aapa Mires and String Bogs," pages 89–91). Tree islands occur within these fens as do isolated trees along the ridges, or strings, that run across the gentle slopes on which they develop, and, just as in Asia, the more southerly and continental regions have mainly basin swamps and fens around the edges of lakes and over flood-plains. Basin fens then merge southward into prairies, again with peat depths of about 10 feet (3 m) similar to those of the Siberian region. The swamps and fens of these eastern parts of the boreal and northern temperate zone are often forested with black spruce (*Picea mariana*) and tamarack (*Larix laricina*).

Valley fens are found in the Pacific oceanic region of North America, and poor fen of an acidic nature is located on many of the hillside slopes where water soaks away. The prairie regions have depressions with both permanent and semipermanent lakes around which marsh and fen systems develop (see "Pothole Mires," pages 100–102). In Wisconsin, "sedge meadows" (a local name for fens) are frequent, often forming a distinct habitat intermediate between the reed-dominated marshes and the forested swamps of the area. These fens have much in common with their European counterparts.

The richness of the organic soils of fens, coupled with an abundant supply of water, has made them attractive to agricultural development. Once excess water is removed, the soils are ideal for arable cultivation, and many of our fenland regions have been developed in this way. The conservation and management of natural fenlands involves the control of water supply from their catchments and the prevention of further successional development to woodland. Mowing, or even grazing cattle, often prove effective methods for holding back tree invasion.

■ AAPA MIRES AND STRING BOGS

The boreal coniferous forest regions of northern Canada, Europe, and Asia present a remarkable picture from the air, not only because of their dense cover of trees but also because of their distinctive wetlands that form extensive, winding vistas in the valleys. Most remarkable is their pattern of elongated ridges and hollows that run along the contours of the slopes, cutting across the flow of water through these mires, giving them a striped appearance. This type of mire is known by the Finnish term *aapa mire* or *aapa fen,* although it is sometimes erroneously called *string bog.* In fact these mires are not bogs at all, since a true bog receives all its water directly from rainfall, while these "string" mires receive water from large areas of catchment. Water thus moves through the ecosystem, often very slowly, for the slopes on which the aapa mires develop are usually shallow. Because of the ridges that run across the mires, the course of water movement is often long and irregular as it meanders slowly through an extensive maze of linear pools and wet hollows, as seen in the photograph and in the diagram.

The ridge systems are referred to as *strings* and the pools as *flarks.* Often the ridges are firm enough to walk on, while the pool systems can be deep and impossible to wade through. This means that it is much easier to walk across the breadth of an aapa mire than it is to walk along the length. The strings are generally narrower than the flarks, though this varies with latitude, and they may support tree growth,

An aapa mire in northern Finland. In the foreground is a linear pool, or flark, and the raised ridge behind it is a string. *(Peter D. Moore)*

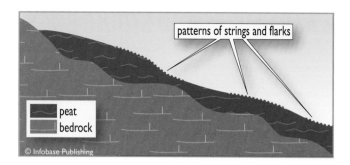

Profile of an aapa mire or string bog. This shallow peatland develops on gentle slopes in northern latitudes. Ridges (strings) alternate with linear pools (flarks) that run across the mire, following the contours of the slope. These mires may extend for many miles.

while the flarks are waterlogged and are rich in floating and emergent aquatic plants such as bog mosses and sedges. Occasionally, beavers may interfere with these developmental patterns by blocking flow channels, and this may result in a general raising of the water table for some distance up the slope on the mire. Such areas are often apparent from a distance because of the dead and dying trees on the submerged and waterlogged strings.

Aapa mires occur to the south of the region where palsa mires abound (see "Palsa Mires," pages 112–114), being found mainly in the boreal region rather than the true Arctic. They are abundant in Finland and northern Sweden and also in the Hudson Bay lowlands of Canada as far south as 46°N, the Lake Agassiz region of Minnesota. In Japan, aapa mires have been recorded as far south as 36°N. They are mainly restricted to regions where the air temperature remains below freezing for much of the winter (December through April in Finland), and where winter snows lie about 16–20 inches (40–50 cm). Aapa mires are also typical of sites with a gentle slope—usually about one percent—down which there is a slow seepage of water.

The peat depth in the string mires is usually quite shallow compared with the raised mires, often less than 16 feet (5 m) but varying with latitude and being generally deeper farther south. Analysis of the content of these peats reveals that this type of mire usually begins its formation as a relatively uniform vegetation of sloping fen, which develops its distinctive raised strings as the peat builds up and impedes the course of water flow, creating an even slower movement of water down the slope as it meanders between the elongated ridges. Studies in Sweden have shown that some of the aapa mires in that country began to form about 6,500 years ago, but that flarks developed only 2,600 years ago. However, this probably varies from one site and region to another.

The question of how the distinctive pattern of string mires develops is still keenly debated. It is very possible that the freeze-thaw pattern of seasonal temperature variation influences the physical development of these patterned mires. Expansion of the wet flarks as they freeze in winter could crush the drier strings and linear patterns could result, much as stones are shifted by frost heaving into lines during the freezing and thawing of arctic and alpine soils. The snow cover, particularly in the Finnish and Japanese wetlands, would be expected to reduce the effect of this action, yet patterning is still produced. So there is still considerable dispute about exactly how their intricate patterns of strings and flarks are formed. A Japanese scientist, Y. Sakaguchi, has pointed out that when water carrying detritus floods across sloping ground, it leaves strings of the detritus in "thatch lines" very like the linear patterns found in aapa fens. This may account for the original development of strings within the shallow, gently sloping mires of the boreal regions, after which the pattern is maintained and emphasized by repeated freezing and thawing. It is also possible that the form of the original ground, surface beneath the peat has an influence in the initiation of the pattern of water flow through these gently sloping mires.

The mineral supply to the surface vegetation depends on a number of factors, including the quality of water draining into the mire, the rate at which it flows through different parts of the system, and the elevation of each particular location. In limestone areas, for example, waters rich in calcium enter the mire and create rich fen vegetation in the wet areas, whereas in acidic rock regions the drainage water will be poor in nutrients and poor fens will result. If the slope is slightly steeper, water flow will be faster and this brings more nutrients to the plants in the flarks. Plants growing on the strings, however, will always receive less input from drainage water and may even become totally dependent on precipitation, like miniature bogs.

The aapa mires, or string bogs, of North America often have regularly spaced elevated areas occupied by black spruce (*Picea mariana*). These form on structures that are sometimes termed *peat plateaus* that interrupt the general pattern of strings and flarks and are thought to develop from *Sphagnum* hummocks that have extended and coalesced to form more extensive raised areas. As temperatures fall with the approach of winter, the initial freezing process takes place in the waterlogged hollows, and the bog moss hummocks are forced upward by the pressure of the ice. This may add impetus to the formation of these peat plateaus and is further exaggerated by the insulating property of the bog mosses on the hummocks, which prevents deep penetration of frost in winter.

More southerly aapa mires have a greater proportion of their surface occupied by strings rather than flarks. The importance of trees, particularly spruce in Eurasia and tamarack in North America, on these strings also becomes greater with increasingly southerly location. The fauna of these mires is typical of the boreal forest regions, with beavers, bears, woodpeckers, and owls regular residents. Some wading birds, from cranes and geese to smaller species, such as broad-billed sandpipers (*Limicola falcinellus*), wood sandpipers (*Tringa glareola*), and golden plovers (*Pluvialis dominicus*), breed in the varied microhabitats of the string mires.

Occupying the continental, boreal regions of Eurasia and North America, where human population densities are relatively low, the string mires have not suffered as badly from interference as many other wetland types. As forestry demands more land, however, the drainage and afforestation of these mires will be demanded. This process is already resulting in the destruction of aapa mires in Finland. Changing climate, especially if this results in warmer winters, could lead to an improvement in the agricultural potential of these regions, and this will place yet more pressure on the aapa mires.

SPRING MIRES

Not all wetlands are large. Nor does a wetland have to be large in order to be important from a conservation and biodiversity point of view. There are some mire ecosystems that are small and yet contain such a diverse assemblage of plants and invertebrate animals that they are worthy of careful conservation. Among these are the mineral flushes and spring mires. Often found in hilly terrain, they usually have a rich supply of calcium carbonate in their waters and

are normally associated with sites at which water emerges from an underground source and saturates the local mineral soil.

Spring mires receive water emerging from the soil at a higher pressure than mineral flushes, which can lead to an accumulation of peaty moss remains interbedded with calcium carbonate. This may build up into a considerable mound of material, several yards thick, covered with vegetation and lime deposits, as shown in the diagram. Mineral flushes do not accumulate bodies of sediment in this way but have a constant percolation of water under relatively low pressure, often trickling down a slope bearing hummocks of vegetation interspersed with channels carrying the water. Both types of mire are usually associated with slopes, and both are more frequent in montane, or moist, cool upland, situations.

Mineral flushes are particularly heterogeneous in their vegetation cover. Usually they consist of small patches of grasses, sedges, and other herbaceous plants that form hummocks, which elevate their surfaces a few inches above the surrounding soil surface. Peaty, organic matter may be present in these hummocks, but more often the soil beneath them is mineral in nature and waterlogged. In areas that have been glaciated during the last ice age, the hummocks may be

Cross section of a spring mire. Groundwater enters the ecosystem as it wells up from the soils and rocks below. As the peat develops, it often contains areas rich in calcium carbonate.

comprised of small piles of glacial detritus, till, or boulder clay that remain uneroded by the passage of streams moving through the mire. The plants on their summits protect them from the erosive forces of the surrounding streams.

The hummocks in these mires are usually small, but the influence of their slight elevation on the chemistry of their surfaces can be considerable, especially if they consist of lumps of acidic boulder clay. The water running around their bases may be effectively neutral or slightly alkaline (pH 7–8), but only two inches (5 cm) above this, on the top of small hummocks, the soil may be distinctly acid (pH 4.0). So the mires possess not only a diverse microtopography but also a very varied mosaic of chemical microhabitats. This is one of the features that gives these mires a relatively high biodiversity.

Another important factor, especially for temperate zone examples of mineral flushes, is their great antiquity. The fact that they are often ancient habitats is not immediately obvious, since it is not something one might expect from a collection of small-statured plants on a wet hillside, but these flushes have often maintained their characteristic open, eroding nature for many thousands of years, often right back to the last ice age. It is ironic that the short-term instability of the soils in such sites has often lent long-term stability to the vegetation.

An example from an intensively studied site in northern England serves to illustrate this point. In a small valley called Teesdale in the north of the Pennine Mountain chain, there are several mineral flush mires of the type described here that are exceptionally rich in rare species of plants and animals. Many of these, like the spring gentian (*Gentiana verna*) and the alpine bartsia (*Bartsia alpina*), together with certain snails that occupy the calcareous runnels between the tussocks, are found in very few locations in the British Isles, and one must go to Norway or to the Alps of central Europe to find them in any abundance. It has proved possible to trace the history of the plants because their pollen grains have been preserved in local geological deposits (see "Microscopic Stratigraphy: Pollen," pages 40–43), and they have been present at the site since the ice retreated some 15,000 years ago. Biogeographers call such species *relicts* of a former widespread distribution, in this case the tundra habitats of the glacial period.

The reason such assemblages of relict species have survived in these mires is that the soil instability has prevented the growth of trees that covered all the surrounding areas and shaded out the lowly alpine plants and their associated fauna. Constant slippage and loss of soils from these slopes under the incessant trickle of water has kept the site clear of forest and has preserved an ancient type of wetland vegetation, together with its relict fauna.

When the release of confined groundwater is more forceful than is the case in these mineral flushes, a spring mire may result, as shown in the diagram. These develop an elevated dome of sediment, but unlike the ombrotrophic (rain-fed) domes of raised bogs, the surface of these mires is fed by the upwelling of groundwater from below. As a result, they are usually rich in nutrients, and if the water is calcareous the surface may have patches of lime deposited. The surface vegetation under such circumstances is equivalent to a rich fen, and snails are usually present, confirming the availability of calcium.

Spring mires can develop to a depth of several yards, depending on the pressure of water emerging from the ground. Peat may develop as a result of the waterlogging, but decomposition is often quite rapid because of the high level of oxygenation, leading to a relatively active microbial population in the sediment. This means that the organic matter is usually broken down into a sloppy, featureless mud, often mixed with calcium carbonate, which may be apparent as layers within the sediment profile. There may even be layers of mineral soil or sand in the profile, resulting from the forceful upwelling of water. Where spring mires occur in acid regions, lime is not present in the sediments, but sandy layers are still apparent.

Acidic spring mires may be vegetated by *Sphagnum* bog mosses, together with plants characteristic of poor fen. Occasionally they may emerge within the lateral acid lawns of valley mires (see "Valley Mires," pages 95–97), in which case they may be identified by a richer growth of vegetation, sometimes even by wetland trees. There are records of upwellings of this kind occurring within raised mires, creating a complex hydrology with upward, downward, and lateral movements of water. This has been described in some peatlands in Minnesota where domed mires occur within general areas of flow-fed, rheotrophic wetlands. These have been regarded as raised bogs (see "Temperate Raised Bogs," pages 103–105), but true raised bogs are fed entirely by rainwater. The Minnesota domes resemble spring mires, especially in dry periods when the flow of water through the peat mass is upward. This may be reversed when rainfall is heavy, but the continental climate in this part of America has low precipitation, only about 24 inches (60 cm) per year.

Most spring mires are relatively small, rarely in excess of 2.5 acres (1 ha) in area, and their sloping nature, combined with a rich flow of water, often results in their being invaded by trees, and sometimes by reed beds, sedge fen, or rushes.

■ MONTANE WETLANDS

Mountains create their own climate. With increasing altitude, temperature falls and precipitation usually increases, so high-altitude sites often bear very different vegetation from lowland sites at the same latitude. It might be expected that wetlands would be favored by increased rainfall and lower rates of evaporation, but there is the additional consideration of steepness of slope, which leads to the drainage of water from many montane habitats. In situations where rainfall is exceptionally high, peatland communities can develop even where the land is sloping. These may extend to cover entire landscapes, forming ombrotrophic blanket mires (see "Blanket Bogs," pages 107–111), but in most mountainous regions of the world, wetland communities are confined to pockets, hollows, and gentle slopes where water can accumulate.

Lakes are often found among mountains, and their fringes may be occupied by a range of wetland habitats, including marshes and swamps, but many mountain ranges in higher latitudes were buried beneath ice only 20,000 years ago, and that ice has scoured out deep valleys in many sites. So mountain lakes are often steep-sided and deep, which leads to the formation of only a narrow band of wetland vegetation around their edges. High altitude plateaus and tablelands, however, such as the Andean Puna of South America, support shallow lakes and wetlands. Perhaps the best known is Lake Titicaca, between Peru and Bolivia. The isolated nature of these upland lakes has led to evolutionary separation of species in some of their basins. Lake Junín in Peru, for example, has its own species of grebe, the Junín grebe (*Podiceps tacznowskii*) that differs from other grebes in the area in its shorter beak length and its different feeding preferences.

Shallow lakes in montane areas can develop from kettle holes, where blocks of ice have melted and left small, wet hollows. These sites become infilled by sediments because erosion of soils and rocks is rapid where slopes are steep and water flow is vigorous. Such sites can become occupied by raised bogs (see "Temperate Raised Bogs," pages 103–105), or by saddle mires if the original wetland occupied a ridge region. Saddle mires are found at the watershed of two valleys, and they extend down the slopes of both valleys to form a saddle shape. Small raised bogs and basin bogs are typical of many mountain regions, including the Rockies of Canada, the Carpathians of eastern Europe, and the Alps of southern Europe. They are usually covered by a vegetation of *Sphagnum* mosses, together with dwarf shrubs and trees, in Europe especially by the mountain pine (*Pinus mugo*), as shown in the photograph.

In the Ural Mountains of Asia, most of the mires have developed on gentle slopes and are rheotrophic (fed by moving groundwater), hence are technically fens. Because the temperature in the northern Urals is low, these mires become patterned into ridges and pools, forming aapa fens (see "Aapa Mires and String Bogs," pages 89–91) and may even contain permanent blisters of ice that develop into palsa mounds (see "Palsa Mires," pages 112–114). Thus we find wetland types that would normally be associated with

High altitude bog in southern Austria covered by dwarf mountain pine (*Pinus mugo*)　(*Peter D. Moore*)

Gentians, such as this blue gentian (*Gentiana clusii*) from southern Germany, grow well in the moist, grazed mires in high alpine regions. The large flowers serve to attract scarce insect pollinators. *(Reportandum)*

arctic and subarctic conditions extending much farther south in the mountains.

Perhaps the most characteristic wetlands of almost all mountainous regions are the small mineral spring systems and fens that are found wherever water finds its way to the surface and saturates the soil. The limited area of these mires does not mean that they are biologically unimportant. They are often remarkably rich in plant life, such as gentians (see photograph), and invertebrate species and sometimes support collections of rare species, all concentrated into small areas. The reason for this is that they are often relict communities dating from earlier times when their area was

more extensive. Open, treeless conditions with soils that were saturated every spring as snow melted were found over much of the temperate zone during the last glaciation. Subsequently, most of these regions have become occupied by forest as trees spread northward during the retreat of the ice, but some wet, steep, and unstable sites have persisted as mire habitats and have remained free from the shade of a tree cover. These sites now continue as small fragments left behind (literally, "relict") in the progress of vegetation development. For this reason, such sites are of high conservation interest. These sloping montane fens may be acidic or basic, depending on the local geology. The calcareous, basic types are generally richer in plant species than the acid mires and have diverse communities of wetland snails.

The instability of these sloping mires is apparent when their profiles are examined, for it is possible to discern peat layers alternating with soil and silt bands that have been eroded from the surrounding slopes and catchment during storm events or sudden snowmelt. Subsequent runoff inundates the mire surface and buries it in mineral material. The diagram illustrates one such layered peat profile from Austria in the European Alps. Only plants and invertebrate animals capable of coping with such instability can survive here.

The East African mountains that lie on the equator have several types of montane mire. The extremely wet climate

Profile of a sloping montane mire from a glaciated region. The peat growth is periodically interrupted by outwash detritus and eroded soils, resulting from episodes of glacial melt or from storms and floods. These form extensive bands in the peat profile.

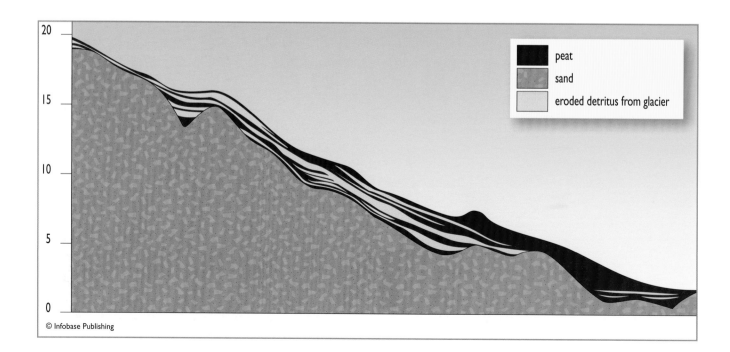

peat
sand
eroded detritus from glacier

of the Ruwenzori range in western Uganda results in the development of blanket mire communities (see "Blanket Bogs," pages 107–111). However, the mountains farther east have mires which have only formed within hollows, often in the summit craters of these ancient volcanoes, as on Mount Elgon on the Uganda/Kenya border (more than 13,000 feet [4,000m] in altitude). The vegetation of these mires is dominated by a sedge, *Carex runssoroensis,* which forms dense tussocks and develops a peaty substrate that is often more than three feet (1 m) in depth. Like the lakes of the South American Andes, these mountaintop mires are so isolated from one another that they have their own endemic species of plants. On the mires of Mount Elgon, for example, are found plants of the giant lobelia (*Lobelia elgonensis*), a species found nowhere else on Earth. Other East African mountains have their own distinctive species of giant lobelia. Despite being on the equator, these high-altitude mires are subject to intense frost action and have some similarities to Arctic mires.

In the Southern Hemisphere, montane mires are also often dominated by cushion-forming plants and are sometimes referred to as *cushion bogs.* These are found in the uplands of Tasmania and New Zealand. The plants that give this mire type its name, such as the genus *Donatia,* have densely packed, spirally arranged, linear leaves, together with an accumulation of dead tissues that create the cushion form. The same genus of plants is also found on the montane and subarctic mires of southern Chile and Argentina in South America. The South American montane mires are in the fortunate position of enjoying protected status as part of the Chilean National Parks system.

Many montane mires in more populated parts of the world suffer from trampling pressures of both domesticated animals and humans. Many are also intensively grazed by cattle and sheep. Although they can withstand grazing reasonably well, the trampling of humans along well-worn trails results in peat compaction and alters the hydrology, aeration, and drainage of the system. Since trails often run along valleys, parallel to streams, they cross these mires at frequent intervals and can create extensive damage. When excessive trampling causes the loss of surface vegetation cover, peat erosion takes place, and the trails then tend to widen and result in yet further destruction. Rock stepping-stones, wooden boards, or even raised walkways may become necessary for the conservation of montane wetlands in heavily used areas.

■ VALLEY MIRES

Valleys act as water-collecting areas, so wetlands are most likely to be found within valley systems. In regions of dry climate, wetland habitats may be entirely confined to valleys. If drainage of water from the valley is impeded, either artificially by a dam or naturally by an outcrop of rock or a deposit of glacial detritus, then a water body may be retained within the valley. If drainage is completely unimpeded and if the gradient of the valley is sufficient, then water may be shed rapidly and no true wetland will develop. Between these two extremes there are situations where drainage water is partially held back or flows more slowly, and in such cases a valley mire develops.

The combination of a low gradient, impermeable underlying rocks or soils, and an ample supply of drainage water can lead to the invasion of plants that tolerate flooding and a lack of oxygen around their roots. Members of the rush (Juncaceae) and sedge (Cyperaceae) families are frequently involved in such colonization. The presence of such plants under wet soil conditions can lead to the formation of organic, mucky soils that themselves hold water because organic materials can act as a sponge in the soil and retain considerable quantities of water, up to 10 times its own dry weight. The formation of spongy, peaty soils, with their tendency to build up because of the poor rates of decomposition in such wet conditions, can lead to changes in the hydrology of the entire valley. The peaty soil may actually impede the flow of drainage water yet further and create conditions in which more and more peat will accumulate. So peat in a valley leads to further peat formation, and a valley mire begins to form.

Valley mires vary in structure and nutrient richness depending on a whole range of conditions, including geology, chemistry, precipitation, valley morphology, and human land use. If there is a rich supply of water draining through rocks and soils that are well provided with calcium, then a rich fen vegetation may develop in the valley. If, however, the calcium supply is poor, then a poor fen system initiates the valley mire development, but valley mires, although they often contain poor fen elements, are more complicated in their arrangement than normal fens. They usually comprise a range of different fen and bog habitats assembled together in a complex pattern, as shown in the profile diagram on page 96. Some ecologists refer to them as a *mire complex* because of their distinctive and varied mix of wetland habitats.

The complexity is due to a number of factors. If the concentration of nutrients in the groundwater is low, then the amount of nutrient received at any point in the mire depends on how fast the water is flowing. The pattern of peat accumulation affects the way in which water drains through the mire and the rate at which it is able to flow, so the peat development actually influences the nutrient movements and, in turn, the vegetation. The outcome, as can be seen from the diagram, is that the location of a stream draining the mire provides the fastest flow of water and therefore the best supply of nutrients. It is around the stream that the

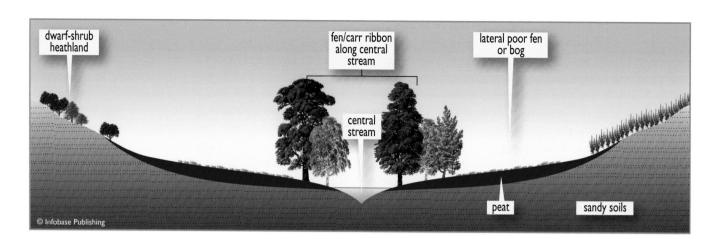

Cross section of a valley mire. A stream runs through the valley, and the flow of water in the vicinity of the stream permits the development of richer vegetation, often with trees. The lateral parts of the mire have slower water flow and therefore poorer nutrient supplies. Consequently, poor fen or bog vegetation develops in the lateral regions. Valley mires can be up to a mile across.

most demanding vegetation is able to grow, often leading to the development of a ribbon of rich fen or even swamp (see "Temperate Swamps," pages 97–98) following the course of the central stream in a valley. In temperate valley mires, willows, alders, cottonwoods, and some birch species are the most frequent components of this *carr*, along with cattails and bur reeds. The flow of the stream often disturbs this bank vegetation, eroding the peaty sediments especially during storms, so there is a turnover.

The sides of the mire remain waterlogged with seepage water flooding over from the stream and draining from the surrounding valley slopes. The movement of this water is often slow, held back by the growing lawn of peat, so the nutrient supply to the plants of these lateral parts of the mire is poor. Poor fen vegetation develops as a consequence, although there may be local flushes with enhanced flow and hence with a richer flora and fauna. These lawns often bear cotton grasses (*Eriophorum* species) and bog mosses (*Sphagnum* species), and the latter can create hummocks that coalesce to form elevated, waterlogged platforms—incipient raised mires. It is possible that truly ombrotrophic, rain-fed conditions can develop on these peat platforms, especially if they are extensive (several hundred yards across), in which case this part of the valley mire is indeed a valley bog, but the term *bog* needs to be used with caution because much of the mire system is actually fen (see "Fens," pages 87–89).

The calcium status of the different parts of a valley mire is best revealed by examining the gastropod mollusks (slugs and snails). The shells of snails require a supply of calcium to make the calcium carbonate from which they are constructed, so the presence of populations of snails indicates a good calcium source. In low calcium conditions, often found on the lateral lawns of valley mires, slugs may be found (they lack shells), but generally no snails. So it is often possible to estimate the acidity (pH) of a location from its invertebrates.

Animal life in a valley mire is often very diverse—a reflection of the diversity of habitats present in a relatively limited area. Since the spatial scale of these mires is generally small, it is the invertebrate life that is normally of greatest interest, ranging from the fast-flowing stream fauna of the central section to the stagnant acid pool fauna of the lateral lawns. Valley mires are often of great importance for dragonfly diversity and conservation because of the range of aquatic microhabitats available. Birds also benefit from the habitat diversity, both in their feeding and breeding activities. Several duck species use this type of mire, and snipe are well adapted for feeding in the mud and peat, probing with their long bills, and breeding in tussocks of sedge and grass.

Human impact on valley mires is often severe, for they are relatively easily drained and reclaimed for pasture or forestry by the construction of a herringbone pattern of drainage ditches. They are also frequently burned, partly because they often have an abundance of very flammable ericaceous dwarf shrubs and partly because landowners use burning as a method of nutrient release from vegetation and peaty soils in times of drought. Fire has often played an important part in the development of valley mires, as in the case of blanket mires (see "Blanket Bogs," pages 107–111). Studies of the peat profiles of many European valley mires, ranging from Norway to Spain, have shown that the basal soils beneath the peat are rich in charcoal fragments, some of which date back more than 4,000 years into prehistoric, Bronze Age times. Early prehistoric burning of vegetation led to charcoal production, and these inert carbon particles blocked the porosity of the underlying soil and led to poor drainage, waterlogging, and the initiation of valley mire formation. In this way, many valley mires may owe their origin to the impact of human beings.

■ TEMPERATE SWAMPS

The invasion of tree species into herbaceous marshes and fens is generally part of a natural sequence, in the absence of any factor that holds back the succession. Most trees have difficulty germinating when the water table is above the soil surface, but many are capable of establishment if there are occasional dry spells, even on a very local scale. Once established, many trees are able to tolerate periodic submergence. The seeds of invasive trees are generally small, so all they need is a microsite above the water table in which they can put out root and shoot. In fact, the initial establishment of trees in a wetland often occurs as a result of the mode of growth of some of the herbaceous species. Tussocky growth among sedges, for example, provides exactly the type of habitat required by trees for their germination, and the elevated tussocks are usually the first spots upon which young trees such as willows, alders, and pines are first found. In this position, the roots are clear of the water and are thus able to avoid total waterlogging.

If this process occurs while the habitat is still regularly submerged by water through much of the year, then a swamp forest becomes established. Here the trunks of the trees are often surrounded by water as they mature, or extensive pools persist among the remaining tussocks. In North America the term *swamp* is often used for this type of ecosystem. The equivalent term in Europe is *carr*. The development of a leaf canopy, associated with a much higher plant biomass than is found in marshes and fens, leads to considerable changes in the entire ecosystem. The microclimate is less extreme and the microhabitats available to animals are more varied, so a whole new range of insect, bird, and mammalian life may take up residence. On the other hand, the lower light intensities beneath the canopy may result in the loss of many open water, marsh, and fen plants that are not able to tolerate the shade, often resulting in a decrease in herbaceous plant diversity. So the arrival of trees in the wetland creates a very different type of habitat.

The forested swamps of western Europe are often dominated by alder (*Alnus glutinosa*), a flood-tolerant tree that has nodules the size of tennis balls on its roots in which reside bacteria that fix nitrogen for the production of proteins, similar to the process found in plants of the pea family. This symbiotic association benefits the tree, which gains proteins, and the bacterium, which gains sugars from the tree's photosynthesis, but the fixation of nitrogen depends on the penetration of air to the roots as nitrogen is not very soluble in water, so some of the alder roots are usually found near the soil surface where air can circulate when the water table falls. In central and eastern Europe, including the mountainous regions, it is the gray alder (*A. incana*) that dominates the forested swamps.

Some swamps may be created by beaver dams. The Eurasian beaver (*Castor fiber*) is much reduced from its former range, and the range of the American beaver (*C. canadensis*) has also contracted as a result of hunting and habitat loss, but both are capable of engineering wetlands by damming valley streams and inundating the land upstream. Many trees die in this process, but wetland trees such as willows, poplars, and alders are favored and encouraged. Since these species are palatable to the beaver, the outcome is very satisfactory, and many other wetland species also benefit.

Forested swamps are widespread in North America but most abundant on the eastern side of the continent. They are found in the valleys of the northeast. The swamp forests of the southerly regions are richer in tree species and tend to be dominated by deciduous rather than coniferous trees, including red maple (*Acer rubrum*), elm (*Ulmus americana*), aspen (*Populus tremuloides*), and various species of ash and oak. These deciduous forests of the swamplands are often referred to as bottomland hardwoods, and they include such famous sites as the Great Dismal Swamp of Virginia and North Carolina and the Okefenokee Swamp of Georgia and Florida. The Great Dismal Swamp is about 80 square miles (200 sq km) in area, and its important trees include Atlantic white cedar (*Chamaecyparis thyoides*) and the tupelo (*Nyssa aquatica*) that can grow in water up to six feet (2 m) in depth.

Some of the most important wetlands of the United States lie in the southern states, particularly in the Mississippi delta region and in Florida. The bottomland hardwoods of the Mississippi delta have been seriously affected by a range of developments in recent years, including flood control measures in the main river channel, drainage and reclamation for agriculture, and development for industry and recreation. Its swamp forests still support large populations of muskrats, but larger mammals such as the jaguar and the black bear are seriously reduced and under threat. It has been estimated that this habitat is being lost at the rate of 40 square miles (100 sq km) per year.

In southern Florida lies the Big Cypress Swamp, 1,500 square miles (4,000 sq km) of swampland dominated by tall bald cypress trees (*Taxodium distichum*), together with dwarf cypress (*T. ascendens*) and slash pine (*Pinus elliotii*). In this swamp the alligator (*Alligator mississippiensis*) plays an important role in the maintenance of diversity by digging channels and deep pools that endure even during periods of drought. These pools provide a source of food and water

Bond Swamp National Wildlife Refuge, Georgia *(John and Karen Hollingsworth, U.S. Fish and Wildlife Service)*

for many species, including herons and egrets, during these times of stress.

In Australia, forested swamps are confined to areas where the water supply is adequate and reliable, most notably in the northernmost region of the Northern Territory at the Kakadu National Park. Important trees include the coolibah (*Eucalyptus microtheca*), river red gum (*E. camaldulensis*), and also paperbark trees (*Melaleuca* species), which are important swamp trees in the tropical regions of Southeast Asia. In the southeastern Australian state of New

South Wales, the floodplains of the Murray River provide an exploitable resource of wetland timber.

Some of the forested swamps of the United States are of great importance in stabilizing water runoff and preventing floods. The Great Dismal Swamp, for example, has a large water storage capacity and absorbs water from excessive rains (as during the catastrophic impact of Hurricane Floyd in 1999). Some swamps have also been used as a renewable supply of energy, as in the case of the alder carrs of southern England during medieval times. The alders were cut to ground level, and their wood was used to fire furnaces for iron smelting. Since alder rapidly sprouts from the cut bases, the forest would recover quickly, so a rotational system of cropping the swamps meant that the supply of biomass energy was sustainable indefinitely. Management by opening clearings in swamp forests can be beneficial to wildlife, allowing light to penetrate to the forest floor and stimulating plant growth within the waters.

Cypress swamp on the Loxahatchee River *(South Florida Wetlands Management Department)*

■ TROPICAL SWAMPS

Uniformly high temperatures and lack of frost provide an ideal growing environment for plant life in general, and trees in particular, as long as the supply of water is adequate. Tropical swamps, in which water is abundant, are therefore highly productive habitats (see "Primary Productivity of Wetlands," pages 48–50) in which the mass of living material, the biomass, accumulates to a greater

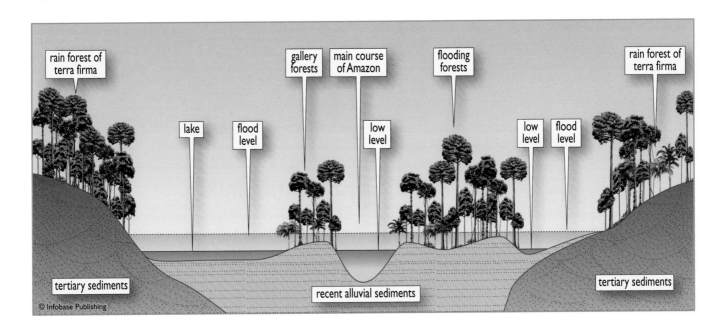

rain forest of terra firma

gallery forests

main course of Amazon

flooding forests

rain forest of terra firma

lake

flood level

low level

low level

flood level

tertiary sediments

recent alluvial sediments

tertiary sediments

© Infobase Publishing

extent than in any other wetland type. Wetland trees that can cope with the waterlogging of their roots overshadow the herbaceous marshes and often assume dominance, leading to the development of the tropical swamp forest.

In the world's greatest mass of tropical forest, the Amazon Basin, wetland systems are found chiefly in the floodplain. This region occupies about a quarter of the total Amazon forest and includes those regions where the many tributary rivers meandered and changed their courses during the last few thousand years. Much of this area remains quite dry during half of the year, although rainfall exceeds evaporation throughout the year, but the area is inundated by up to 30 feet (10 m) of water during the annual floods (April to October); see the diagram above. At this time, the spread of the floodwaters brings aquatic animals deep into the forest, and many trees have seeds that are eaten and dispersed by fish during the floods.

Several types of swamp forest occur. In the regions that border the Andean mountain chain, the floods bring large quantities of suspended silt, much of which is deposited in the upper swamps. These are called white-water swamps, or *varzea,* and the trees of these fertile regions are tall with buttressed roots that help support their great height. Among the trees are the native rubber tree (*Hevea brasiliensis*) and the sandbox tree (*Hura crepitans*) that can reach heights of 200 feet (60 m).

There are also permanently waterlogged sections of the forest, often developed where the river once meandered into a wide curve but has now straightened its course once again, leaving a crescent-shaped section as an isolated "oxbow" lake. These are important wetland habitats that rapidly become colonized by plants as soon as the scour of moving water ceases, leading to the formation of floating grass meadows and shrub vegetation. Trees then follow to form a swamp forest.

Floodplain wetlands of the Amazon forest. The Amazon River periodically floods as water enters it from the Andes Mountains. The floodplain forests, therefore, are regularly converted to swamps.

Among the first and most important trees to establish itself here is the palm *Mauritia flexuosa.* Once the palm cover is dense, a deciduous tree, *Luehopsis hoehnei,* related to the lime, is able to invade. This type of swamp forest is locally called *aguajal* and provides an important habitat for a number of bird, reptile, and mammal species, including the zigzag heron (*Zebrilus undulatus*), sun bittern (*Eurypgya helias*), and rufescent tiger heron (*Tigrisoma lineatum*). The black caiman (*Melanosuchus niger*) also uses these forested oxbow lakes, as does the giant river otter (*Pteronura braziliensis*), because their permanent waterlogging and relative stability supply ideal breeding locations. The hollow stems of the *Mauritia* palms are also used as nesting holes by blue-and-yellow macaws (*Ara ararauna*), a species that has become threatened by vigorous collecting for the caged-bird trade. Permanently waterlogged swamp forest of this type, dominated by *Mauritia* palms, becomes more abundant near the mouth of the river because the push of the ocean tide holds back the drainage of the freshwater.

Some of the rivers that enter the Amazon Basin have passed through acidic, nutrient-poor sands, and the drainage waters (pH about 4.0 to 5.0) are said to be among the least fertile on Earth. The low level of nutrients leads to the development of swamp areas that are poorer in productivity, biomass, and species diversity than the *varzeas*. These are the black-water swamps, or *igapos.* The waters are dark and have only a slow decomposition rate resulting in high levels of suspended organic detritus, much of which is carried away by the floodwater, thus draining the forest of what little nutrient capital it possesses.

Swamp forests also occur in the coastal regions of South America north of the Amazon Basin, in Suriname and Guyana. These are similar to the flooded swamps of the Amazon and share many of the same tree species, such as palms and the kapok tree (*Carapa guianensis*), but they generally accumulate more peaty materials in their sediments. They blend into mangrove forest at the coastal fringe as the salinity of the water rises.

African swamp forests are found scattered throughout the continent south of the Sahara. As in the Amazon forests, the palm family figures among the important tree species, with different types of palm featured in different parts of the continent. In west African swamps, such as those of Nigeria, the raffia palms are important (genus *Raphia*, the source of raffia for matting and basket weaving), while in the more southern swamps, such as the Okavango, trees of the date palm genus *Phoenix* are found. These tall, spiny palms fringe the waters of the swamp and are often accompanied by strangler figs (*Ficus burkei*) that begin life high in the canopy of other trees, grow by extending their tangled root system down the trunks to the peaty soil below, and eventually smother their hosts in the mass of woody growth. Also frequent in the Okavango swamp forest is the African mangosteen tree (*Garcinia livingstonei*), which is the favorite perch of Pel's fishing owl (*Scotopelia ussheri*), a remarkable fish-eating owl whose presence can often be detected by the piles of fish scales in the droppings below its perches. Fan palms and acacias are also present but usually on the more elevated islands that only suffer flooding very occasionally.

In East Africa, the date palm genus *Phoenix* is important in the swamps rather than the raffia palms. Like most African swamp forests, these accumulate a deep peaty substrate, which can be used by peatland scientists to determine how the forests developed and how long they have been here. An upland swamp forest in Rwanda, the Kamiranzovu Swamp, has been determined to be 38,000 years old from analysis of its basal peats. The upland swamp forests of Africa have forest cover that is shorter and less dense than their lowland equivalents; in many ways they resemble the swamp and coastal types of wetland found in temperate regions.

It has been estimated that 60 percent of the tropical peatlands of the world are found in Indonesia, and the coastal regions of the southeast Asian islands are also rich in peat-forming swamp forests (see "Tropical Raised Mires," pages 106–107). Many of these are so unique that they demand particular attention, for they are elevated into domes, just like the raised mires of the temperate zone.

The tropical swamp forests are exploited for their fishery resources, with the peoples of the Amazon Basin deriving about 60 percent of their protein intake from subsistence fishing. The expansion of urban populations in the Amazon, however, has led to the development of a fishing industry that is placing an increasing strain on these resources, and there is a need for greater planning and conservation of fish populations in the expansion of this industry. Forest exploitation is not a major problem as few of the wetland trees are of value to commercial timber interests.

■ POTHOLE MIRES

Certain areas of the continental interiors of both North America and Russia have landscapes that are littered with small wetland hollows, each with an area of only a few acres or less. Most are less than 10 acres (4 ha) in area and are generally only about three feet (1 m) in depth. From the air, both their density and pattern is apparent, but they are far less evident on the ground because of the flatness of the landscape in which they lie. The density of these small wetland sites can be as high as 150 per square mile. In North America these wetlands are called *prairie potholes* and are most abundant in southern Canada (Alberta, Saskatchewan, and Manitoba, which contain about 65 percent of the prairie potholes) and the northern edge of the prairie region of the United States (Iowa, Minnesota, the Dakotas, and Montana).

Strictly defined, these wetlands consist of a series of shallow-water wetlands together with fringing marshes and fens, but their distinctiveness lies in their small size, their density, and the fact that they expand and contract at certain times of the year, drying in summer and becoming flooded during the spring snowmelt. As a result, some of the hollows may become virtually dry during the summer, depending entirely on occasional rainstorms to provide their water. The underlying soils are heavy and nonporous, and the potholes have no outlet streams, so evaporation is the only source of water loss and this can lead to increased salinity in some of them.

Two factors are involved in the location and development of these pothole mires. The first is a relatively flat landscape lacking major drainage rivers, within which there are hummocks and hollows, permitting the accumulation of small water bodies. The second factor required is a distinctly continental climate with hot, dry summers and cold winters. The general landscape conditions, both in North America and Russia, were produced by the activities of glaciers during the last ice age. The pothole region lies within the limits of the last major glaciation (about 110,000 to 10,000 years ago) and lay beneath a mass of moving ice that scoured the basal rock. As the climate warmed, the ice mass began to decompose, leaving behind large volumes of mineral detritus that had been accumulated during the time of ice expansion. Within this heavy "boulder clay" were blocks of ice that became separated from the main glacier during its retreat, almost like icebergs, and these slowly melted within their matrix of rock debris. As they decayed, they formed hollows within the boulder clays, sometimes called *kettle holes,* and these provided the landscape within which the pothole mires developed.

Because the pothole wetlands lose water during the summer, peat formation is impaired. A falling water table allows oxygen to enter the sediments, and this stimulates the activity of decomposing bacteria, so the organic matter is broken down and does not accumulate. Silts from the surrounding basin, however, are washed in during snowmelt, so the potholes do gradually fill up with sediment. The waters of the prairie potholes are generally neutral to alkaline in reaction, but some of the northern basins have acidic waters.

The summer drop in water table in the prairie pothole mires affects the vegetation because it causes the marginal wetland to dry out and so does not permit the growth of plants that are drought-sensitive. The farther one goes from the center of the pothole, the drier the summer conditions, so a concentric series of zones often develops, reflecting the bands of increasing dryness. Where water is reliably present, even during the dry season, cattail (*Typha*) and bulrush (*Scirpus*) species predominate, but they are replaced by salt marsh grasses and glassworts (*Puccinellia* and *Salicornia*) in sites that are saline. In the less permanent water zones, fine-leaved grasses are dominant rather than these coarse-leaved marsh species. Trees, such as alder and willow, inhabit some of the wetlands in what is otherwise an almost treeless landscape.

These wetlands have long been an important habitat for the animals, both vertebrate and invertebrate, of the prairie region. The wetlands provide a source of water at a time and in a place where water can become scarce. Animals such as otters, raccoons, beavers, muskrats, and mink are very dependent on the potholes as a habitat, and in the past, before the intensive development of agriculture in the region, bison and deer undoubtedly used the wetlands. Above all, however, the potholes have provided a resource for wildfowl, particularly ducks. The pothole wetlands have been termed "the duck factory" of North America because of the numbers of breeding wildfowl in these reed-fringed pools. Although the potholes represent only about 10 percent of the wetlands of the United States, it has been estimated that 50 percent of the duck population breed within them. The ducks breed at precisely the time in the spring when the water supply is at its best and the growth of the fringes of marsh plants provide cover for nesting. Mallard (*Anas platyrhynchos*), pintail (*Anas acuta*), shoveler (*Anas clypeata*), blue-winged teal (*Anas discors*), and redhead (*Aythea americana*) all use these wetlands. The potholes also provide important stopover sites for a range of birds on migration between the Canadian north and the Gulf of Mexico, using the Mississippi flyway.

Human populations have long been established in the area and have exploited the pothole wetlands for hunting and fishing. Native American peoples also burned the prairies regularly for driving game and regenerating grass growth, and this custom would undoubtedly have had an impact on the vegetation of the potholes during the dry season. To the European colonists, the wetlands were an unwelcome source of mosquitoes and other blood-feeding insects that made the already difficult prairie life even more so. Perhaps this is one of the reasons why so much of the wetland has been drained, although agricultural development was the major incentive for reclamation. In the 1850s about 18 percent of Minnesota, the Dakotas, and Iowa was covered by wetland; now the figure is about 8 percent, the remainder having been drained for agriculture. Even the wetlands that remain are in danger of receiving excessive inputs of fertilizers and pesticides from the surrounding land, where the intensity of agriculture continues to increase. Since the 1980s, government-funded incentives to drain and develop the area have ceased and wetland restoration has been increasingly undertaken.

There are evident economic advantages in restoring the potholes, for these water bodies can store excess storm water, helping to prevent flooding, can act as irrigation reservoirs in drought, and are a source of water for livestock.

There is similar concern for the conservation of the mires of the Russian steppes, where agricultural development is depleting the habitats available in this summer-arid environment. As in the United States, these mires are particularly important to birds as breeding areas.

Mires of the pothole type are also found in Europe, particularly where dry, continental conditions exist in areas once glaciated and now covered in boulder clays or other glacial deposits. Even in the relatively oceanic British Isles, small wetlands, often temporary in nature, are found that resemble pothole mires. In the west of England, small mires have been formed in deep hollows created by underlying salt deposits becoming soluble and collapsing. The resulting pools, often called *meres,* are deep and steep-sided. They often become colonized by floating rafts of vegetation and develop into quaking mires (see "Quaking Bogs," pages 102–103).

In eastern England the climate is much more continental in nature (hot, dry summers and cold winters). Here are found some different meres that have developed over glacial deposits and fit more closely to the North American pattern of pothole mires. They are filled with water through the winter and gradually dry out during the summer, leaving a characteristic concentric pattern of rings caused by the invading vegetation. In the case of these English mires (such as Ringmere in East Anglia) the annual celery-leaved buttercup (*Ranunculus sceleratus*) is the main invading plant, and the rings are formed by different stages in the life cycle being represented within each ring. The innermost is the youngest, with newly germinated seedlings, while the outer ring is the oldest, and it is here that the plants first begin to produce their yellow flowers, making the rings even more prominent.

■ QUAKING BOGS

Most peatlands have soft surfaces, and walking across them can be a difficult process because of the lack of support from the uncompacted peat, but some mires feel particularly unstable because their surface vegetation is merely a thin layer floating on water—not only does the surface sink, it also rocks under the weight of a person. Some marshes and swamps are of this type, such as the papyrus swamps of the tropics in which the dominant plants colonize open water areas by extending their rhizomes out over the surface. But there is also an acidic bog habitat that occurs in temperate regions and that floats in this way; the Germans call it *Schwingmoor* ("quaking bog"), which is an apt and graphic description.

The process of succession in shallow lakes is driven by the infilling of sediment, eventually passing through marshes and swamps and perhaps leading on to raised bog formation (see "Temperate Raised Bogs," pages 103–105). If the lake is small and steep-sided and lies within a limited catchment with a poor inflow of water, sedimentation can be very slow and marginal vegetation will contribute little to the buildup of organic material on the lake bottom or to the trapping of suspended sediment entering the lake in the streams. This type of lake often remains in that condition for a very long time as a result. Because the water is deep and stable, the layers become firmly stratified, and there may not even be regular mixing of bottom and surface water. This reduces the occurrence of nutrient enrichment by water mixing that would enhance the lake's productivity.

Lakes of this type (termed *meromictic*) often result from kettle holes, depressions in glacial drift where an isolated block of ice has gradually melted to produce a small, deep lake. The sediments that accumulate are undisturbed by winds on the surface and build up in a laminated form, rather like the growth rings in a tree. Each season is recognizable in a core taken from such a site, often with a dark organic layer produced in summer when plankton are productive, and a pale, mineral layer in winter when soil washed in from the catchment predominates.

The steepness of the sides of the lake may make it impossible for marginal vegetation to invade because the water is too deep for them to take root in the bottom sediments, but it is possible for invasion to take place as a result of the extension of floating rafts from the lake edges. These gradually spread over the lake surface, eventually leaving just a single pool of open water at the center of the lake. In time, even this disappears, and the raft of vegetation thickens and begins to support additional plants and even hummocks of mosses and woody plants. The plants involved in the initial colonization of lakes in this manner include reeds and cattails, but often it is the smaller aquatic species that are important, such as bottle sedge (*Carex rostrata*) and bogbean (*Menyanthes trifoliata*).

The raft at the edges of the lake is older than that at the center and is usually the thickest part, and trees frequently establish themselves on the floating mat, as shown in the profile diagram. In Europe it is often the Scotch pine (*Pinus sylvestris*) that invades, and in eastern North America, the Atlantic white cedar (*Chamaecyparis thyoides*) in acidic sites and northern white cedar (*Thuja occidentalis*) in lime-rich regions. Farther out, the carpet is thinner, but it may still support small trees or shrubs such as heather (*Calluna vulgaris*) in Europe or leatherleaf (*Chamaedaphne calyculata*) in North America. The presence of trees on the floating mire surface can create eerie effects when a person walks over the peat, for as the surface rocks under human weight, the tall trees sway violently. Hummocks of bog mosses (*Sphagnum*) also develop on the raft, particularly on the open, sunny areas, and eventually spread and merge into elevated mounds, effectively raised above the influence of the water below. The surface vegetation can thus develop into a small raised bog system floating upon an enclosed water body.

Where tree development extends all the way into the center of the bog, the decrease in the height of trees as one approaches the middle is often apparent. This is especially true when the floating peat layer is thin, for any tree growing on thin peat over water will gradually sink and may kill its own roots (especially in the case of trees such as Scotch pine that cannot live when submerged). Only very slow-growing trees survive, and the dead remains of the fast-growing individuals demonstrate the disadvantages of rapid extension in such a habitat. It is effectively a natural form of bonsai.

Quaking mires of this type have been studied in the eastern part of the United States (from Minnesota eastward) and in southeastern Canada. They are also present, but rare, in the British Isles and western Europe. Their status in Europe is complicated by the long history of human activity in wetlands, many of which have been harvested for peat, resulting in deep, steep-sided depressions within peat bodies. These old cuttings, some of which are very extensive, may become water-filled and then recolonized by floating rafts of vegetation to form human-induced quaking mires.

The impact of human land use on quaking mire development may be more widespread than is currently appreciated. Scientists have examined the question of why, at a particular stage in a lake's history, it becomes colonized by floating marginal vegetation, and some surprising facts have emerged. A study of sediments from one quaking mire in

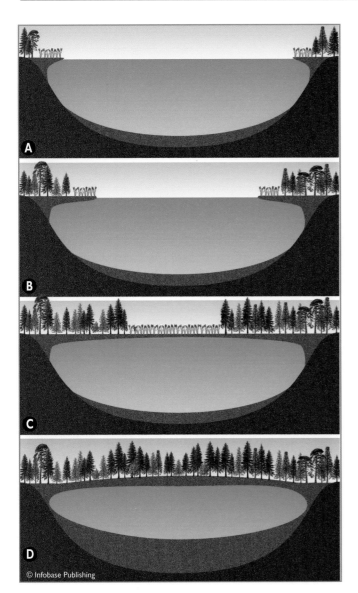

The development of a quaking mire. (A) Small, steep-sided basins are colonized by floating vegetation around their edges. (B) These floating mats extend outward over the lake. (C) Eventually even the central pool is covered, and the entire lake surface bears a floating carpet of peat on which swamp trees become established. (D) Accumulation of peat leads to increased thickness, and even tall trees can be supported.

Wales, for example, showed that immediately prior to the formation of the floating carpet, the surrounding catchment had been cleared of trees and the soils had been cultivated for growing hemp (*Cannabis sativa*) during the 16th century, when hemp was used extensively to make rope for the English navy. The sudden input of plant nutrients from the locally disturbed soils stimulated the marginal vegetation of the lake into its raft development. In Ontario, Canada, very similar processes have been described in which the impact of European pioneer farmers led to the nutrient enrichment

and consequent eutrophication of lakes and the subsequent development of quaking mires.

This is not the only explanation of quaking mire formation, however. In some sites in northwest England, very thick masses of peat are found overlying deep bodies of water, and it is thought that these have developed as a result of basin sinking. This area of Britain is rich in underground deposits of salt and as these dissolve, the basin deepens and fills with water below the peat. So salt solution may account for the development of some types of quaking bog.

◼ TEMPERATE RAISED BOGS

Some of the most impressive of the world's peatlands are the extensive domed masses of peat that can be seen in parts of North America, northern Europe, Russia extending into Siberia and the maritime Pacific provinces of the Far East, and also the southernmost region of South America. There are also mires in New Zealand that have many features in common with the raised bogs of the Northern Hemisphere. The domed mires or raised bogs are so called because their surfaces are actually elevated above the surrounding land surface, forming a convex dome often several miles across and with a central height of 30 feet (10 m) or more above its margins. Often such bogs occupy a single basin, but in more rolling country they may overspill their containment and join with other expanding bogs to form extensive units, called *ridge-raised bogs.*

The process by which these mires grow is slow and complex. They often originate from lake sites, or from estuarine habitats, and usually pass through a swamp forest stage during their infilling and upward progress. Eventually the vegetation becomes dominated by various species of bog moss, *Sphagnum,* which has remarkable properties as a peat former (see "Plants Life in Bogs: Bog Mosses and Lichens," pages 147–151) and which is largely responsible for the ultimate elevation of the peat dome. The dead remains of the bog moss become compacted under the weight of the growing vegetation, but the rate of decay is so slow that the surface is gradually raised above the water table. A cross section of a raised bog shows the layers of different kinds of peat that have formed in the course of its development, as shown in the diagram. In this way the bog leaves a record of its own developmental history.

In New Zealand there are raised bogs in which *Sphagnum* has not played the leading role. Instead, plants belonging to the higher plant family Restionaceae have been implicated in building up large masses of peat, sometimes to a depth of 40 feet (12 m), and have elevated the peat surface above the influence of groundwater. These mires, particularly the Kopuatai Peat Dome in the South Auckland region, have developed under warmer conditions than their northern

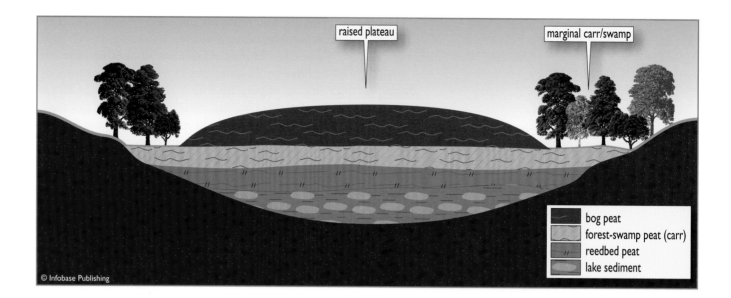

raised plateau

marginal carr/swamp

bog peat
forest-swamp peat (carr)
reedbed peat
lake sediment

© Infobase Publishing

Cross section of a raised bog. The sequence of different sediments reflects the course of successional development that this complex habitat has undergone over a period of several thousand years.

counterparts, but under high precipitation (about 50 inches [127 cm] per year).

Precisely how a waterlogged ecosystem can maintain its water when elevated above its surroundings has been a matter of debate for many years, but the explanation seems to be that the bulk of the peat dome is almost impermeable to water, so the rainfall that lands upon the dome is held there until it drains off laterally (see "Hydrology of Peat," pages 24–27). This type of mire, then, is truly "ombrotrophic," or rain-fed, so such mires can develop only under conditions of adequate rainfall; otherwise, their surfaces would dry out. In fact it is more complicated than that because evaporation depends on temperature, so even a high precipitation will not allow the growth of raised bogs if the temperature is too high. In temperate regions, however, it seems that a minimum of 19 inches (48 cm) per year is needed to sustain a raised bog. In southern Chile they develop only in areas with more than 24 inches (61 cm), and in New Zealand they seem to need more than 50 inches (127 cm) to maintain healthy growth.

The vegetation of raised bogs varies considerably in different parts of the world, but they have in common this elevated structure that makes them into bogs. In oceanic areas, such as Ireland and other parts of western Europe, they are normally treeless, or with relatively few trees, as shown in the photograph on page 105 of a Finnish raised bog. They are vegetated with dwarf shrubs such as heather (*Calluna vulgaris*), together with cotton sedges (*Eriophorum* species) and, of course, expanses of *Sphagnum* bog mosses. This low vegetation permits an extensive view of the dome of peat, which

is often flattened as a plateau on the summit. The sloping sides lead down into surrounding wetlands of swamp and fen. Such a view is rarely possible if the bog is forested.

Trees are more abundant on raised bogs in continental areas, that is, those that are further removed from the oceanic influence of the sea, as for example in central and eastern Europe and in northeastern North America. Any trees present must, of course, be able to cope with waterlogging, but some pines and spruces succeed in growing in these wet, acid habitats. In interior Alaska and in the region south of Anchorage, raised bogs occur that have stunted and contorted trees, usually of lodgepole pine (*Pinus contorta*). In eastern North America, black spruce (*Picea mariana*) and tamarack (*Larix laricina*) are the usual invaders of domed mires.

The surface of the raised bog is far from uniform, as can be seen from the photograph. Pools, sometimes several feet deep, alternate with drier hummocks of vegetation, and these collections of pools and hummocks are often arranged in patterns, although this may only be apparent when viewed from the air. If the central part of the raised bog forms a plateau and is relatively flat, then the pool arrangements may be random and ragged, but if the slope of the dome is appreciable, the pools arrange themselves in a series of concentric rings. If the entire mire is on a gentle slope, the pools may take a crescentic form, curving along the contours. Raised mires with these crescent patterns of pools are called *eccentric bogs*. These are found in the Atlantic region of eastern Canada and are frequent in Scandinavia, but they become scarce farther south in America and Europe. There are some eccentric bogs, however, in the cool montane microclimate of the European Alps.

There has been considerable debate among ecologists about how pools are formed and whether they last for a long or a short period. The question was first raised when people began to cut away at the bogs to harvest peat, leaving exposed peat faces behind. Often it was possible to detect dark and pale alternating layers in the peat, and similar patterns were discovered when corings were taken in the middle of bogs. These bands were interpreted as alternating pools and hummocks, the supposition being that pools might deposit more peat, grow more quickly, and eventually overtake in height the neighboring hummocks, leaving them to become the next generation of pools. This idea of a so-called *regeneration complex* was extremely popular through much of the 20th century, but more detailed analyses of peat exposures in Ireland led to a very different conclusion. Pools actually seem to stay in roughly the same place through time but occasionally expand or contract according to the conditions of wetness on the bog surface at any given time. It was these periods of pool spreading that gave a banding effect to the peat.

This raises a new question. Why does a pool on a bog surface tend to remain a pool rather than fill in to become a level area or grow into a hummock? The answer to this seems to lie in the process of decomposition. The dead leaves and twigs of vegetation that fall into bog pools actually decompose faster than those that accumulate in moss lawns and hummocks. Why this should be is still unclear, but it is possible that algae in the pools generate more oxygen, and this encourages bacteria and fungi to break down any plant lit-

An extensive raised bog in Finland. The oceanic raised bogs of western Europe are usually poor in tree cover.   *(Peter D. Moore)*

ter. Whatever the cause, the outcome is that pools do not become filled with detritus but retain their depth and simply expand or contract their area according to the current surface wetness.

Just as vegetation and topography vary over the surface of a raised bog, so it varies around the edges of the dome. The sloping sides of the dome (called the *rand*) can be quite steep, and because of this water drains from these areas quickly, leaving them drier than other regions. This is often reflected in the nature of the vegetation. Toward the base of the rand, the peat surface is considerably lower than that of the plateau, and hence receives drainage water from surrounding regions in addition to rainfall. Such areas are rheotrophic, or flow-fed, because much of the mineral nutrient content of the water is derived from surrounding soils and is brought to this low-lying region (the *lagg*) in drainage water. This supply of water with its much richer content of dissolved minerals permits higher productivity in this zone, and as a result trees are much more abundant, including willows, birches, and alders.

Human interaction with raised mires has long taken on a destructive form. The value of peat as an energy supply has led to its extensive exploitation by cutting, drying, and burning. More recently, the horticultural use of peat as a soil conditioner has resulted in even more vigorous destruction of the raised bogs. These scarce and fragile wetlands have often taken 7,000 years or more to develop, so they are effectively irreplaceable. This in itself should be ample justification for their high-priority status for conservation.

◼ TROPICAL RAISED MIRES

Although many of the forested tropical swamps, such as those of northern South America and East Africa, accumulate peat in their sediments, the process of peat buildup is most apparent in the forested mires of Southeast Asia, particularly Sarawak, Borneo, New Guinea, and Sumatra. In the low-lying coastal and riverine wetlands of this region, large areas of the swamp have produced such large quantities of undecomposed vegetable detritus that the accumulating peat has now elevated the ground surface above the level of water draining through the catchments. These mires have become truly ombrotrophic, or fed entirely by rainwater rather than groundwater, and they are the tropical equivalent to the temperate raised bogs (see "Temperate Raised Bogs," pages 103–105).

Another aspect of the great scientific interest of these wetlands is their similarity to the coal-forming swamps of the Carboniferous period. Although the vegetation contains very different tree species, the general structure and form of the vegetation is very similar to that of the coal swamps, and their extensive nature and deep deposits of organic sediments with very low silt content also make them the closest thing we currently have on Earth to the coal-generating mires that have played such an important part in climatic and human history (see "Coal-Forming Mires," pages 174–175).

It is also in these Southeast Asian swamps that some of the richest biodiversity of vegetation and animal life is found, comparable to that of the Amazon, and this, when coupled with the vulnerability of these regions to exploitation, is a strong reason for concentrating conservation efforts on this unique area of the Earth. Of particular importance is the occurrence of many endemic species on these raised peatlands. These are species that are found only in this type of habitat and only in this region, so their survival is specifically linked to the protection of these peatlands.

In Papua New Guinea the coastal mangroves grade into flooded forests of nipa palm and then spread through sago palm into the peat-swamp forest itself. These peat forests consist of elevated domes of peat developed between the main drainage rivers and tributaries. In the more western mires, the tree family Dipterocarpaceae has become dominant, especially the species *Shorea albida*, which has massive buttress roots and is valued for its timber. This type of forest is easily spotted from the air because of the generally pale tops to the tree canopies, often marked by the holes from lightning strikes. Clambering plants, such as *Pandanus*, are found in the lower stories of the forest, especially in the dense, low forest of the central dome area, and the insectivorous pitcher plants of the genus *Nepenthes* are also frequently present (see "Plants in Bogs: Carnivorous Plants," pages 151–154). The bog mosses (*Sphagnum* species), so typical of the temperate raised bogs, may also be found on the rain-fed bog plain, the plateau occupying the center of these raised mires.

There is a definable zonation of forest types from the riverine and the marginal swamp forests that are flooded by groundwater, especially during the monsoonal rains, through the sloping edges of the elevated dome (often about 1–2 miles [2–3 km] from the rivers)—what would be termed the *rand* in temperate raised bogs—and onto the bog plateau itself. These features are shown in the diagram on page 107. Not only do the species of tree vary along this gradient, but the size and the density of trunks change. The peripheral swamps have tall trees, up to 130–160 feet (40–50 m) in height; the density of large trunks is relatively low and the water table is permanently high, leading to the presence of "breathing roots" or pneumatophores in some species. On the rand area, however, the water table can fall below the peat surface during the dry season and in this region is a transitional forest of lower stature, perhaps only 80–100 feet (25–30 m) in height. Moving onto the peat dome, the size of the trees (including *Shorea albida*) becomes yet smaller—maximum about 65 feet (20 m)—but their density is greater, forming what is referred to as *pole forest*. Apart from being smaller in height, the diversity of tree species also declines from about 30–50 species in a half-acre (0.2 ha) area around the edges of the swamp to only 12–25 species in a similar area of the bog plain. Even farther from the river, a taller interior forest is found, reaching heights of 150 feet (45 m) with a poorer ground flora.

The acidity of the peat and the water is greater on the dome area of the bog plain, usually less than pH4. This is because the dome, which is usually elevated 17–35 feet (5–10 m) above the surrounding water table, receives only acid rainwater with no groundwater influence. The peat itself also has a high capacity for retaining nutrients, chemically trapped on the surface of dead vegetable matter, so the infertility and acidity of the water becomes increasingly pronounced as it resides on the peat mass. In some respects, this water resembles the black water input of the Amazon Basin (see "Tropical Swamps," pages 97–98). The water received in this way remains perched on the top of the impervious peat dome and may gradually drain laterally over the surface of the rand. In the same way, nutrient elements, such as phosphorus and potassium, are scarcer in the center of the bog plain.

Among the vertebrate animals of these swamp forests are several primate species, including the orangutan (*Pan pygmaeus*). These animals achieve their highest densities in the marginal swamp forest, one survey of Kalimantan showing approximately five individuals per square mile (2.5 per sq km). They occur also within the pole forests of the inner bog, but at densities only about half those of the mixed marginal forest. Since so much of the drier forest of this region has been heavily degraded by logging, the swamp forests may well prove the most appropriate habitat for the conservation of the orangutan in the islands of Southeast Asia. The same survey of Kalimantan also found several Red Data Book bird species (i.e., species recognized by the International Union

Profile of a tropical raised mire from Indonesia. The deepest peat domes (up to 65 feet; 20 m deep) lie on the raised areas between drainage rivers, and may be several miles in diameter.

for the Conservation of Nature [IUCN] as in need of special conservation concern) using the swamp forest extensively, including wrinkled hornbill (*Aceros undulatus*), helmeted hornbill (*Rhinoplax vigil*), Storm's stork (*Ciconia episcopus*), short-toed coucal (*Centropus rectunguis*), and Wallace's hawk eagle (*Spizaetus nanus*).

There are many threats to the future of the bog forests. The demand for timber extraction is considerable because several of the trees of this habitat are good timber producers, among them *Shorea albida*. The possibility of controlled and managed logging must be considered, and some studies suggest that opening areas of the forest by selective logging could actually increase the diversity of small mammals present. A more variable range of microhabitats, including denser ground cover, an abundance of detritivore invertebrates, and the presence of more dead logs, may all add to the heterogeneity of habitats and increase small mammal diversity, but there also are destructive effects, and these may be particularly important for the rarer species that are restricted to the primary bog forest habitat.

Some mammals and birds are under threat because of collection for the pet trade worldwide. Illegal logging and tree extraction can lead to easier access to the bog forests and may allow trappers easier access to the interior of these mires. Commercial peat extraction for horticulture or energy production is not yet a problem in Southeast Asia, but the resource is there and all that is needed is the political will and the finance to permit the exploitation of the large volumes of peat. Combined with timber extraction and subsequent reclamation for agriculture, this would result in irremediable damage to these unique ecosystems.

The massive tsunami of 2004 reminded the world that human communities and coastal ecosystems are very vulnerable to incursions of the sea. Natural disasters of this kind may prove even more damaging to the tropical raised mires than human exploitation. This will be made even worse if global sea levels continue to rise as a consequence of global warming and the melting of the ice sheets and glaciers of the world.

■ BLANKET BOGS

Most wetlands are confined by the topography of the landscape. They depend on water, and water is most easily available in valleys and hollows where it accumulates, having drained from surrounding slopes and soaked through the soil as groundwater. Even raised bogs (see "Temperate Raised Bogs," pages 103–105), whose surfaces are elevated above this groundwater, begin their development in basins, valleys, and estuaries where water collects. This is not the case with the blanket mires; they are exceptional among wetlands in that they occur not only among the hollow localities of a landscape but also develop over the hilltops, slopes, and plateaus, forming a complete peat cover like a blanket over an entire region, as shown in the following photograph.

Blanket mire in the Inverpolly National Nature Reserve, Northwest Scotland *(Peter D. Moore)*

The main climatic requirement for the development of this type of wetland is rainfall. Without persistent wetness it would be impossible to maintain high water tables on slopes and hilltops, so blanket mires are only found in regions of high precipitation, low evaporation, and a lack of any prolonged dry season. Although equatorial regions are wet, high temperature means that decomposition rates are rapid and that peat formation is unlikely outside the waterlogged valley bottoms. On certain mountainous sites, however, such as the Ruwenzori range in western Uganda, altitude leads to cooler conditions and lower evaporation, and there blanket mires have developed.

Most blanket mires, however, are found in the temperate zone, mainly at high altitude, but can be found even at low elevations if the conditions are sufficiently cool and oceanic (as in the western islands of Ireland, Scotland, and Norway). Western Europe, from southwest England to Norway, western Iceland, Newfoundland, parts of the Pacific Northwest and Alaska, Kamchatka and Japan, the southernmost tip of South America and some oceanic islands, such as the Falklands, Tasmania, and New Zealand, are the main centers of blanket peat formation. Beyond these, there are isolated mountainous sites where patches of blanket peats develop, such as the European Alps and South Africa.

The vegetation of the blanket mires resembles tundra in its constituent species and in its open, treeless appearance. It is poor in terms of species diversity, often dominated by bog mosses (*Sphagnum*) and cotton grasses (*Eriophorum*), together with other members of the sedge family, such as the tussocky deer sedge, *Tricophorum caespitosum*. The treeless, open landscapes of blanket mire are important breed-

ing locations for many bird species, especially waders and some raptors, such as harriers.

Blanket mires are essentially ombrotrophic, that is, they are fed entirely by rainfall. This is particularly true of those parts of the wetland that occupy the high ridges, plateaus, and summits of the hills, but since the mire covers other parts of the landscape, the movement of water through the peat can be complex and can profoundly affect the distribution of chemical nutrients and hence the vegetation. Water draining from the upper parts of a mire collects in small channels and finally in more extensive wet areas through which water soaks and periodically emerges, where its movement brings a richer supply of nutrient elements. These wet areas contain more varied assemblages of mosses and other plants and hence diversify the habitat. Similarly, the valley areas contain sites where the water flows rapidly or where pools develop and these create distinctive microhabitats for animals and plants. So, although the blanket mire system is strictly a bog, being rain-fed, it also contains minor habitats that more closely resemble poor fens or valley mire wetlands (see "Valley Mires," pages 95–97). Pool systems are also often found in the flatter regions of the mires, as shown in the photograph, forming concentric patterns around the highest points. Sometimes the upper regions are dissected by erosion channels, and it is still disputed whether the erosion of blanket mires is a consequence of the natural development of peat instability on hilltops and slopes or whether it has resulted from human-induced air pollution.

The peat profiles of blanket mires also vary with the landscape, as shown in the diagram. Hollows in the bedrock beneath the peat may contain sediments that once accumulated in small lakes and swamps and in which a succession has taken place leading to bog formation, often through a wooded stage of development. These sites frequently contain the oldest deposits, so they were evidently the initial sites of wetland development in the landscape. After the valley hollows, it is the flat plateau regions that have the deepest peats (in western European blanket mires, often 6–12 feet [2–4 m] in depth), but their basal sediments often contain wood remains, sometimes also with an abundance of charcoal present. These brushwood peats are then superseded by a uniform, homogeneous bog moss/cotton grass peat that continues to the surface. The slopes over which peat develops vary in steepness according to the quantity of rainfall at the location. In Wales and western Ireland, for example, blanket peats may develop even on slopes of 20° to 25° in steepness because the climate is so consistently wet (about 250 rain days per year). Again, these slope peats often have evidence of former scrub woodland at their bases, but sometimes the soils have been eroded from the slope before the commencement of peat formation.

One of the most interesting questions relating to these distinctive wetlands is precisely how they first formed and spread. It is evident from the profile sections through blanket mires (see the diagram on page 110) that peat was initially developed in hollows, later began formation on flatter hilltops, and finally spread both up and down the slopes. This spread of the peatland was evidently at the expense of former forest cover, but was this major landscape change a result of a climatic shift toward colder and wetter conditions, or was it a consequence of some other factor that reduced the tree cover and allowed the bogs to spread? Evidence suggests that in western Europe, where most of the work on blanket mire has been centered, both factors were involved. There is much evidence from a range of sources that conditions became wetter in this region after about 7,000 years ago, and there were further changes toward cooler, wetter conditions about 5,000 and 2,500 years ago, as well as at other times. Blanket bog initiation and growth is often associated with these wetter periods.

There is also evidence from the fossils in the peat profiles, however, that prehistoric human activity played a part in the spread of many blanket mires. Charcoal layers tell of past fires and are often accompanied by evidence in the fossil pollen record of human activities. Perhaps the prehistoric pastoral peoples of these times cleared scrub woodland in the marginal uplands for grazing and used fire to assist their manual clearance work. This had many effects on the hydrology of the landscape. The removal of trees means that more rainfall reaches the ground instead of being intercepted by the canopy and evaporated back into the air. It also means that less water is removed from the soil through tree roots to supply the tree's leaves with water for transpiration. Thus, more water enters the soil and less water is removed from it. The soil water retention would also have been assisted by deposits of charcoal from the burning trees, for these inert carbon particles effectively block the soil pores and prevent water from draining away.

The outcome is that the original oceanic woodlands in these hilly areas were removed, solid ground became wetter, and peat began to form over the waterlogged soils. Grazing, as well as continued burning by the pastoralists, prevented reinvasion of trees and the blanket bogs spread out to form the extensive treeless vistas now so typical of western Ireland

Blanket bogs of the Silver Flowe Peatland in southern Scotland *(Peter D. Moore)*

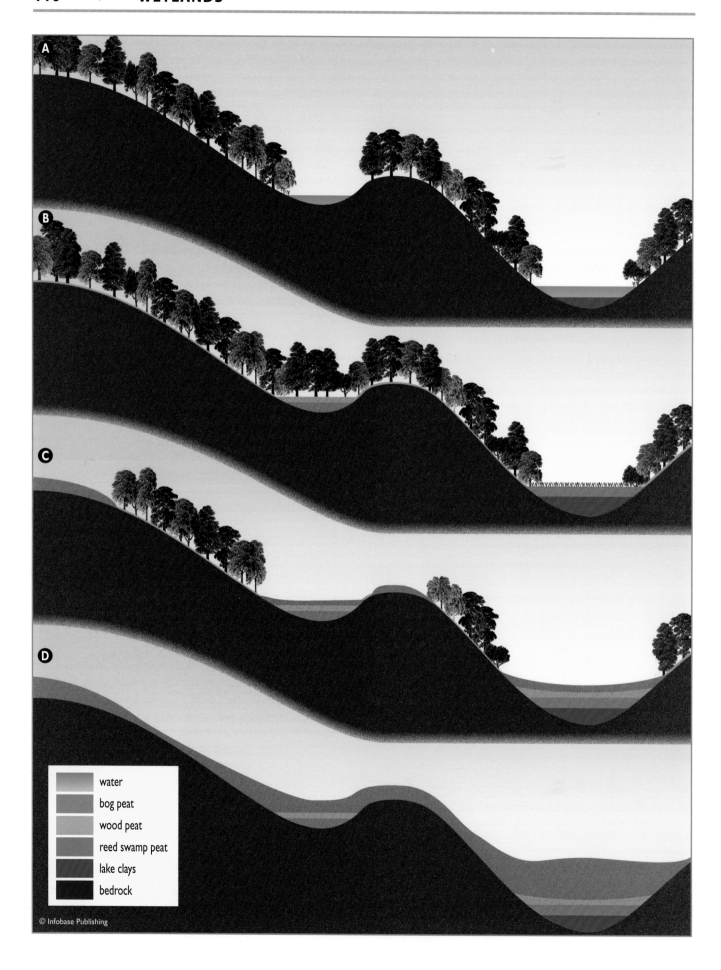

water
bog peat
wood peat
reed swamp peat
lake clays
bedrock

© Infobase Publishing

(opposite page) Series of profile diagrams showing the development of a blanket mire landscape. (A) Hilly, wooded country in an oceanic climate has topogenous mires developing only in hollows. (B) Over the course of centuries these low-lying mires undergo succession and become colonized by vegetation. (C) Prehistoric clearance of forest by people, together with increasing climatic wetness leads to the development of blanket mires, initiating on hilltops and plateaus. (D) Further forest clearance, burning, and grazing by domestic animals, together with further climatic deterioration, leads to a complete blanket of peatland, covering hilltops, slopes and valleys.

and Scotland. Further research from other blanket mire areas of the world is needed to establish whether human activity was important elsewhere and just how much blanket bog would have developed under the influence of climate alone.

One of the major problems in determining the relative impact of climate and prehistoric peoples in the initiation of the blanket bogs of western Europe has been the fact that both factors are known to have been operative at the same time. Many of the blanket bogs of Ireland, Wales, and Scotland began their formation about 5,000 years ago, which is precisely when the first major impact of prehistoric farming (Neolithic—New Stone Age) began in these areas. The fact that one can detect evidence of landscape change at the time of bog initiation (using such evidence as pollen and charcoal) may simply be coincidental. The climate at this time was also becoming cooler and wetter in northern Europe, so this undoubtedly also had an effect, but which impact was greater is now impossible to say. There was another wave of blanket bog spread in the Bronze Age, especially in parts of Wales and Northern Ireland, but again there is evidence for increased wetness about 3,000 years ago when these people began their pastoral activities in the oceanic uplands of northwest Europe. The circumstantial evidence associating human activity with peat initiation is now extremely strong, apparent even in the blanket bogs that cover the coastal islands in the vicinity of Bergen in western Norway.

The most probable compromise explanation is that the cooler and wetter climate of the last 5,000 years in these areas has made the formation of blanket bogs more probable. The removal of the natural forest and the imposition of burning and grazing land management practices by local peoples often tipped the hydrological balance in these cool, wet environments and stimulated the bogs into spreading out of hollows and over surrounding slopes. The right climate (cool and wet) is clearly necessary for blanket bog formation, and many of the blanket bogs of the world (such as those in Tierra del Fuego in South America, the Falkland Islands in the South Atlantic, and the Ruwenzori Mountains in Uganda) may well have developed under climatic influence alone. In those regions of the world where prehistoric human impact resulted in widespread forest destruction and where the climate was marginal for blanket bog formation (such as Wales and Ireland), the human factor is likely to have been a necessary contributory factor for their development.

The bleak, hilly topography, with its deep peat cover of blanket bog, presents an impressive landscape. Until very recently, however, these bogs have been regarded generally as a kind of wet wasteland, fit only to be "reclaimed" and made more productive from the human point of view. Fortunately for the bogs, they have not been particularly useful for arable farming, although there is evidence in Europe that periods of warm climate (such as 1800–1815) were accompanied by attempts to grow cereals on drained blanket bogs even at moderate altitudes. In Britain, the very high grain prices associated with the Napoleonic Wars at that time undoubtedly provided an economic incentive to such developments, but the bogs' wetness, cool climate, and nutrient poverty have generally prevented heavy arable exploitation.

Pastoral farming, however, has been more realistic, and sheep farming on blanket bogs has long been practiced in western Europe. Even cattle were grazed on them during the Middle Ages. Lowering the water table in these bogs makes them more suitable for grazing, so there have been increasing attempts to drain them using channels. Also, the fertilization of their surfaces using nitrates and phosphates increased their productivity from the agricultural point of view but usually destroys their natural vegetation. The greatest danger for the future of blanket bogs, however, comes from forestry. Drainage, fertilization, and the removal of grazing animals (domesticated sheep and wild deer) have improved the productivity of softwood forests when planted on these bogs. North American trees, including Sitka spruce (*Picea sitchensis*) and lodgepole pine (*Pinus contorta*), have been very successful crops on the European blanket bogs.

Fortunately, conservation policies are being established that ensure that some of these bogs are preserved and are protected from commercial exploitation. It is ironic that this habitat, which humans converted from forest to bog many thousands of years ago, is now being preserved from a reversal of this change—the plantation of forest and the destruction of the bogs. The justification for this conservation is the distinctive flora and fauna that have developed in response to the establishment of these bogs. Some wading birds, such as dunlin (*Calidris alpina*) and greenshank (*Tringa nebularia*), breed mainly in this type of habitat within the temperate zone. Predatory birds, such as northern harrier (*Circus cyaneus*) and merlin (*Falco columbarius*), also breed mainly in the blanket bog regions of northern Europe, and one bird, the red kite (*Milvus milvus*), may owe its survival in Britain to the presence of these remote wilderness areas. Even though many areas of blanket bog may be human-created, they are now part of the world's biodiversity and demand our assistance in their survival.

■ PALSA MIRES

Among the most northerly of the peat-forming wetlands, lying beneath the midnight sun and grazed by caribou, are the palsa mires of the Arctic regions. These wet areas of erupting peat mounds and open pools develop in areas of discontinuous permafrost, that is, where part of the subsoil remains frozen throughout the year while other areas melt each summer to produce a saturated soil surface. The average yearly temperature in these regions is less than 32°F (0°C), and the summer growing season is less than 120 days. In the low-lying areas of these cold regions, where water accumulates through the summer, the vegetation becomes dominated by cotton grasses and bog mosses forming extensive and uniform lawns, but these are interspersed with protuberant masses of lichen-covered mounds, and black, eroding hummocks of peat, often 6–10 feet (2–3 m) in height and about 150 feet (45 m) across. These are the palsa mounds. When excavated, they reveal a solid core of ice that lasts right through the Arctic summer.

If it were possible to observe a palsa mound over a number of centuries, one could see that it grows gradually from year to year as its ice core expands, but the precise mechanism that underlies palsa formation has only recently been discovered. The original lawn of sedges and moss from which the palsa erupts is not entirely level; some areas are very slightly higher than others due to the uneven growth of the mosses and the peat that accumulates below them. In winter, when the entire landscape is frozen, the surface of the wetland becomes covered with snow, but the harsh arctic wind causes drifting and the raised areas have a thinner snow cover than the hollows. Snow cover actually protects the vegetation and the soil from frost, so it acts as an insulating blanket. When it is blown away, the frost penetrates more deeply into the ground, and this deeply frozen soil thaws more slowly during the brief summer.

Over the years, the ice mass in the soil becomes increasingly permanent and greater in volume, for water expands on freezing so that the ice occupies a larger volume (approximately 10 percent larger) than the equivalent mass of water. As a result the ground begins to swell and the palsa mound starts to rise.

The story does not end there, however, because the elevation of the surface causes changes to the water balance of the soil. Water drains more easily from the summit of the palsa in summer, leading to drier conditions on the mound. The Arctic actually receives very low rainfall, especially during the summer, and this can lead to serious drought in well-drained situations such as those at the top of a palsa mound. This in turn leads to changes in vegetation on the palsa summit. Instead of being covered by deep-rooted sedges, the convex surface of the palsa is invaded by short arctic herbs and lichens. Many of these lichens, the dry, crisp "reindeer mosses" (which are actually intimate combinations of photosynthetic algae and protective fungi) are pale gray or white in color, and this further affects the temperature balance of the palsa. This pale surface sunlight reflects much of the sunlight (rather than absorbing its warmth), and this keeps the palsa mound cool even during times of radiant sunlight, so the ice core of the mound persists through the summer and grows yet faster through the winter.

Eventually, depending on the size of the mound, the direction it faces, and the steepness of its sides, the surface vegetation begins to break up and erode. There are physical stretching forces on the "skin" of the expanding palsa that cause it to split, and the thin cover of vegetation is also easily disturbed by such grazing animals as reindeer (caribou). Once the turf splits, the impact of rain, snow, and ice soon enlarges small crevices into gullies and erosion begins in earnest. The underlying peat that is exposed in this way is usually dark brown or black in color, so the reflective properties of the lichens give way to the heat-absorptive capacity of peat, since dark surfaces absorb heat more efficiently than light ones. Once a large area of peat is uncovered, the palsa can experience a relatively rapid meltdown in the summer months that can result in its total collapse and even the creation of a water-filled crater in its place. These crater pools are surrounded by the remains of ramparts derived from the fragmented rim of the degraded palsa.

Palsa mounds thus undergo a cyclic development in which the full cycle is believed to last many centuries or even thousands of years, depending on local conditions. Of course, no one has been able to watch or record the complete sequence over this long time span, but the existence of palsas at different stages in their developmental history allows ecologists to piece together the story of their rise and fall, which is summarized in the diagram (opposite). The resulting landscape of these wetlands is a complex mosaic of elevated hillocks dry enough to walk over, interspersed with wet cotton grass lawns and deep, water-

(opposite page) The rise and fall of a palsa mound. (1) The Arctic wetland surface is relatively flat. (2) Any slight irregularities, such as slightly higher ground, results in less snow cover and poorer insulation on the raised areas, so ground ice persists through the summer and swells as more ice forms. (3) The ice core continues to grow and raises the mire surface above the surroundings, as a result of which it becomes drier and clothed with dwarf shrubs and light-reflecting white lichens. (4) Eventually, the top of the palsa mound begins to erode as a result of water runoff, and bare, black peat is consequently exposed, which rapidly absorbs summer sunlight and instigates an ice-core meltdown. (5) The palsa mound collapses, leaving a pool surrounded by a circular rampart.

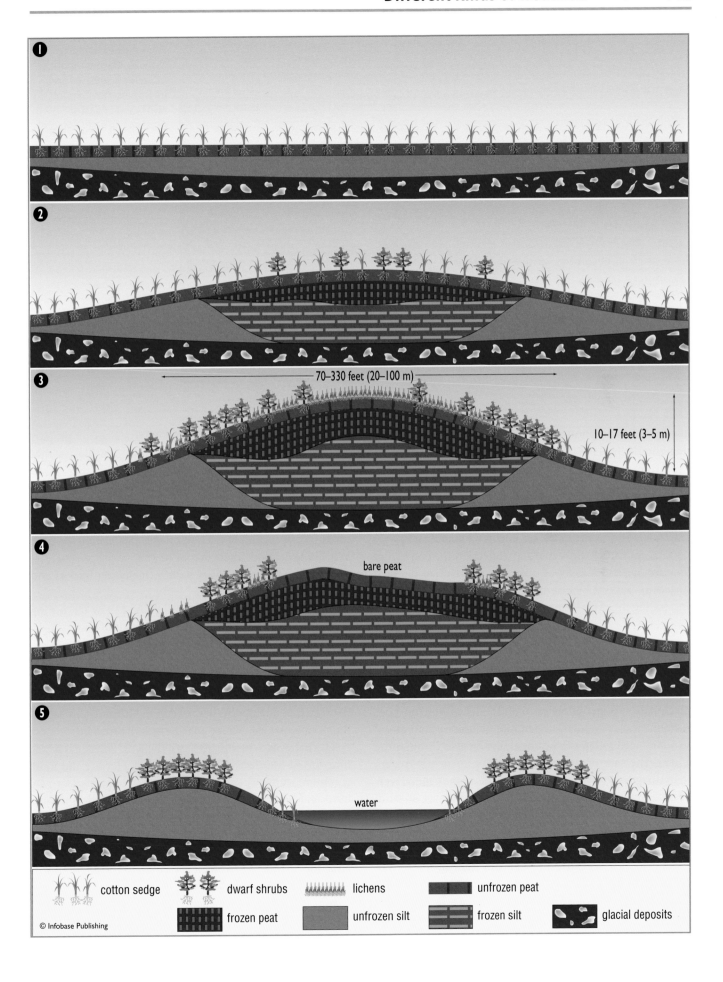

70–330 feet (20–100 m)

10–17 feet (3–5 m)

bare peat

water

cotton sedge   dwarf shrubs   lichens   unfrozen peat

frozen peat   unfrozen silt   frozen silt   glacial deposits

© Infobase Publishing

Palsa mire in northern Norway. Mature mounds of peat-covered ice are interspersed with pools and lawns of cotton grasses.   *(Peter D. Moore)*

filled pools. The photograph above shows these different stages in the palsa cycle at a site in Arctic Norway.

An experimental demonstration of the process by which palsa mounds begin to form has been conducted by scientists from Oulu University in Finland, who braved the Arctic elements, sweeping the snow from an area of cotton-grass wetland at a site in the far north of Finland and keeping it clear of snow through three winters. By the end of that time the ice had developed sufficiently to elevate the peat surface by several inches, thus initiating the palsa mound and setting the cycle in motion.

ARCTIC POLYGON MIRES

Wetlands are abundant in the Arctic, despite the fact that precipitation is low, often less than 16 inches (400 mm) per annum. The very low temperature reduces evaporation, so water accumulates in summer and freezes through the winter. The subsoil, or permafrost, is permanently frozen, so drainage is always poor, and this adds to the general wetness of summer conditions. Any ridges of gravel or sand, however, may become quite dry. It is a peculiar feature of the tundra regions that arid and wet ecosystems can exist alongside one another.

As the frozen landscape begins to thaw in the spring, water moves over the surface of the ground, and wetland vegetation develops in depressions, seepage hollows, and drainage channels. These wetlands are technically fens (see "Fens," pages 87–89) as they are fed by flowing water rather

than solely by rainfall (as in the case of true bogs). A proportion of the dead plant material accumulates as peat within the waterlogged soils, but the peat depth beneath these mires is generally shallow, usually less than 20 inches (50 cm). The vegetation is dominated by sedges (*Carex* species) and cotton grasses (*Eriophorum* species), together with a range of moss species including the bog mosses (*Sphagnum* species). Where the bog mosses form mounds above the general level of the flat fens, some dwarf shrubs, including birches and willows, may survive.

Only the surface layers of the ground thaw out during the brief summer weeks, usually to a depth of about 12–15 inches (30–40 cm). Below that lies the permanent ice of the *permafrost*. This upper horizon of the soil that defrosts each summer is called the *active layer*. The presence of peat in this soil layer acts as an insulating blanket, preventing the deep thawing that might otherwise occur and helping to preserve the permafrost. As winter approaches, the freezing process creates patterns over the surface of the fens. Polygonal structures of 30–100 feet (10–30 m) in diameter are the most common, with ice penetrating deepest in large angular patterns and breaking up the landscape into a kind of mosaic form. This is particularly apparent in aerial view. Flying over the Arctic, even at the high altitude of an intercontinental jet, reveals an extraordinary, blotched pattern almost resembling the texture of marble in appearance. The depth (down to 3 feet [1 m]) and the width (up to 10 feet [3 m]) of the ice channels become greater each year, so the pattern becomes more marked as time passes. As the ice forms in the channels, it expands and displaces soil from below the peat layer, heaving it upward to the surface where it is deposited as banks along the sides of the ice cracks. So the linear ice wedges become embanked with soil, creating poorly drained pools in the central parts of some polygons. These add to the diverse topography and to the range of habitats available to the insects, birds, and mammals that inhabit the tundra through the summer.

The polygons themselves can be divided into low-center and high-center types, as illustrated in the diagram. In the former, the wet ditches have a margin of raised banks, and the center of the polygon is depressed so that sedge-meadow wetlands or shallow pools develop there in summer. In the high-center types, the mid-region of the polygon is elevated and often forms a series of cracked, eroding, peaty hummocks. These drier regions are important for ground-breeding birds.

The polygon mires are restricted to the High Arctic region of Alaska and Canada, together with the most north-

erly regions of Siberia. In the Antarctic region, there is no precise equivalent to these mires, but the islands to the south of Tierra del Fuego bear a grass-dominated vegetation that grows over peaty soils and within which penguins and albatrosses breed. Unlike the Arctic, however, these areas of shallow peatland receive a rich supply of rainfall, from 40 to 160 inches (1,000–4,000 m) per year. In many respects they are more comparable to the blanket mires (see "Blanket Bogs," pages 107–111).

In some areas, particularly on estuarine and floodplain deltas, raised domes are interspersed with open pools. These are pingo fields. The domes (sometimes as high as 330 feet [100 m]) are created by water welling up from saturated layers below the ground and then freezing as they near the surface, forming a subterranean ice mass similar to, but usually bigger than, palsas (see "Palsa Mires," pages 112–114). Like palsas, these structures periodically collapse to form open water bodies that add further wetland habitats to the tundra. In general, smaller ponds have a longer ice-free period in summer than the larger lakes, so they are important to the wildlife during the arctic summer.

The low temperature of the Arctic regions reduces the rate of microbial decomposition in the soils, so the organic litter produced by plant growth may accumulate

Profile of arctic polygonal mires. (A) A low-centered polygon mire in which the central region of the polygon is depressed and holds water. (B) A high-centered polygon mire in which the mid-region of the polygon is elevated by the ice in the frozen soil (permafrost) and is covered by draining and eroding peat.

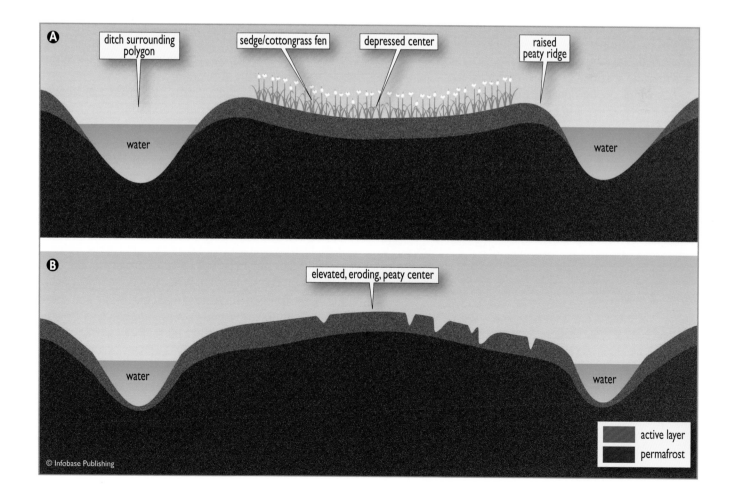

as a peaty material especially where soil wetness creates conditions of low oxygen supply and the bacteria and fungi are even more restricted in their activities. Ponds often have dark-colored bottoms because of the accumulation of dead remains of the aquatic plants, which include aquatic buttercups (*Ranunculus* species), mare's tails (*Hippuris vulgaris*), and water sedge (*Carex aquatilis*). The drier, raised rims of polygons are occupied by dwarf shrubs, such as bearberry (*Arctostaphyllos uva-ursi*) and dwarf birch (*Betula nana*), as shown in the photograph, but the diversity of plants is very low, as is the productivity of these communities. The polygon wetlands are generally dominated by the cotton grasses, such as *Eriophorum angustifolium,* especially in the Prudhoe Bay region of Canada where much arctic wetland research has been carried out, and this vegetation is particularly important in conservation terms because it is the main habitat of the breeding herds of caribou during the summer. These habitats are also important feeding grounds for lemmings. Farther south, in the Low Arctic, more peaty communities dominated by *Sphagnum* bog mosses and the cloudberry (*Rubus chamaemorus*) are increasingly frequent. Here they merge into the palsa mire regions.

Ecologists have carried out some work in Russia concerning the age and development of the polygon mires of the High Arctic, and these wetland structures have proved to be surprisingly recent in origin, often less than 500 years. There are some older peat deposits, however, dating back as far as 5,000 years, but, in geological terms, the Arctic wetlands must be regarded as a fairly recent habitat.

Despite the low productivity of the arctic wetland vegetation, it becomes rich in invertebrates for the short period of summer freedom from ice and snow. This abundance,

Bearberry and dwarf birch growing on the surface of an arctic mire *(Jennifer Culbertson, U.S. Fish and Wildlife Service)*

coupled with the permanent daylight, provides suitable conditions for many birds to breed, including waterfowl (geese and ducks), waders, gulls and terns, and some smaller, perching birds, such as the Lapland longspur (*Calcarius lapponicus*) and the snow bunting (*Plectrophenax nivalis*). These, in turn, support predators including arctic fox (*Alopex lagopus*), gyrfalcon (*Falco rusticolus*), and snowy owl (*Nyctea scandica*).

Despite the remoteness and the low human population density, the arctic wetlands are nevertheless under threat from human activities. The exploitation of oil in Alaska, for example, has brought problems of damage to permafrost soils by vehicles and the disruption of caribou migration routes by the building of pipelines. Where decomposition is slow, even the seemingly simple problem of trash disposal can create unforeseen difficulties in the arctic environment.

■ INLAND SALINE WETLANDS

Salt water is normally associated with the oceans, and saline wetlands are often found along the coasts. When water evaporates it leaves any dissolved chemicals behind and enters the air in an almost pure (distilled) form. Hence the water that falls in rain is relatively pure, being contaminated only by the dust that it picks up on its movement through the atmosphere, together with acidic pollutants and any salts that may be suspended when whole droplets of water enter the atmosphere from the ocean water that is whipped up by high winds. As the water percolates through vegetation canopies, through soils and via rocks into underground water bodies (aquifers), it accumulates dissolved chemicals once more, so the water that enters wetlands from ground drainage is not pure but contains varying quantities of chemical elements (see "Soil Chemistry in Wetlands," pages 65–67).

In most wetlands water drains through the system, so dissolved chemicals can leave as well as enter, and this ensures that there is no buildup of salts, but some wetland ecosystems lack any exit streams or drainage and the only loss of water is by evaporation. When this occurs, the evaporating water leaves behind its load of chemicals and these accumulate in the water body over long periods of time in a process termed *salinization*. This leads to an increasing concentration of salts in the water, and here *salts* means not only common salt, sodium chloride, but also sodium sulfate, magnesium chloride, magnesium sulfate, calcium carbonate (lime), and calcium sulfate (gypsum). Closed basins that suffer salinization of this sort are most frequent in hot climates where water supply is restricted both in quantity and duration of supply. Spring snowmelt in the highlands of Iran, for example, leads to a flush of water into closed desert

Inland saline wetland at Wadi el Natrun, northern Egypt. The purple color is caused by a high population density of microscopic dinoflagellates. *(Peter D. Moore)*

basins where it evaporates during the hot, dry summer to produce temporary saline wetlands.

Inland saline wetlands are found in greatest abundance in the interior of South America, in North Africa and parts of East Africa (such as the Rift Valley), in the Near East and central Asia, and in Australia. Salty wetlands are also found in southwestern parts of the United States, such as the Salton Sea in southern California.

The degree of salinity that develops in such a wetland depends on a number of factors, such as the length of time over which the site has been accumulating salts, the chemistry of the catchment rocks, the rate of water flow into the site, and the intensity of evaporation. In extreme situations, such as that of the Dead Sea, which separates Jordan and Israel, and in some of the salt lakes of northern Egypt, the concentration of salt can rise to a level where only highly adapted, microscopic forms of life are able to survive, as shown in the photograph of Wadi el Natrun in Egypt. This is the extreme situation, however. More usually, a range of plant and animal life persists, and these wetlands consequently form important resources for wading birds and wildfowl. Perhaps the best known and most important crustacean of the saline wetlands is the brine shrimp (*Artemia* species), which is important as a food resource for many birds, including flamingos.

In South America, the high tablelands of the Andes (the Puna) in Peru, Bolivia, Chile, and Argentina contain many lakes and wetlands, some of which have high salinity. The most distinctive birds in these saline wetlands, as in those of many other parts of the world, are the flamingos. Three species are found in the Andes, two of which are found nowhere else on Earth (endemic species). Flamingos feed by holding their heads upside down and filtering water through their bent beaks to extract small crustaceans and algae, as shown in the photograph above. They are confined to saline lakes where there are no fish, for in such situations there is no competition for the planktonic organisms that form their prey. They even nest in the center of shallow saline lakes, where they build up mounds of salty mud in which they lay their eggs. One of the hazards of these hot, salty locations, however, is that the young birds may become crusted with salt on their feet and legs, which eventually develops into encircling anklets that can so incapacitate the young birds that they starve or are killed by predators.

Flamingos are remarkable because they feed by filtering water, holding their bills in an upside-down position. *(Pierre Janssen)*

In the Mediterranean regions and North Africa, salt lakes are mainly seasonal, becoming waterlogged during the winter and drying out in dry summers. Winter is therefore their most productive season. Then, they can provide feeding grounds and staging posts for large numbers of migratory and wintering waders and wildfowl. Some of these wetlands, like the salt pans of the Camargue in France, and Lake Ichkeul in Tunisia, are near the coast and occasionally receive an input of salt water from the sea. Others, such as Lake Tuz in Turkey, lie deep inland in closed basins. The Camargue has more than 8,000 breeding pairs of Eurasian greater flamingo (*Phoenicopterus roseus*). In East Africa, the soda lakes of the Rift Valley are also famed for their flamingos, with Lake Nakuru in Kenya supporting more than a million lesser flamingos (*Phoeniconaias minor*). In these soda lakes, the main fish of the region, *Tilapia,* is absent, so the waters are often colored by abundant blooms of algae and certain zooplankton.

Many of the saline wetlands of Asia, including those of the Indian subcontinent, are largely seasonal, depending on the input of water, which may or may not be regular. The occasional water bodies formed in wet periods in the deserts of eastern Iran and the Rajasthan Desert of India are vital staging posts for migratory waders traveling between their northern breeding grounds and the sea coasts to the south. Some saline wetlands lie at high altitudes in Afghanistan and Tibet, and these tend to be more permanent. The demand for water for agriculture in these arid regions, however, presents a particular threat to these wetlands.

In the interior of Australia, closed basins in the middle of arid regions receive occasional rainfall and flooding, which lead to the development of temporary saline and brackish wetlands, one of the most famous being Lake Eyre in South Australia. This basin is more than 620,000 square miles (1 million sq km) in area, and its water supply is both low and erratic. Plants and animals (including fish and amphibians) need to survive long periods of drought between wet episodes.

The conservation of inland saline wetlands involves overcoming many varied difficulties. In some regions, such as the Puna of South America, saline wetlands are threatened by pollution from agricultural runoff and toxic waste from mining activities. These chemicals also accumulate in the ecosystems, along with the salts, and contaminate the waters, causing the death of the organisms that lie at the base of the food chain. In France, the Camargue has long been exploited for salt production by evaporation in shallow lagoons, while the Dead Sea coastline is mined for its mineral resources, particularly phosphates. The spectacular scenery and wildlife often associated with inland saline wetlands are also a great tourist attraction, but this can present particular problems since the output from a tourist industry (including sewage and detergents) can easily enter the closed basin waters and cause pollution.

◾ SALT MARSHES

Coastal wetlands generally differ from inland wetlands in two important respects: The water supply is mainly saline (although this can also be the case in certain inland closed basins), and the water table changes with the rhythm of the tides. As is the case at inland sites, coastal wetlands become vegetated only where the erosive power of water flow and wave action are not too strong, and this usually means that they develop on coastlines with gentle land/sea gradients—what geomorphologists term *low energy coasts.* It is under such conditions, especially in the sheltered parts of estuaries, in enclosed bays, and in the lee of shingle spits and barrier islands, that salt marshes and mangrove swamps develop.

Salt marshes are characteristic of the temperate and high latitudes, and they differ from the tropical mangrove swamps in that they are dominated by herbaceous plants rather than trees. Like inland wetlands, they undergo a process of succession that is driven by the accumulation of sediments, but they differ from inland wetlands in that sediments are delivered in pulses with each flooding of the tide. The process begins with the development of extensive tidal flats of fine-grained sediments in the intertidal zone of sheltered coastal sites. The deposited sediments contain both inorganic particles of weathered rocks and seashells, and also organic materials derived from terrestrial ecosystems and arriving in these coastal areas through river erosion (allochthonous material, that is, brought in from outside the ecosystem). This means that the coastal mudflats are rich in energy that has originated in other ecosystems but is now available to an army of detritivores (detritus-feeding invertebrate animals) and decomposers that inhabit these waterlogged mudflat deposits. Small animals, such as crustaceans, and microbes involved in tapping these vast energy reserves need to be adapted to very low oxygen conditions (see "Animal Life in Water," pages 137–139). The abundance of small detritivorous animals in these rich, organic sediments attracts predators, particularly wading birds with long beaks that can probe the mud in search of them.

Some plant life exists in the mud, but the instability and constant movement of mud makes anchorage difficult, so there are very few large seaweeds present in the mudflats, except where they can gain stable anchorage on a rock. Microscopic algae, however, such as the diatoms, are able to occupy the surface layers of the mud and photosynthesize to accumulate yet more energy. There are also some flowering plants that can occupy mudflats, the most widespread and important being the sea grasses, especially the genus *Zostera*—an important food reserve for some vegetarian birds, including such geese as the black brant (*Branta bernicla*). In warm regions, other genera of sea grass plants are also important, such as *Posidonia* (the dead leaves of which cling together in balls that are washed up on shorelines and are used in the Mediterranean region as field fertilizers), *Thalassia* (which has nitrogen-fixing nodules), and *Syringodium*.

Salt marshes are the temperate equivalents to tropical mangroves. They occupy the coastal fringes of protected shores, and some are dominated by the cord grasses (*Spartina* species). *(George Gentry, U.S. Fish and Wildlife Service)*

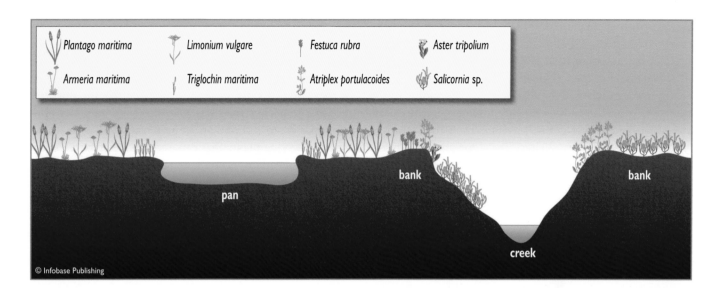

Profile of the topography and vegetation of a typical European salt marsh at low tide. Pans remain filled with water and are without vegetation. Creeks drain at low tide and their banks become covered with salt-tolerant plants, depending upon the forces of erosion on them. Seawater enters the ecosystem along the creeks and overflows during spring tides, depositing suspended matter along their edges to form elevated banks.

The presence of rooted plants in the mud stabilizes it and the development of leafy fronds above the mud surface reduces the speed of flow of the current, which in turn leads to more of the bedload (the material carried along the surface of the substrate) and the suspended sediment being deposited around the plants. So the presence of plant life increases the rate of sediment deposition and alters the nature of the environment. The most important change that results from raising the surface of the sediment in these tidal environments is that longer periods are spent above the water during the low tide periods, and this can affect the degree to which air can penetrate the mud.

Even more influential than the sea grasses in this process of mud colonization and the encouragement of sedimentation are the glassworts (*Salicornia* species) and the cord grasses (*Spartina* species), shown in the photograph (opposite). These two genera of plants are found throughout the world in salt marshes and, like the sea grasses, occupy the zone that is submerged during each high tide. Both are more erect and robust than the sea grasses, so their influence on sedimentation is even greater. Most *Salicornia* species are annual, so they depend on seed survival and dispersal for the maintenance of their populations, but *Spartina* species are perennial and spread locally and even disperse more widely by vegetative means as well as by seeds. As a result, *Spartina* often forms very extensive and uniform beds in the lower regions of salt marshes. The resulting stability of the substrate may allow the development of dense mats of dwarf forms of seaweeds around their stem bases. The high productivity of this community encourages a range of invertebrate herbivores, particularly mollusks.

Even from the early stages of salt marsh formation it is evident that the surface of the mud is uneven. Runnels, or creeks, begin to form and carry the load of water into and out of the marsh during each tidal movement. Hollows form that are waterlogged throughout the low tide period—mud

pools in which there is no episode of soil drainage and aeration between tides. These pools usually remain unvegetated and can persist as pools ("pans") even late into the marsh development. So each marsh begins to assume a distinctive pattern of topography as the succession proceeds.

As the level of the sediment rises (a result of the constant deposition of detritus from tidal flooding), the less extreme high-water events (neap tides) may not immerse the entire surface. Instead of 600–700 tides a year, a surface may be submerged only 150–500 times. These new conditions permit different species of plants to invade and this zone of vegetation is often termed *Middle Marsh* to distinguish it from the *Low Marsh* below it and the *High Marsh* above it. Salt-marsh grasses (e.g., *Puccinellia* and *Distichlis*) are able to invade, together with sea lavenders (*Limonium*), asters (*Aster*) and plantains (*Plantago*), and the original pioneer species of *Salicornia* and *Spartina* become less prominent. The vegetation height at this stage of succession is lower than formerly because of the loss of vigor in the growth of *Spartina*. This plant does not appreciate the lower frequency of flooding and the greater soil stability associated with the elevated substrate. Further plant species invade as the sediment surface continues to rise, but the fact that flooding is now less frequent means that the source of sediment is less rich and reliable. Also, all but the finest suspension of particles will have been deposited lower down the shore, so the High Marsh (flooded less than 100 times a year) has a very slow accretion rate. The

The upper part of a mature salt marsh, as seen here in eastern England, is a complex pattern of salt pans and creeks. *(Peter D. Moore)*

range of microhabitats present in the High Marsh is shown in the diagram, and illustrated in the photograph.

Many of the plants of the Middle and High Marsh bear flowers on long stalks that carry nectar and are pollinated by bees and butterflies. For this reason, apiarists (beekeepers) may place hives along the drier edges of the marsh and harvest a particularly fine-tasting honey. The animal component of the community is inflated by the arrival of species that are normally regarded as terrestrial but can cope with the new set of conditions. Even ground-dwelling invertebrates can survive infrequent flooding in air pockets among the turf. Vertebrate grazers, such as rabbits and hares, may venture out onto the marsh when it is uncovered and take advantage of its productivity, and many bird species find raised tussocks where they are able to nest.

The pattern of creeks, which begin to become apparent even in the early stages of development, persists into the High Marsh, tending to be more complex and branched in the large marshes with very low slopes and straighter with a parallel arrangement in steeper sloping marshes. The creeks are carved out by the erosive force of the tidal water that moves through them, and one might be inclined to suppose that it is the water draining from the marsh as the tide recedes that creates the greater erosive force, just as rivers erode their courses as they move from the land to the sea, but this is not the case. The movement of water into the creeks as the tide floods is often twice the speed of the ebb tide waters that leave the marsh. So the main erosive force is inward rather than outward. This does mean that some of the scoured material from the lower creeks moves on into the upper marsh with the tide and is often deposited as the waters first overflow the banks of the creeks, forming ridges of new sediments along their edges and to some extent damming the drainage of water from the flat areas of High Marsh beyond. The creek edges, because of their slight elevation and better drainage, often have a distinctive vegetation that consists of species that need better aeration around their roots. One of these is the sea purslane (*Atriplex portulacoides*), found along the coasts of western Asia and Europe and as an introduced species in North America. It is often restricted to creek edges in its distribution because of its need for oxygenated soil. These raised creek edges, especially in the higher parts of the marsh, are also the locations where the more flood-sensitive plants first gain a foothold in the succession.

The flat lawns of turf between the High Marsh creeks, together with the relatively stagnant pans that have no drainage outlet, create distinctive aquatic microhabitats. These water bodies are occasionally flooded by seawater, but between these events they may suffer evaporation in hot, dry weather, thus raising their salinity, or may be subjected to rainwater dilution during wet weather. The animals and plants inhabiting these pans need to be able to tolerate rapid changes in salinity as well as extremes of salt concentrations. Not surprisingly, there are few organisms that can fulfill these demands, so life in the pans is sparse and low in diversity. The low productivity of the pans as a result of the sparsity of plant life also means that there is little infilling and that the pans tend to persist.

The upper edge of a salt marsh usually occurs where the marsh sediments abut upon the shoreline. Usually there is a change in gradient at this point as dry land is reached. This final zone is normally occupied by more robust herbaceous plants, such as rushes (*Juncus* species), and some sedges (*Scirpus* and *Carex* species), together with short turf plants such as sea heath (*Frankenia* species) and sea milkwort (*Glaux maritima*) Only the very highest of spring tides covers this band of vegetation. As a consequence of these occasional floods, however, it does receive shoreline detritus that adds to its nutrient input and also brings the seeds of certain strandline plants. Effectively, this is the end point of succession because flooding is too infrequent to bring significant amounts of new sediment, so the soil surface will rise no further, unless there is a change in sea level. At the same time, the occasional flooding means that truly terrestrial plants, including trees, are unable to maintain populations in this zone.

In the case of estuaries, which are often lined with salt marshes, the tidal waters mix with freshwater draining from the land to create a brackish region in the river mouth. Here, there is often a gradation of vegetation from salt to fresh marsh. In the higher reaches, the marsh becomes invaded by trees such as alder, many species of which are relatively tolerant of brackish conditions and can extend toward the mouth of the estuary. Fish and aquatic crustaceans also occupy different zones of the river depending on the intensity and frequency of saline influence. The detritivorous amphipods, for example, are represented by different species in an estuary at different distances from its mouth, depending on the penetration of salt water upstream.

Salt marshes grow and extend outward into mudflat areas and commence a process of terrestrialization but are finally limited in this process by periodic sea flooding and limited sedimentation. A marsh's extension outward, if not confined by shingle spits or barrier islands, can proceed at a mile per century or more. This may be unwelcome as far as humans are concerned, especially if harbors become silted and invaded by such plants as *Spartina*, interfering with navigation. Salt marshes do offer agricultural opportunities, especially for pastoral farmers, as sheep can be grazed very successfully in their higher reaches, as shown in the photograph from northern France. Artificial bridges may become

This salt marsh near Mont-Saint-Michel in Normandy, France, is heavily grazed by sheep. Grazing reduces the diversity of plant life on salt marshes. *(Peter D. Moore)*

necessary to allow the animals access to areas isolated by creeks and to give them a means of escape during exceptionally high tides. The excess of salt in their diet does not seem to be harmful, and certain parasites of sheep, such as liver fluke, are absent from salt marshes.

The extraction of minerals, including salt, from salt marshes has been one of the major human uses of this habitat. In Scotland, the silts of the Hebridean marshes are rich in lime derived from seashells, and these sediments have long been harvested to add to the acid soils of surrounding hills in order to improve their agricultural productivity. In the Spanish estuary of the Río Tinto at Huelva and at the mouth of the Rhône in southern France, salt has been harvested from marshes by the creation of extensive shallow pans in which salt becomes crystallized as the water evaporates in the hot, dry summer. In the case of the Rhône delta, known as the Camargue, the saline wetlands have become world famous for their breeding flocks of flamingos, which build elevated nests of mud out in the center of the natural salty lagoons.

The acquisition of new agricultural land by the reclamation of salt marshes has long occupied the ingenuity of engineers. Much of the world's salt marsh area has been enclosed by barriers, usually earth walls, and the isolated marsh has been mechanically drained and irrigated to remove salt. In warm climates these reclaimed, low-lying regions have proved suitable for rice cultivation. Such areas remain below the high-water level of spring tides and so are always in danger of flooding as a result of exceptional tidal surges.

The Earth is currently experiencing a period of global warming that many climatologists blame on human injection of "greenhouse gases" into the atmosphere, and if this continues it will have some impact on sea levels and, hence, on salt marsh development. Some estimates suggest that the global sea level could rise by more than three feet (1 m) in the next hundred years, but more recent calculations based on detailed models indicate that the rise may be only about one foot (30 cm) or less. Even a rise of this order, however, could lead to the erosion of many salt marshes. In the past, the consequence would be that saline waters would move farther into estuaries and the salt-marsh zone would shift landward, but the human-induced reclamation by building seawalls means that the salt marsh can no longer do this. The marsh is likely to become constricted between the advancing sea and the immovable coastal defenses, which will result in a major threat to the future conservation of salt marshes. With rising tides, however, the maintenance of coastal defenses is likely to become much more costly, and this may lead governments to abandon some reclaimed areas and allow the sea access once more—a policy of controlled retreat. For once, economic stresses may actually favor wildlife conservation measures.

MANGROVE SWAMPS

Tropical, and some subtropical, coastlines with *low energy* regimes (that is, gentle wave action and low current velocities) bear a tree-dominated type of wetland ecosystem that has been variously called *mangrove* and *mangal,* which is illustrated in the photograph. The accompanying map shows the global distribution of mangrove, together with its temperate equivalent, salt marsh. Objections to the use of the term *mangrove* are largely based on the fact that it can be applied either to the habitat or to the trees themselves, thus causing confusion. Like salt marshes, mangrove swamps typically occupy gentle inclines, are based upon fine-particle, silty, muddy sediments (usually much softer than those in salt marshes), and are subjected to regular tidal movements of saline water. Also, like salt marshes, they show a zonation of vegetation from the sea to the shore, occur particularly in estuarine conditions, and develop in a successional progression, which is shown in the diagram. They differ from salt marshes in their greater biomass, higher productivity, greater structural complexity, and much higher biodiversity.

Mangrove swamps are found around the coasts of tropical South America, the whole of Central America and the Caribbean, and around the Gulf of Mexico and Florida. They also occur in tropical West Africa and East Africa northward to the limit of the Red Sea, around the coasts of the Arabian Peninsula, India, and Southeast Asia as far north as the southern tip of Japan, through the tropical Pacific Islands, around the northern and eastern Australian coast, and into the northern tip of New Zealand. One of the greatest areas of mangrove swamp on Earth is at the mouth of the Ganges in Bangladesh—the Sunderbans. In Southeast Asia, mangroves occupy an estimated 1 percent of the total land area of the Earth.

Mangrove swamps are wetlands that occupy the coastal fringes of many tropical and subtropical regions. They provide protection for coastlines against tidal surges, storms, and tsunamis.   *(Bill Wilen, U.S. Fish and Wildlife Service)*

The mangrove trees themselves are not particularly diverse. Only 34 species truly rate as mangroves, and most of these species are found in the eastern part of the mangrove range. Although the Caribbean has fewer than 10 mangrove tree species, the east coast of Queensland, Australia, has more than 30 species. The most common mangrove trees come from five different families of flowering plant; they are not uniform as far as their classification and evolutionary origins are concerned. The term *mangrove* even includes a group of palms of the genus *Nypa*, whose fruits have been found in a fossil state in the 60-million-year-old geological deposits east of London, England, showing that at that time a tropical climate extended even into these high latitudes. Perhaps the most widespread types of mangrove tree are those belonging to the genera *Avicennia* and *Rhizophora*, which are found in both the New World and the Old World mangal wetlands.

On the seaward side of the mangal fringe lie mudflats, and these are often colonized by sea grasses, just as in the case of their temperate zone equivalent, the salt marshes. Among these sea grasses, the seeds of mangrove trees fall, stick into the mud, and germinate, eventually growing and stabilizing the mobile sediments. Precisely which species of mangrove tree is the first to establish itself varies from one site to another, but in the relatively simple mangal ecosystems of the Caribbean, *Rhizophora* is the first colonist. *Rhizophora* has a remarkable system of arching roots, whose primary function is to permit oxygen exchange in the water-logged habitat but which also serve to bind mud and trap sediments. It is possible that the accumulation of sediment eventually allows the invasion of *Avicennia* species, but geomorphologists argue about just how efficient the sediment-trapping system is. The mangal zones close to the shore experience higher levels of salinity due to the evaporation of water from those soils that spend longer amounts of time out of the seawater, exposed to the heat of the sun, and in which salt consequently concentrates.

In northwest Australia, in contrast, it is *Avicennia*, together with *Sonneratia* species, that first colonize the open muds, so there is evidently no simple sequence that operates in all parts of the world. Zonation is certainly apparent, but whether the

World distribution of mangroves and salt marshes. On the whole, mangroves are restricted to the Tropics, and salt marshes replace them in the temperate regions. Only in southern North America and in the south of Australia are mangroves found well outside the Tropics.

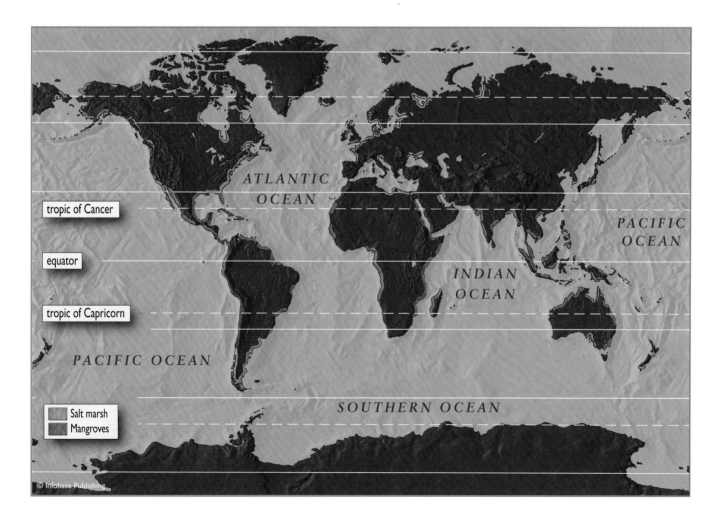

tropic of Cancer

equator

tropic of Capricorn

ATLANTIC OCEAN

PACIFIC OCEAN

INDIAN OCEAN

PACIFIC OCEAN

SOUTHERN OCEAN

Salt marsh
Mangroves

© Infobase Publishing

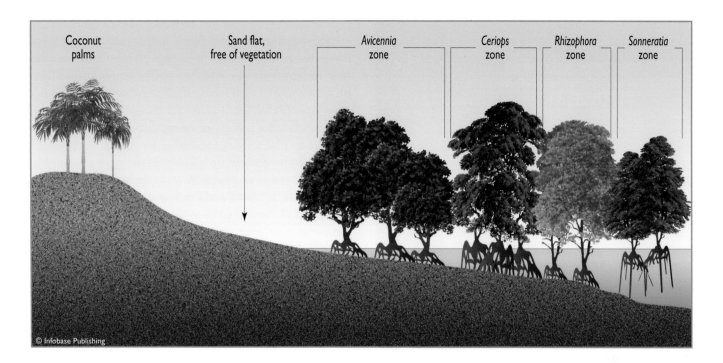

Coconut palms | Sand flat, free of vegetation | *Avicennia* zone | *Ceriops* zone | *Rhizophora* zone | *Sonneratia* zone

Profile of tropical mangrove swamp in East Africa, showing a series of vegetation zones. Different species of tree are best suited to different depths of water and different salinities, hence this zonation pattern develops. The precise pattern of the zonation varies in different parts of the world. The total distance represented diagrammatically here is approximately one mile.

zonation that occurs in mangrove swamps relates to successional processes, or whether it simply reflects the different preferences for water depth and salinity tolerance among the different tree species is still a matter of ecological debate.

Where mangal swamp occupies the seaward edge of estuaries and deltas, it may line the creeks and rivers for some distance inland, depending on the penetration of saline waters upriver. This condition is found in the delta of the Niger River and of the Rio Gêba in West Africa, where the extensive mangrove swamps are a home to hippopotamus and two species of crocodile, both living in salt water.

An impressive feature of mangal wetland both above and below the high-water mark is the complexity of its architecture. Above the mud surface, in the layer periodically flooded by seawater, the arching roots of *Rhizophora* are accompanied by the upright roots of *Avicennia* (termed *pneumatophores,* or *pneumorrhizae,* these erect roots take up oxygen for respiration) and the knotted "knee roots" of *Ceriops* that twist in and out of the mud (see "The Biology of Coastal Wetlands," pages 166–168). These diverse structures combine to form a complicated, periodically submerged, landscape, as shown in the photograph. The cover and protection afforded by this assortment of tangled roots is the haunt of a very wide range of mollusks, crustaceans,

and worms and also forms a valuable habitat for fish, which breed in the submerged forest of roots and also use the cover to hide from predators. One particularly interesting type of fish, from the biological point of view, is the mudskipper (*Periophthalmus* species), an air-breathing fish that occupies burrows in the mud and emerges onto the mud surface at low tide (see "The Biology of Coastal Wetlands," pages 166–168). Because of the value of mangal swamps in sheltering small fish, the mangals are extremely important in tropical coastal fisheries. Wading birds, particularly herons, egrets, and ibises, including the scarlet ibis shown in the photograph, use these regions at low tide to feed upon the

Black mangrove trees in a Florida mangrove swamp are surrounded by breathing roots, or pneumorrhizae, emerging from the mud.   *(South Florida Wetlands Management Department)*

The scarlet ibis is a wading bird of the tropical swamps of the Caribbean region. *(Birmingham Zoo Press Room)*

rich range of organisms available. Populations of crocodiles are also resident in the creek systems of mangal swamps.

The complexity of the upper architecture of the trees offers opportunity for a range of canopy-dwelling animals and plants, including a wide range of epiphytes—plants that grow on the branches of trees. Insect- and fruit-eating birds and bats inhabit the canopy, together with one very characteristic monkey from the mangals of Borneo, the proboscis monkey, with its prominent nose. Fish-eating herons and ibises breed in the canopy.

Mangals, like many tropical forests, are constantly under threat from humans as sources of timber and pulpwood. In the Sunderbans, for example, the extraction of wood for pulp production and newsprint, timber for cheap scaffolding, charcoal and fuel wood for the support of local populations have all resulted in the clearance of perhaps as much as half of the original forest. Reclamation of mangal-dominated landscapes for rice production is also a major threat.

Coastal forest loss in the Sunderbans, however, is not simply a matter of concern for wildlife conservationists but also for those involved in coastal protection. Without the mangal strip, the entire area becomes less well protected against tidal surges associated with typhoons and tsunamis. In Thailand, government funds are available for mangrove rehabilitation, and in Indonesia, mangrove swamps have been given a new legal status that forbids removal of trees from the swamps and tree harvesting in the vicinity of the coastal strip, but this is not true of all mangrove areas. In Malaysia, there are plans for major coastal reclamation, and in the Philippines, 75 percent of the mangrove swamp has been removed during the last 60 years. Where the mangroves represent an important local resource and potential source of revenue, as in Costa Rica and Nicaragua, there is the possibility of exploiting the wetland as a renewable resource and harvesting trees for telegraph poles and charcoal, and bark for leather tanning, in a sustainable manner.

NORTH AMERICAN WETLANDS

Climate, particularly the ratio of precipitation to evaporation, plays an important part in determining the kind of wetland that can develop in any region (see "The Water Cycle," pages 1–4). Climate varies strongly with latitude, so the distribution of the different kinds of wetland also has a distinct latitudinal component, but other factors also underlie patterns of wetland distribution, including altitude, air mass movements, and distance from the ocean. The global pattern of wetlands is therefore the outcome of several combined factors, and this is clearly illustrated by the distribution of wetland types in the North American continent, as shown in the accompanying map.

Northern Canada and Alaska are occupied by the Arctic polygon mires and palsa mires, merging in the south with the string bogs, or aapa fens, of the boreal forest region. In the far west at these high latitudes, however, the rainfall that results from proximity to the Bering Sea and the Pacific Ocean leads to the development of ombrotrophic raised bogs. The eastern regions of the northern temperate zone also bear raised bogs, together with floating quaking bogs that develop over steep-sided lakes. Southern Ontario and the New England region are particularly rich in quaking bogs. Blanket bogs are also found in some northern regions of the east coast, especially in Newfoundland, where the precipitation is high because of the effect of the Atlantic Ocean.

The Rocky Mountains contain a variety of mires, mainly in their valleys, consisting of fens, spring mires, and valley mires, with marshes and swamps surrounding lakes and other areas of open water. These are rheotrophic mires, dependent on catchment drainage to maintain their water supply. The prairie regions of the Midwest of North America are rich in

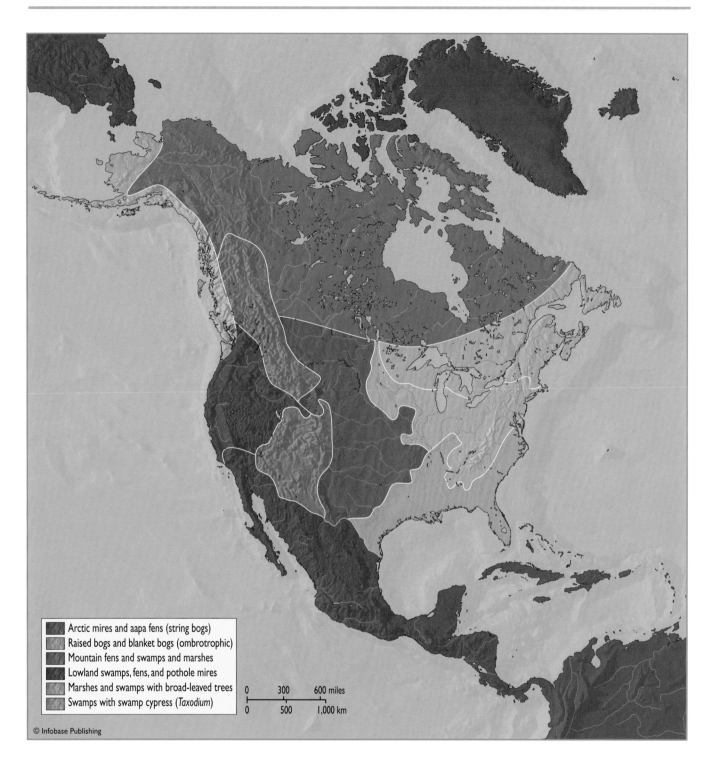

Arctic mires and aapa fens (string bogs)
Raised bogs and blanket bogs (ombrotrophic)
Mountain fens and swamps and marshes
Lowland swamps, fens, and pothole mires
Marshes and swamps with broad-leaved trees
Swamps with swamp cypress (*Taxodium*)

0   300   600 miles
0   500   1,000 km

© Infobase Publishing

North American wetland types. The distribution of some types of mire, such as the Arctic mires and the aapa fens (string bogs) are clearly related to climate and are found only in the cool temperate zone and the Arctic tundra. Similarly, the swamp cypress forests are restricted to the warmer regions of the south. Raised bogs and blanket bogs require high levels of precipitation, so they are found mainly along the Pacific Northwest coast and Alaska or in the maritime regions of the East and the Great Lakes. The intervening regions have a range of marshes, fens, and swamps, depending on altitude and water supply.

pothole mires, rheotrophic wetlands developed around small lakes. Marshes and swamps are found wherever water accumulates in the southern states, especially in river floodplains. The swamps are usually dominated by broad leaved trees through much of the area, but the more southerly regions have increasing proportions of coniferous swamp cypress trees, which assume dominance in the far south.

Southwestern areas of North America are largely dry and so contain few wetlands, although some marshlands persist locally. Inland saline wetlands occur in some of the desert regions, such as the Salton Sea in California. Coastal saline wetlands are found wherever conditions are suitable on both the east and west coasts. Salt marshes are the most widespread coastal wetland type throughout the temperate latitudes, but the most southern regions, especially along the coasts of Baja California, the Gulf of Mexico, and Florida, have a coastal fringe of mangrove swamps.

The zonation pattern of wetlands through North America thus illustrates the ways in which the major wetland types of the world are distributed in response to climatic factors, a pattern that is reflected in the other major continents.

CONCLUSIONS

There are many different types of wetland in the world, and these can be classified according to water sources, water chemistry, depth of water table, developmental history (as recorded in stratigraphy), and vegetation.

Shallow freshwater wetlands are very widespread over the globe, as are marshes, dominated by emergent reeds, sedges, and cattails in the temperate zone and by papyrus in some tropical areas. While marshes have a summer water level above the soil surface, fens have a water table that falls below the soil surface. They are rich in plant and invertebrate species but generally poorer than marshes in their bird life. Aapa mires, or "string bogs," are found in the cool temperate regions, especially the boreal forests of North America and Eurasia, and are distinctive in their pattern of ridges and linear pools that run at right angles to the slope of the ground. Spring mires, as their name suggests, develop where water emerges under pressure from the ground, and in mountain districts springs and seepage zones can lead to distinctive, sloping montane mires. Valley mires form in low-lying topography, where drainage water moves along a valley floor, but where the lateral parts of the stream become waterlogged. Water movement and nutrient supply are richest on either side of the central stream.

All of these wetland types are dominated by herbaceous vegetation, but this is not true of all wetlands. Forested wetlands in the temperate zone are usually referred to as swamp or carr and contain trees that are capable of survival even though rooted in water. Forested swamps are also found in the Tropics, some of them being flooded periodically, such as those in the floodplain of the Amazon River and its tributaries in Brazil. These tropical forested swamps are among the most biodiverse wetlands in the world.

Pothole mires are characteristic of temperate regions that have been glaciated in the past, leaving landscapes pockmarked by the retreating ice. The resulting wetlands are ideal regions for breeding wildfowl. Steep-sided lakes formed by melting ice can sometimes develop into quak-

ing bogs as floating vegetation expands from the margins and forms a raft strong enough to bear developing masses of peat. More frequently, however, temperate lakes pass through a process of infilling, leading to increasingly shallow conditions as marsh gives way to swamp and carr. In the course of time temperate forested wetlands may be invaded by bog mosses of the genus *Sphagnum,* resulting in the death of trees and the development of domes of moss peat, forming raised bogs. These are ombrotrophic in their hydrology, receiving water entirely from rainfall rather than land drainage. Raised bogs can also develop in the Tropics, especially in the coastal regions of Southeast Asia, but here they remain dominated by forests rather than mosses. In many respects, such tropical raised bogs are the closest modern equivalents to the ancient coal-forming wetlands of the geological past.

Areas of high rainfall, especially in oceanic regions of the temperate zone but also on some low latitude mountains, may develop blanket bogs, ombrotrophic peat-forming ecosystems that extend over entire landscapes covering hill slopes, plateaus, and valleys. The development of blanket bogs has often been accelerated or even initiated by the tree-felling activities of prehistoric people, usually followed by burning and grazing by domestic animals. In the high latitudes these blanket bogs merge into palsa mires, in which the landscape consists of a mosaic of shallow pools, sedge marshes and fens, and raised peat masses with ice cores. Palsas develop only under conditions of extreme cold when the frozen peat fails to melt in the summer, leading to the development of massive ice blisters that eventually break down and collapse to form new pools. In the High Arctic, however, conditions are too cold and dry for palsa formation, and here the ground becomes split by ice into large polygons. The centers of the polygons and the deep crevices that separate them bear wetland vegetation, forming polygon mires.

Finally, there are the saline wetlands, which include inland lakes and marshes that have no outlet streams and which therefore become increasingly saline as water is lost by evaporation and the salts are left behind. Most of the saline wetlands, however, are coastal in their distribution. In temperate regions, salt marshes are dominated by herbaceous vegetation, and these are characterized by a zonation pattern that results from different periods of immersion by seawater and then exposure to the atmosphere. Coastal wetlands in the tropical zone, however, are mainly occupied by a specialized group of saline-tolerant trees, the mangroves, and these develop into a protective strip around many tropical coasts.

The wetlands of the world, though varied in their vegetation, structure, and development, all have in common the need for an abundance of water. Their resident plants and animals must all, therefore, be adequately equipped to survive in this humid environment.

5

Biology and Biodiversity of Wetlands

The Earth is currently the location of all the known living things in the universe. It is possible that there is life elsewhere, but this is a matter of speculation rather than proven fact. The vastness of space and all the other stars and planets it contains are largely, perhaps entirely, a bleak and lifeless desert, while the Earth teems with life. Life makes the Earth special, and it is reasonable that people should have due regard and respect for the richness of living things that exist here, the Earth's biodiversity. Among the world's biomes, wetlands rank as one of the most diverse.

■ THE MEANING OF BIODIVERSITY

The word *biodiversity* is currently much used both by scientists and journalists, but it is not easy to define. The main component of biodiversity, and the one that immediately comes to mind, is the richness of species present in an area. Assembling species lists is therefore the first step in determining the biodiversity of an area, but diversity involves more than the number of species present. Consider a collection of 100 colored balls and suppose that they come in 10 different colors. The collection could have 91 balls of one color and one each of all the remaining nine colors, or it could have 10 balls of each color so that there was an even distribution of colors among the balls. The same could be true of a natural ecosystem with 10 species and 100 individuals. The individuals may be evenly spread among the different species, or one species might dominate the system and take up the bulk of the community. Faced with this situation, the ecosystem with the more even and equitable distribution of individuals among the species should be regarded as more diverse than the ecosystem that is dominated by one species. Ecologists therefore make a distinction between richness (number of species in an area) and diversity (a combination of richness and evenness).

Biodiversity, however, is even more complex than this. All the individuals in a population, unless they are vegetative clones, differ slightly in their genetic constitution. In human societies only identical twins have exactly the same genetic makeup, while all others differ to some extent. This variety makes society much more interesting and diverse; the thought of identical people behaving in precisely the same way is the basis for disturbing science fiction, as in Aldous Huxley's *Brave New World*. Sexual reproduction involves the shuffling and recombination of the genetic attributes of parents, leading to totally new arrangements in the offspring, making them into distinctive and unique individuals. This is true of humans and plants, mammals and microbes.

The outcome of genetic recombination, therefore, is genetic diversity. A population of organisms in which breeding takes place between unrelated individuals is said to exhibit *outbreeding*. New genetic characters are constantly being added to the population, and new combinations lead to great genetic variability. This, in turn, equips the population to cope with new challenges, including adaptation to changing climate, pollution pressures, predation, disease, and so on. A small population in which breeding takes place between related individuals is deprived of new genetic input and is said to be *inbred*. There is a danger that genes leading to physical weakness, normally recessive to more dominant healthy genes, consequently accumulate and make their presence felt in the population. Hemophilia in humans is an example of the type of recessive gene that shows itself under such circumstances. An organism that has outbreeding populations and a diverse genetic constitution is therefore more likely to survive in a changing world. It is said to have a wide *gene pool*.

Biodiversity, then, is a concept that includes the breadth of the gene pool within the populations of organisms that constitute its species lists. Genetic diversity is a part of biodiversity because it is a component of the range of variation found within a particular habitat or ecosystem.

Biodiversity also includes the wealth of different microhabitats found within a landscape. This is closely related to species diversity because a diverse landscape, consisting of patches of many different habitats and microhabitats, will contain more species than a uniform landscape that lacks variety. A varied geology and topography in a region develops a range of different soil types, which, in turn, may support a variety of vegetation types with different species compositions and physical structures. Even one wetland area may contain patches of open water, marshes dominated by reeds, some swamp areas with trees, and perhaps some slightly drier sites, elevated above the water table and bearing bog mosses. Within these vegetation types many different animals and microbes are able to survive, leading to high biodiversity.

Conservationists lay much stress on biodiversity. They regard each species as an irreplaceable product of genetic material, sorted by the long and painful process of natural selection and honed to a high degree of fitness. Regions rich in biodiversity are therefore ranked highly in terms of conservation priority. Some conservationists lay stress on the "hotspots of biodiversity" around the world, mainly concentrated in the Tropics, where large numbers of species reside. It is reasonable to rate such sites very highly, but it is dangerous to neglect less diverse areas which nevertheless contain species of plants and animals found nowhere else in the world. Wetlands are extremely variable and contain many species of plants and animals that are confined to these moist habitats. Wetlands occur in many different parts of the world, so the range of organisms that contribute to wetland communities is extremely wide. Consequently, wetlands are extremely biodiverse.

■ SPECIES REPLACEMENT IN SUCCESSION

The biodiversity of wetlands changes in the course of time. Wetlands are dynamic ecosystems—they are in a constant state of change. This is particularly true of wetlands in an early stage of development, including open water areas, reed beds,

Cattail marsh, Outer Banks, North Carolina *(U.S. Fish and Wildlife Service)*

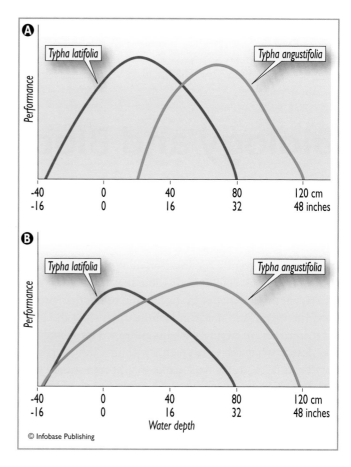

The relative performance of two wetland cattail species, *Typha angustifolia* and *T. latifolia,* in relation to water depth. (A) The natural distribution of the two species in the field when they are found together in competition. (B) The range of the two species when they grow alone, in the absence of competition. When there is no competition, both species are able to grow in a wider range of conditions.

marshes, and so on. Wetland types often replace one another in a reasonably predictable sequence (see "Succession: Wetland Changes in Time," pages 14–15), and this is recorded in the sediment profile. These directional changes in the composition and the nature of ecosystems are called succession, and they occur in all types of habitat, both wetland and dryland.

The presence of reeds slows the movement of water through a wetland, and the particles of silt and clay that remained suspended in fast-moving water become sedimented and are deposited around the reed bases. How much material accumulates in this manner depends on many variables, including the density of the reeds, the speed of water flow, and the sediment load of the moving water. The latter will also vary with *allogenic* (external) factors such as the erosion rate of the catchment and the proximity of an input stream.

A consequence of this sedimentation is that the water becomes shallower, and this gives an opportunity for the invasion of plants that could not have grown in deeper water because of the problems of reaching the surface with their stems and leaves. In a study of reed beds and fens in Europe, for example, the true reed (*Phragmites australis*) is usually dominant in waters on average 22 inches (56 cm) deep (ranging from three to 43 inches [8 to 110 cm]). Tussock sedge (*Carex paniculata*) communities are most productive in sites where the water depth is zero (summer water table at ground level), and meadowsweet (*Filipendula ulmaria*) grows best where the summer groundwater table is on average 5.5 inches (14 cm) below the soil surface. As the water becomes shallower (or, strictly, as the surface of the soil rises), different species of plants find themselves at an advantage and are able to compete for a place in the community, and this fuels the development of the succession.

In this successional sequence, the pioneer species appear to pave the way for their successors. They are said to *facilitate* the invasion of new species. This does not imply that there is any planning or altruism on the part of the pioneers; their very presence makes it inevitable that conditions should change in favor of the newcomers. It often happens that as conditions continue to change, the pioneers are completely "out-competed" by the new arrivals because the latter are bet-

ter suited to the changed environment. This *autogenic* (internal) aspect of succession and species replacement, therefore, is the outcome of facilitation followed by competition.

Competition, in biological terms, occurs when two individuals demand the same resource from their environment and that resource is in short supply. In the case of plants, the resource is often the light needed for photosynthesis, so the plants actually compete for space because this gives them access to the light resource. Competition is at work in wetland vegetation, as can be seen by observing the difference in the performance of species when they grow in the absence of their competitor and when they grow in its presence. Cattail marshes (see photograph opposite) are ideal habitats in which to examine these interactions. In experiments with the cattails, *Typha latifolia* and *T. angustifolia*, for example, it was found that when growing without competition, *T. angustifolia* could occupy wetlands with water 47 inches (120 cm) deep, right through to fens where the water table was 8 inches (20 cm) below the soil surface. Its optimum water depth for growth was about 23 inches (60 cm). The related, but more robust, species, *T. latifolia*, however, grows less well in deeper water and has its optimum for growth at about 8 inches (20 cm) water depth. When they

Ecological Niche

Ecologists have long pondered the question of how so many different species of plants and animals manage to live alongside one another in a reasonable degree of harmony and stability. Some species clearly have very different ways of living, so the problem does not arise. Blue jays, oak trees, and deer can all live together because they are all engaged in very different modes of life, but how can two species of grass or two species of vole coexist in an area when they appear to be in competition for the same resources? Usually, when the grasses or the voles are examined closely they reveal subtle differences in their precise requirements. As in the cattails shown in the diagram, they may have slightly different environmental limits and preferences, or they might obtain their needs in slightly different ways. Hawks and owls can live together because one hunts in the day and the other at night. The sum of all an organism's requirements, together with its manner of obtaining its needs, it called its *ecological niche*. The niche can be considered as the role played by an organism in its community; it describes the way in which an organism makes its living.

The ecologist G. Evelyn Hutchinson developed the concept of the ecological niche by considering two

possible forms that it can take. If an organism is unconstrained by other competitive species seeking the same resources, then it expands to its full potential. In the absence of *Typha latifolia*, for example, *T. angustifolia* would expand its population throughout the deep and shallow water, as shown in part B of the diagram on page 130. Hutchinson called this potential range its *fundamental niche*. If its competitor is present, then the potential is never fully achieved, and the species occupies a more limited range that Hutchinson termed its *realized niche*. The realized niche is always smaller than the fundamental niche because of the competition that all species experience in natural conditions. If two species are to coexist in the same area, then they must differ in some aspect of their fundamental niches. Even in a situation where two species consume the same food, each species will differ from its competitors in the way it exploits that food, as in the case of hawks and owls. In this manner, the species in a community seem to divide the resources among themselves in a process called *resource partitioning*. In the course of succession, the pattern of resource partitioning is constantly changing as new species enter the community and others are extinguished.

grow together in conditions of competition, *T. latifolia* is able to exclude *T. angustifolia* from the drier end of its range because it grows so much more vigorously. So the more delicate *T. angustifolia* becomes restricted to the deeper water (between 8 and 47 inches [20 and 120 cm] deep), where it grows more efficiently. These interactions are shown graphically in the diagram.

The two cattails thus differ in their individual requirements and are said to occupy different *ecological niches* (see sidebar on page 131). Each wetland species, then, has its place in the course of succession. The adaptations possessed by each plant and animal species fit it for a particular location in space and time where it will make a living by carrying out a specific function (primary producer, herbivorous consumer, detritivore, etc.). As time passes, species will replace one another because conditions have changed, and new plants and animals are better fitted for the new circumstances. Eventually, one might expect the system to reach a kind of equilibrium (sometimes called *climax*), when further development ceases. In practice, however, changes in climate and other environmental factors, together with internal changes, such as peat erosion and fire, mean that such stability is an unlikely ideal.

The composition of communities thus changes in the course of wetland succession, and the overall biodiversity tends to increase, although this is not always the case. When a raised bog replaces swamps and fens, for example, the structural diversity of the habitat decreases and the plant diversity also falls, so overall biodiversity is often lower. Most wetlands are complex, however, and contain areas in different stages of succession, which means that the species found in all of the different stages may well occur in close proximity within a wetland area. In such a case, the habitat diversity ensures that overall biodiversity is enhanced. Succession is thus an important contributor to biodiversity.

■ LIVING IN THE WET

Living in wetlands presents certain problems to living organisms, whether animal, plant, or microbe. The type of problem encountered depends largely on the chemistry of the wetland (particularly whether it is salty) and partly on its hydrology (pattern of water distribution and movement). Living in a tree surrounded by water may lead to problems for an animal, but living submerged in the waterlogged mud around that tree's roots presents very different problems. All of the living things associated with wetlands have had to face difficulties in their evolutionary history, and all show distinctive adaptations to the habitats in which they are found.

The most general feature of wetlands, of course, is water, and an adaptation to life in or near water is the main char-

acteristic of all wetland organisms. Most of the plants of wetlands, and many of the animals, spend some time with at least part of their bodies submerged in water. Some may spend their entire life fully submerged. Waterlogging affects the process of decay in wetlands (see "Decomposition and Peat Growth," pages 54–56) as a result of its impact on microbes and also how the chemistry of waters and sediments is influenced by waterlogging. It also has a profound effect on the plants and animals that find themselves permanently immersed in water. The main problem is precisely that which is encountered by the bacteria and fungi involved in decomposition, namely a lack of oxygen availability.

Submerged aquatic plants photosynthesize successfully as long as light penetrates to them through the water (which depends on the turbidity, or clarity, of the water and the degree of shading from the canopies of the plants above the water) and as long as there is an adequate supply of carbon for fixation into sugars. The carbon comes in the form of dissolved carbon dioxide gas in the water or as hydrogen carbonate (bicarbonate) ions. While photosynthesis is taking place, oxygen is generated and is released into the surrounding water, where it dissolves. This oxygen, together with that dissolved from the atmosphere at the air/water interface, is available for the respiration of wetland plants, animals, and microbes.

In the absence of photosynthetic aquatics (perhaps because of dense shading by overhanging vegetation), in situations where little oxygen is dissolved from the surface (for example, in stagnant waters) and in conditions of high microbial activity (where organic matter is being decomposed), oxygen may be in very short supply. When this happens, all organisms, except those microbes with alternative biochemical pathways for respiration, experience problems in maintaining their energy-producing respiratory cycle.

Many of the adaptations of submerged wetland organisms, therefore, are concerned with obtaining sufficient oxygen. There are two main ways in which this problem has been solved in the course of evolution. One solution is for the organism to develop a means of transporting oxygen from the atmosphere to its location in submerged, anoxic conditions. Mobile animals may carry air from above the water and retain it in a thin film around their bodies while beneath the surface. Others may carry the air and store it beneath the surface of the mud in tunnels or chambers. Plants do not have this option but can develop air passageways in their tissues so that emergent parts of the plant body can take in air that will subsequently diffuse through the stems or leafstalks into the submerged organs beneath the mud. The presence of precious stores of oxygen in this airless environment has inevitably led to many animals developing techniques for "stealing" this air and gaining oxygen at second hand.

A second solution to the anaerobic problem is a biochemical one. Even terrestrial animals experience a lack of oxygen occasionally, so the problem is not confined to wetlands. For example, a sprinting human cannot breathe fast enough to keep up an adequate supply of oxygen to the muscles, and fatigue soon sets in. This tiredness is an outcome of the buildup of lactic acid in the muscle tissues, which itself occurs because the blood sugars are not fully oxidized to carbon dioxide. A similar process occurs in plant roots when they are submerged in water for long periods. The sugars used in respiration are not fully converted to carbon dioxide but are turned into ethanol in a process called anaerobic respiration. (It is the anaerobic respiration of yeast that is used in the commercial production of alcohol.) Ethanol is a toxic compound, and its accumulation leads to the death of roots, which is the immediate cause of plants unused to long-term submergence being killed by the experience. Those wetland plants that have a long evolutionary history of flooding, however, have developed biochemical pathways that avoid the accumulation of ethanol in their roots or maintain tolerance to its presence. They may divert the product of anaerobic respiration into malic acid or into shikimik acid (harmless organic acid), which do not have the toxic effects of ethanol. In this way, they avoid the problems of flooding damage.

A secondary effect of the lack of oxygen in submerged conditions is the presence of toxic ions, such as Fe^{2+} and Mn^{2+}. Those plants with air pathways through their root tissues seem to deal with this problem by leaking some oxygen into their soil environment and oxidizing these ions into a harmless form (Fe^{3+} and Mn^{3+}). Sometimes this can be observed by extracting plant roots from anaerobic muds and observing the brown stains caused by ferric (oxidized) compounds around them. It is also possible that the malic acid formed during anaerobic respiration may combine with toxic ions and render them harmless.

Many wetland invertebrate animals have developed compounds with a high affinity for oxygen so that they can store and move this commodity around their bodies. Even hemoglobin, the oxygen-carrying compound in the blood of mammals, is found in many anaerobic-dwelling invertebrates, including wetland lumbricid earthworms, some mollusks, and midge larvae. The African swamp catfish (*Clarias mossambicus*) has a form of hemoglobin that has an unusually strong affinity for oxygen, and this animal, like the lungfish, is also able to tolerate high concentrations of carbon dioxide in the blood. Eels, which are often associated with oxygen-poor conditions, are able to conduct some gaseous exchange through their skins to supplement the intake of oxygen through the gills.

As in plants, anaerobic respiration occurs in the tissues of animals subjected to anoxia, and, since this is an inefficient respiratory process that fails to make full use of the energy available, large stores of food (usually in the form of glycogen) are often found in animals of the mud (such as

bivalve mollusks). This, of course, can make them attractive as a food source for predators, including human beings.

■ MICROBES IN WETLANDS

Biologists divide the living organisms of the world into five kingdoms. One of these kingdoms, the bacteria, is distinct from all other kingdoms by lacking a true nucleus in the single cell that forms the body of its members. These are called *prokaryotic* cells. All of the other four kingdoms contain organisms with nuclei in their cells, called *eukaryotic* cells. These four kingdoms consist of animals, plants, fungi, and Protoctista (or protists). The first three of these are familiar to most people, and the fourth contains an assortment of organisms that have variously been regarded as plants and animals in earlier systems of classification. The protoctista consists of the algae (red, green, and brown), various unicellular photosynthetic organisms (including diatoms, chrysomonads, dinoflagellates, and euglenids), slime molds, amoebae, and various other organisms formerly grouped within the protozoa. The term *microbe* is usually used to cover the bacteria and the fungi, so it encompasses two kingdoms of organisms.

In all ecosystems, microbes are most abundant in the soil. An ounce of soil can contain as many as 3 million bacteria (100,000 per gram). The identification of bacteria, however, poses immense difficulties, so it is extremely difficult to estimate the richness of species in soil. Individual bacteria need to be cultured and developed into large colonies if they are to be identified effectively, and at present up to 95 percent of soil bacteria are regarded as unculturable, so they defy identification. In recent years, microbiologists have approached the problem by extracting DNA from soils and amplifying this by a culturing system, but the DNA of bacteria is variable, and the bulk of the DNA extracted from a soil cannot be assigned to particular species, so researchers have to be content with arbitrary labels for the "species" they extract. Bacteria are present and active in wetland soils, especially those that occasionally become aerated as the soil water table falls. The mud that accumulates at the bottom of ponds and bog pools is permanently waterlogged, and oxygen can become scarce. Sometimes, however, even in a stagnant pool, the photosynthetic activity of algae and other unicellular protists, such as the euglenoids, generates oxygen at the mud/water interface and allows the aerobic bacteria and fungi to respire and decompose the accumulating organic matter.

Some bacteria are photosynthetic, and these need light for long-term survival. The most abundant of the photosynthetic microbes in tundra environments are the blue-green bacteria, the cyanobacteria (see sidebar). There are other photosynthetic bacteria found in the surface of wetland mud

Cyanobacteria

The blue-green bacteria were once regarded as algae, and their system of photosynthesis is very similar to that of the algae and of higher plants. They generate ATP and a reducing agent (both needed for the fixation of carbon in photosynthesis) by using the energy of sunlight, which they absorb with the blue-green pigment phycocyanin together with chlorophyll a. Most are aerobic and, like higher plants, use water as an electron donor, releasing oxygen as a waste product. Some are able to operate by using hydrogen sulfide as an electron donor, however, releasing sulfur instead of oxygen, like the purple bacteria. Many species are filamentous, forming long chains of individual cells linked together at their ends. These bacteria were the main photosynthetic organisms of the early Earth, dominating the planet from 2,500 million years ago to 600 million years ago. They formed dense mounds called *stromatolites,* which can still be found in the shallow seas of the west coast of Mexico, the Bahamas, Australia, the Persian Gulf, and even beneath the ice in Antarctica.

Apart from their ability to photosynthesize, the cyanobacteria are remarkable biologically because they can fix atmospheric nitrogen. One of the photosynthetic cells in a chain enlarges and loses its pigments, forming a heterocyst that houses the enzyme nitrogenase. Most cyanobacteria are terrestrial or freshwater, though some have been found in seawater. They are particularly abundant in tropical wetlands, including agricultural systems, such as rice paddies, where they are important as a source of nitrates in the growth of the crop. In natural wetlands, their capacity to fix nitrogen can be a problem because excessive fertilization of water, called *eutrophication,* leads to problems of overproduction of organic matter, high microbial activity, oxygen depletion, and the loss of aerobic life.

that have green or purple pigments. The purple bacteria are unusual because they do not use water, H_2O, as a hydrogen donor, which is the process used by all green plants. Instead the purple bacteria use hydrogen sulfide, H_2S. Whereas green plants produce oxygen as a waste product in their photosynthesis, purple bacteria produce sulfur.

The photosynthetic bacteria are not easy to detect in a pond, but there is a simple experiment that can be used to discover their presence. Mix the surface organic mud from a stagnant pond with calcium sulfate and pack it in a glass test tube. Then fill the tube with water and place it for a few weeks in the light. If the photosynthetic bacteria are present in the mud, green and purple patches appear in the tube as colonies of the bacteria develop. The hydrogen sulfide they need for their photosynthesis is produced by another anaerobic bacterial group, the sulfur bacteria, including *Desulphovibrio* species, which take the oxygen they need for respiration and decomposition from the sulfate ion, SO_4^{2-}. In the process, they reduce sulfate to sulfide, S^{2-}, which combines with hydrogen to form hydrogen sulfide. This is then used as a source of electrons by the photosynthetic bacteria. It is thus possible to construct a simple microbial ecosystem resembling that of the surface of natural anaerobic pond mud.

Photosynthetic bacteria need light for their activities, so they occur only in the surface layers of pond mud. In the dark regions below the mud surface, anaerobic, dark-living bacteria, including the sulfur bacteria, continue to operate, breaking down organic matter. The hydrogen sulfide produced accumulates as bubbles of gas that periodically erupt through the mud and water and burst into the atmosphere above, resulting in a distinctive and unpleasant smell that is characteristic of stagnant mud, similar to that of rotten eggs. Wading through mud compresses it and encourages the release of this distasteful gas, providing evidence of the anaerobic decomposition that is proceeding below.

■ SUBMERGED PLANT LIFE

Compared with plant life on land, a submerged photosynthetic plant has some advantages and some disadvantages. Its problems, in other words, are quite different from those of its terrestrial relations. Water, for example, is not in short supply. Whereas a land plant expends a great deal of energy on the production of roots that can forage in the soil for vital water supplies, an aquatic plant is surrounded by water. All of the adaptations associated with drought resistance in land plants (small, inrolled or leathery leaves, waxy cuticles, stout cylindrical stems, and so on) are not needed in the water. Water is also a denser medium than air, and this means that plants find it easier to support their leafy canopies without resort to tough stems and branches. There is no equivalent to woody trunks among aquatic plants; the nearest approximation is the rubbery stem of some of the wave-washed, low-tide brown seaweeds. Water supports the plants, and those parts that need to be at, or close to, the surface can be provided with air-filled flotation structures, thus avoiding any need for the kind of scaffolding required by land plants.

Water hyacinth (*Eichornia crassipes*) is a tropical aquatic plant that has become a weed because of its high rate of vegetative reproduction. Many tropical lakes, including Lake Victoria in Africa, are badly affected by local growth of the weed, causing difficulties for boat traffic. *(Peter D. Moore)*

Hardened, woody materials are, therefore, uncommon in submerged aquatic plants. They are sometimes needed for the protection of seeds, especially if these are likely to become trapped in mud or desiccated for long periods or if they are adapted to pass through the gut of an animal and survive to enjoy the consequences of wider dispersal. For support purposes, however, the natural buoyancy of the water is adequate.

The supply of elements needed by a plant for constructing its component parts is also freely available. Whereas a land plant gains most of its elements (with the obvious exception of carbon) from roots embedded in the soil and often relies on microbes around those roots to liberate nutrients from organic matter and clay particles, aquatic plants are constantly bathed in nutrient solution. Sometimes that nutrient solution is very dilute in vital elements, such as phosphates and nitrates, and the plant needs to be adept at concentrating them (as in ombrotrophic bog pools and in water bodies within poor fens).

Roots in submerged aquatic plants have some absorptive function, but this need not be restricted to the substrate. In the Canadian pondweed (*Elodea canadensis*) and in the bulbous rush (*Juncus bulbosus*), for example, stems proliferate and produce dense bunches of roots at specific growth points (nodes). The roots contain no chlorophyll, so they do not photosynthesize, but they absorb elements from the water and, if they extend down to the basal mud, they also serve to anchor the plant, thus preventing its being washed away in the flow of water.

Photosynthesis requires a supply of light (from which energy is trapped) and carbon dioxide (or bicarbonate ions), which is used as a source of carbon for the building of sugars and, subsequently, more complex organic molecules. Photosynthesis under water is perfectly possible as long as these two resources are available. Light is absorbed by water, but in the shallow water bodies included in the definition of wetlands, this does not deplete the resource significantly. If, however, the water contains suspended particles of material (either inorganic or organic), light is absorbed more effectively and may be in short supply even in a shallow water body. When this occurs, submerged aquatic plants can exist only close to the surface, or they may be replaced by aquatics with floating leaves, such as the water lilies, or by plants with their entire body floating in a mat, such as water hyacinth (*Eichornia crassipes*), as shown in the photograph. Water lilies, with their tuberous stems in the substrate and their leaf blades on the surface, are limited in the depth of water they can occupy by the length of the leaf stalk that needs to be developed to link these stems to the floating leaves. They seldom occupy water bodies deeper than 12 feet (4 m).

The availability of carbon in the water depends on its being dissolved as carbon dioxide from the atmosphere or, alternatively, on the supply of calcium (or magnesium) carbonate from the rocks of the catchment. Carbon dioxide is also generated by plant and animal respiration in the water, and by the decay processes taking place mainly in the organic-rich substrates on the bottom. Dissolved carbon dioxide, like dissolved oxygen, does not diffuse as rapidly in water as it does in air, so, at any given time, its availability will depend on how rapidly the aquatic plants use up the supply in their photosynthetic activities and how quickly it is replenished. At night, when no photosynthesis takes place, carbon builds up in the water, but in the daytime, when the plants are actively photosynthesizing, it may become locally scarce, forcing plants to compete for the limited supply. Some submerged aquatic plants, such as the quillwort (*Isoetes* species), a primitive group of plants related to the ferns, have evolved a distinctive biochemical pathway of carbon accumulation using an enzyme for fixation (PEP carboxylase), which is more efficient in its carbon-trapping abilities than the normal carbon-fixation enzyme (rubisco). This gives them an advantage in the struggle for the limited carbon resource (see sidebar).

In many ways, the floating and emergent aquatic plants get the best of both worlds. They are supported by water, have their mineral nutrients supplied by the flow, but use the atmosphere to provide carbon dioxide for photosynthesis and the oxygen needed for respiration. Hollow stems and leafstalks may allow the penetration of these gases down into the submerged parts of the plants, and the respiratory carbon dioxide generated in the roots can diffuse upward. There are some emergent plants, however, in which the roots have developed a very different function and grow upward out of the water to tap the atmospheric oxygen for their respiration. The trees of mangrove swamps (see "Mangrove Swamps," pages 123–126) have these "breathing roots," or pneumorrhizae, and they come in a variety of forms. Some are peg roots, sticking up out of the mud; others form loops, rising above the surface then plunging down again; and some take the form of stilts, propping up the tree trunk while taking in oxygen from the air. A similar adaptation is found in some freshwater swamp species where flooding causes the development of masses of roots from trunks and stems above the new water level. It is frequent in the black-

Aquatic Photosynthesis

Most aquatic plants and algae adopt the same photosynthetic system as that used by terrestrial plants. They use sunlight to generate an energy-rich molecule, ATP, and to form a reducing agent, NADPH. The reducing agent and the energy are then used to reduce carbon from carbon dioxide to organic forms of carbon, starting with simple sugars. The only difference is that a submerged aquatic plant has to obtain its carbon dioxide in a dissolved form or may resort to using the hydrogen carbonate ion, HCO_3^-. On the whole, this system works perfectly well, but dissolved carbon dioxide, like oxygen, diffuses very slowly in water, and this important resource can become severely depleted. This is most likely to happen in shallow water where the temperature in the middle of the day may rise and the solubility of gases, including carbon dioxide, is reduced. Intense competition between all the photosynthetic members of the phytoplankton also reduces the raw material of photosynthesis in the waters. Any mechanism that can provide a plant with an advantage in harvesting CO_2 at low concentrations is therefore at a premium.

One such mechanism is called *Crassulacean acid metabolism* (CAM). This biochemical pathway, as its name reveals, was first found among the plant family Crassulaceae, which consists largely of succulent species from arid environments. These plants face problems of desiccation, especially in the daytime if they open their stomata to absorb gaseous carbon dioxide from the atmosphere and consequently lose water from their internal tissues. Instead, they open their pores at night and fix the carbon in a temporary form as organic acids, using the enzyme phosphoenolpyruvate carboxylase (PEP carboxylase). Sealed within the plant tissues, the organic acids then decompose to release carbon dioxide again that can be refixed using the conventional photosynthetic system involving the enzyme rubisco (ribulose bisphosphate carboxylase/oxidase). The problems of aquatic plants may seem far removed from those of the succulent species of the desert Crassulaceae, but they have come to use precisely the same enzyme, PEP carboxylase, to overcome their difficulties in accumulating carbon from low environmental concentrations. PEP carboxylase has a much higher affinity for CO_2 than rubisco, so it can operate at much lower CO_2 concentrations. This is precisely what is needed by plants growing in shallow, warm wetlands, and several species have evolved the CAM mechanism for temporary carbon accumulation. Like the desert succulents, they take up carbon dioxide from the water during the night, when the respiration of all the aquatic organisms replaces the carbon that has been lost from the water during the day. They store this fixed carbon as organic acids in the night and then release the CO_2 into their tissues in the daylight, fixing it finally by using the conventional rubisco system. They thus gain a competitive advantage over their neighbors in the struggle to take up a very limited carbon supply.

gum (*Nyssa sylvatica*), in many willows (*Salix* species), and in the marsh-dwelling species of willowherbs (*Epilobium*).

ANIMAL LIFE IN WATER

The animals of aquatic habitats may spend their entire life below water, as in the case of most fish, or they may spend one part of their life cycle below water, as in the case of mosquitoes, or they may spend time both above and below water according to their immediate needs and inclinations, as in the case of frogs, crocodiles, and hippopotamuses. The most aquatic habitats of the wetlands, therefore, contain a wide range of animals with many different ways of life and adaptations.

Many fish breed within the protection offered by submerged and emergent aquatic plants. The complexity of the habitat provides opportunities for them to lay their eggs in locations where they may be less easily found by predators. Among the cichlid fishes of central Africa, some, such as *Tilapia sparrmanii,* build nursery areas where they dig several holes in the mud among the roots of aquatic plants, in which they lay their eggs. They then guard these against predators and even continue their "brooding" activities when the small fish fry hatch. The three-spined stickleback (*Gasterosteus aculeatus*), common in the temperate waters of Europe, Asia, and North America, even builds a nest among the rooted plants of shallow-water wetlands, using plant material that it glues together using a sticky secretion from its kidneys. The male cares for its eggs by fanning them with its fins to provide a good supply of oxygen, and after hatching it continues to guard the young.

Those wetlands that expand their area during wet flood periods offer many opportunities for fish to exploit resources that are unavailable for the remainder of the year. In June, July, and August, much of the forest region along the Amazon River in Brazil becomes flooded, following heavy rain and snowmelt in the Andes. Many species of fish (possibly between 2,000 and 3,000 species), including many cichlid fishes, spread over new areas and consume the fruits and seeds dropped into the water by the surrounding trees. Although piranhas are renowned for their flesh-eating activities, there are species of piranha that are entirely vegetarian, and these benefit from the abundance of seed. They also contribute to the dispersal of these periodically inundated trees. Some fish (known locally as *aruanas*) are able to leap out of the water and take resting insects or even hummingbirds from low branches.

Some aquatic mammals live out their lives in the waters of the Amazon, such as the river dolphin (*Inia geoffrensis*), which also ranges widely through the flooded forest. Navigation in the complex tangle of submerged branches is difficult, but these animals use a form of sonar, rather like

underwater bats, to ensure that they do not collide or become entangled in the vegetation. They use the same adaptation for locating their fish and crustacean prey. The sea cow, also called manatee (*Trichechus inunguis*), is another mammal that spends its life submerged, usually in the estuaries of the great tropical and subtropical rivers or along coastal wetlands. In the Amazon, the manatee is entirely freshwater in its preferences but, unlike the river dolphin, it is herbivorous. Its prodigious capacity for consuming aquatic plants has led to the use of the manatee for the clearance of waterways in Guyana.

When the Amazon floods recede, some aquatic animals use the opportunity for laying their eggs upon dry land. This is the case with the river turtles (Suborder: Cryptodira), which lay up to a hundred eggs each along the mud and sandbanks exposed by the falling water level. The very high egg production of such species is indicative of the intensity of predation on both the eggs and the young turtles, for very few survive to maturity. The same applies to the caimans (a tropical equivalent of the alligator), for these reptiles also lay their eggs in the dry season and suffer similarly large losses to birds, rodents, and indeed to turtles. Caimans, however, actually build nests and protect them from predators. Being cold-blooded, the caimans cannot, of course, actually incubate eggs; the heat from the decomposing vegetation in the nest construction aids embryo development within the eggs. Those caimans that survive to maturity may live up to 40 years. Alligators are distinguished from crocodiles by the arrangement of their teeth. In alligators (see photograph) only the teeth of the upper jaw are visible when the animal's mouth is closed, which gives it the appearance of a smile. In crocodiles there is an enlarged fourth tooth on the lower jaw, which projects upward outside the upper jaw, giving the

Alligator basking on a log   *(Bailey Dick, U.S. Wildlife Service)*

appearance of a crafty leer. Alligators are found only in the southeastern states of the United States, and in the Yangtze River valley in China.

There are 13 different species of crocodile inhabiting the world's wetlands, the most familiar being the Nile crocodile (*Crocodylus niloticus*), now extinct in the wetlands of lower Egypt but still found in Lake Nasser above Aswan and through much of East Africa. When young, the Nile crocodile feeds on aquatic worms, beetles, and crustaceans but graduates to turtles, waterfowl, and mammals as it matures. Their speed and agility depend on their temperature, which they raise by basking out of the water, and this may also aid their digestive enzymes to operate efficiently. Spending the night in water helps to ensure that they cool less rapidly.

The hippopotamus (*Hippopotamus amphibius*), being warm-blooded, is not limited by such considerations. Indeed, it tends to spend the daytime in the water and emerge at night to graze along riverbanks and wetlands (see sidebar).

Much less spectacular than these large aquatic animals, but by no means less important, are the aquatic invertebrates of wetlands. Mosquitoes, for example, are found in practically all aquatic wetland habitats, acid and alkaline, arctic and tropical, saline pools, and even the warm pools of volcanic craters. The larvae are fully aquatic, although they do breathe air through pores, or spiracles, at the end of their bodies in addition to absorbing oxygen from the water over their body surface. They swim with a jerky twisting movement, and they are covered with stiff hairs, which may help to protect them from predators. The mature insect, of course, lives above the water surface and feeds on sugars from flowers, but the female will also take vertebrate blood, and almost all vertebrates are affected by mosquitoes, with the obvious exception of fish. Even among fish, the mudskipper (a fish that emerges from the

---

Hippopotamus

The hippopotamus (*Hippopotamus amphibius*) is one of the largest of all wetland creatures. It can grow to 15 feet (4.5 m) in length and can weigh up to 6,000 pounds (2,700 kg). Despite its size and its short legs, it can run fast on land, easily outpacing a man. It spends most of its time in the water, however, and its great bulk is supported within an aquatic medium so that it swims well both on the surface and below. The nostrils are situated on top of its snout, so it can float close to the surface and almost totally submerged yet still be able to breathe through the projecting apertures. When it submerges the hippopotamus closes its nostrils, and it can then remain underwater for five minutes or more. Its diet is entirely vegetarian, but it has enormous teeth, especially the lower canine teeth, which it uses for fighting. The hippopotamus can be aggressive in behavior and is greatly feared in its native Africa because it sometimes attacks fishing boats and people in the water, especially if the animals have young with them. They also attack on land when humans come between them and their wetland habitats, readily charging when they are nervous. More people are killed each year by the hippopotamus than by any other animal in Africa.

At night the hippopotamus leaves the water and grazes along riverbanks, often invading agricultural land around villages. It returns to the water for the day and defecates mainly in the water, thus transferring nutrients from the land to the aquatic habitat and encouraging water eutrophication. These animals have been hunted and persecuted extensively in Africa, partly for meat, partly because they represent a danger to people, and partly because of their pest status in destroying crops. Hippopotamus grazing, however, does diversify the wetland habitat, creating open water, holding back the successional development of reed and papyrus beds, and opening trails through swamps. They have very few enemies, and their populations can become high in those localities where they are protected. In the Nile Valley north of Lake Victoria, for example, densities of up to 80 animals per square mile (30 per square km) have been recorded, but at this level the hippopotamus population is likely to overgraze and cause serious damage to the local environment. Conservationists regard 20 animals per square mile (8 per km²) as the optimum number for maintaining a diverse and sustainable wetland habitat.

A single hippo can consume at least 150 pounds (70 kg) of herbage in one night's grazing, which means that in areas of high population density the plant biomass removed from a square mile each night can be as high as six tons. When the hippopotamus grazes, it uses its muscular lips rather than teeth to pluck the vegetation, and it prefers relatively short grass that it is able to tear up by the roots. The hippopotamus is selective in its grazing, so the animal can alter the composition of wetland vegetation by taking some species and leaving the less palatable ones behind.

water) is a potential host. Many viruses and parasites are transmitted by mosquitoes.

There are many other insects that spend part of their life below the surface of the water and then emerge in their adult form. Mayflies (Ephemeroptera), for example, are familiar because of their swarms flying above water bodies in the springtime. They are easily recognized when at rest because they hold their wings vertically above their backs and they have three prominent bristles projecting from their hind parts. These bristles are also found in their aquatic larval stage. Their eggs are laid on the water surface, although some species can submerge themselves and lay eggs on the stones beneath the surface. When the young hatch, they have gills and so are fully aquatic. They feed mainly on vegetable detritus on the bottom of rivers and ponds, and they in turn are fed on by carnivores such as fish, dragonfly larvae, and water beetles. Another predator that is particularly fond of mayflies is the dipper (*Cinclus mexicanus*), a bird that has the remarkable ability to walk along the bottom of streams and rivers, turning over stones and extracting its invertebrate prey.

When the mayfly nymph is mature, it comes to the surface of the water, its skin ruptures, and the adult insect emerges. This emergent insect is usually quite dull in color, but within an hour or so it molts once again to produce the final mature mayfly. It is brightly iridescent, enjoying a very short life of dancing flight, mating, and egg laying before dying and donating its body to the aquatic detritus that will supply other organisms with their required energy.

Even some vertebrates have life histories that involve a juvenile stage beneath the surface of water. These are the amphibians—frogs, toads, newts, and so on. They spawn beneath the surface of the water, and their young (tadpoles) spend their early life feeding on both vegetable and animal materials and respiring through gills. The abundance of the spawn is an indication of the high risk of predation suffered by many amphibians. The likelihood of any individual tadpole surviving to maturity is very low, so large numbers have to be produced in order to ensure that the adult population is at least maintained at a stable level. There are few more graphic illustrations of Darwin's proposal that young are overproduced, thus ensuring that only the very fittest will emerge as breeding adults from the pressures of predation, disease, and competition for food.

Aquatic ecosystems provide numerous feeding opportunities for animals, so there is a very high diversity of species, ranging from the very small to the very large. Some minute animals (the zooplankton) feed on the microscopic plants (the phytoplankton) that live and photosynthesize while suspended in the waters. Meanwhile, a range of aquatic bacteria, viruses, and fungi attack these and larger animals, or simply feed on their dead remains if they manage to survive predation and die a natural death. Other animals filter feed.

This means that they draw water through their bodies and sift out the energy-rich materials, the bodies (both living and dead), or the fragments of tissue that are suspended in the water. Some of these filter feeders are very small, such as the microscopic *Vorticella* that attaches itself to submerged surfaces and waves its filter apparatus on stalks. Other filter feeders are much larger, such as the bivalve mollusks, including freshwater mussels that draw water through their siphons and extract their food.

Other animals are scrapers. These progress over surfaces and rasp away at the coating of microscopic plants and bacteria. The gastropod mollusks, such as water snails, are typical of this group. They have a tough, barbed, tongue-like structure, the radula, with which they scrape at surfaces of stones or plants. Many animals belong to the detritivore feeding group, consuming dead material (mainly plant matter) that is constantly being produced beneath the water surface, whether from local aquatic plants or brought in from other ecosystems (such as forest leaf litter). The segmented worms and the unsegmented nematode worms are among these detritivores, together with mites and a range of insect larval stages.

All of these animals are preyed on by a range of predatory organisms. Water beetles are voracious predators, the larger ones even taking fish. Some of the insect larvae, such as dragonflies (Odonata), are also predatory on small animals. Vertebrate predators include truly aquatic animals, such as fish, together with the amphibious frogs and newts. Reptiles, including snakes and crocodiles, are predatory on these vertebrates, as are some mammals, including otters and freshwater dolphins. There are also the many birds that exploit the underwater world in their search for food, including kingfishers, grebes, herons, egrets, dippers, and such specialist feeders as the Florida snail kite (*Rostrhamus sociabilis*), which feeds almost exclusively on gastropods.

The abundance and variety of underwater animal life indicates the richness of opportunity for making a living in the submerged environment. This has resulted in a high level of complexity in ecological interactions, particularly in the way organisms feed upon each other (see "Food Webs of Wetlands," pages 50–54).

■ ANIMAL LIFE ABOVE THE WATER

Reed beds and marshes are among the most productive habitats on Earth, so it is not surprising that many animals, both vertebrate and invertebrate, inhabit them. Herbivores find in these wetlands an ample, constantly replaced supply of food. The tropical marshes of sedge and papyrus are productive throughout the year, while the temperate wetlands have a seasonal productivity. This seasonality is even more pronounced in the Arctic, where the productive period lasts only a few months, and herbivore populations are transitory.

Arctic mires are grazed in summer by caribou (*Rangifer tarandus*). This is the same species, but a separate subspecies, as that found in Europe and Asia, where it is called the reindeer. Migratory herds of these animals, which are semidomesticated in Lapland, move into the sedge marshes of the tundra during the summer and graze on its herbs and also take lichen, or "reindeer mosses" (*Cladonia rangiferina*), from drier parts of mires. These animals have broad hooves with concave undersides, which are ideal for walking across wet, soft surfaces. The short vegetation cover on the mires is often tunneled by lemmings, small rodents that reach high population densities in certain years.

The boreal wetlands to the south are grazed by other deer, including the moose (*Alces alces*). This animal spends much of its time during the summer wading in shallow wetlands and grazing on aquatic and emergent plants (see the photograph on page 57), possibly because of their high sodium content. This element is often scarce in the continental interiors and it is needed by this large deer.

Farther south still, the subtropical marshes of Asia are grazed by semiaquatic water buffalo (*Bubalus bubalis*). Although wild buffalo are still present in northern India, Sri Lanka, and Burma, the domesticated water buffalo is the most common large grazing animal of the marshes. This beast has a particularly placid disposition and, although it cannot be herded, it is easily led and has been domesticated in India for at least 2,500 years. Both wild and domesticated water buffalo are often preyed on by tigers. Unlike lions and leopards, tigers prefer ambush to open-country stalking, and they have no dislike of water, so reed beds and marshes are ideal hunting country for them.

The marshlands of the South American Tropics are grazed by a giant rodent, the capybara (*Hydrochoerus hydrochoeris*), which is related to the guinea pig but is much larger, about the size of a sheep (about 4 feet [1.3 m] in length). They are remarkable among rodents in having webbed feet, which are ideal for walking over the floating carpets of the marshes. They are sociable animals, often found in groups of up to 20 along waterways. Like all rodents, their long incisor teeth keep growing indefinitely and are worn down by constant grazing. Their predators are many, including jaguars and foxes out of the water, and caimans within it.

Another South American rodent that occupies reed bed and marsh habitats is the coypu (*Myocastor coypus*). Although smaller than the capybara, it is still quite large for a rodent, often reaching a length of 2 feet (.7 m). It is distinctively aquatic in its habits, swimming readily and feeding on the shoots of wetland plants, particularly the reeds that form floating platforms at the water's edge. The underfur of these animals, below their bristly outer pelts, is particularly soft and keeps the skin dry and warm in the water. This has

encouraged a trade in the pelts of these animals, and fur farms have been established in North America, Europe, and Russia. In places, coypu farming has led to the accidental introductions of the rodent into local ecosystems. The animals have proved to be pests in many wetlands, destroying reed beds and undermining riverbanks. In eastern England, an intensive trapping campaign has succeeded in eliminating the feral coypu population, despite its initial success in the wild.

The tropical marshlands of Africa are inhabited by several types of grazing antelopes, including reedbuck, kob, and waterbuck. The waterbuck (*Kobus ellipsiprymnus*) is distinctive in its greasy appearance and penetrating, musky smell that is said to be detectable from more than a quarter of a mile. They never stray far from water and prefer reeds and other grasses as their main diet. The high productivity of these tropical marshes can support high densities of waterbuck population; up to 250 per square mile (1,000 per km²) have been recorded at Lake Nakuru National Park in Kenya. Some males (about 7 percent of the population) occupy territories into which they attract females and mate with them, while many of the remainder ("satellite") males hang around the edges of the territories in hope of becoming a favored turf-holder one day.

Perhaps the most highly adapted antelope of the southern African wetlands is the lechwe (*Kobus leche*). These animals, despite their relatively small stature, can wade into waters up to 3 feet (1 m) deep. They actually use the wetlands as a means of avoiding predators, leading their young into deeper water when in danger. Lions will sometimes follow them into water, and there is always the possibility that crocodiles will be waiting for them farther out. Unfortunately, they have not adapted as well to human predation. Their tendency to herd and to be easily driven has led to extensive overhunting by humans, who find them relatively easy prey. The erection of dams in wetland areas for water conservation has also resulted in the flooding of many areas once used by lechwe, and their population in many regions has dwindled by half.

Marshes are generally favorable habitats for amphibians, and some frogs, such as the African sedge frogs, tree frogs (see photograph), and bush frogs, are able to climb the shafts of reeds to hunt insects in the herbaceous canopy. Frogs are preyed upon by various snakes, particularly aquatic water snakes (*Nerodia* species), shown in the photograph, and ribbon snakes (*Thamnophis sauritus*) in the temperate marshes of North America. Of the wetland snakes, the most remarkable for their size are the South American anacondas, the largest on record being a green anaconda (*Eunectes marinus*) killed in Colombia, that measured more than 37 feet (11 m) in length. These enormous snakes inhabit the margins of marshes and feed on caimans, birds, and mammals such as the capybara.

Amphibians, such as this green tree frog, are common members of wetland communities, some preying upon insects in the swamp canopy but returning to the water to breed.   *(South Florida Wetlands Management Department)*

Dragonflies and damselflies are important invertebrate predators of wetlands.   *(Aldra)*

Invertebrates of many kinds inhabit the marshland habitat. Among the most prominent are dragonflies and butterflies. Dragonflies (Odonata), shown in the photograph, spend their larval stages below the water, but the nymph (the immature insect) crawls up the stems of emergent plants to pupate, so that the adult insect emerges into the air above the water level. After mating, eggs must be deposited below water, and the female again uses emergent plant stalks to enable her to keep her wings dry while placing the eggs under the water surface. Although dragonflies are at their most diverse in the tropics, they inhabit almost all types of wetland with standing water and emergent plants.

The southern water snake (*Nerodia fasciata*) is a very variable species and is found in both freshwater and saltwater wetlands. The subspecies shown here is the Atlantic salt marsh variety found in coastal Florida.   *(Robert S. Simmons, U.S. Fish and Wildlife Service)*

One species, belonging to the genus *Zygonyx*, lives as a larva in the spray zone of the Victoria Falls in Zimbabwe. Many butterflies are also associated with marshes, some because their larval food plant is a wetland species, and many others because they come to take water and mineral salts from the muddy banks of waterways. Sodium is particularly in demand by butterflies because the males constantly use up their sodium reserves in the production of sperm sacs that they donate to the females. This is the likely reason why the muddy sides of the Amazon and many tropical rivers are often covered with butterflies.

■ AQUATIC BIRDS: DABBLERS AND DIVERS

For the naturalist, one of the most distinctive features of wetlands is their abundant bird life. Birds are often the most conspicuous of animals in the wetlands, although this is not true of all birds, for several species are secretive and rarely seen. Shallow water, reed bed, and marsh offer many habitats and many food resources for birds, including both vegetarian and carnivorous species. For this reason, many different feeding strategies are found among wetland birds.

Many ducks and geese feed upon vegetable matter, either above or below the water. Mallard (*Anas platyrhynchos*), green-winged teal (*A. crecca*), northern pintail (*A. acuta*), cinnamon teal (*A. cyanoptera*), wigeon (*A. americana*), and many other ducks are classified as dabblers. They feed on aquatic plants and associated small animals in shallow water by dipping from a floating position, or, in some species, by upending and stretching their necks to reach the bottom. The depth at which they can feed is largely determined by their neck length, but free-floating aquatic plants may be accessible for them even in deeper water. Young birds, on the other hand, need rich supplies of protein and consequently eat more insects than the adults. For the first 25 days of their lives, mallard ducklings do not put their heads under water but feed entirely on surface materials, especially floating insects.

Like most species of geese, the dabbling ducks will also feed on land, especially on seeds sown on arable land. The feeding action on land is fast and also very selective, taking only the preferred plants or grain. Recognition of the favored food is visual, and the feeding action must be extremely efficient to permit selection when working at high speed. The selectivity of geese and ducks on riverbanks can lead to a highly specialized vegetation in which avoided species predominate—species such as the European daisy (*Bellis perennis*), for example. In Germany, this plant is called *Gänseblümchen,* literally "little goose flower," because of its prevalence in areas grazed intensively by geese. Thus, the grazing of herbivorous geese and ducks can affect the species composition of wetland vegetation.

Not all ducks are dabblers, however. Other ducks, together with coots, dive to obtain food, whether plant or animal. Coots (*Fulica americana*) are omnivorous (but largely vegetarian) and feed both on the surface, by upending, and underwater, by diving. They leap clear of the water prior to diving and tighten their feathers to exclude the air layer that would otherwise give them unwanted buoyancy. Even so, their dives are mainly quite shallow, down to about six feet (2 m), but dives of 20 feet (6 m) have been recorded. Goldeneye ducks (*Bucephala clangula*) are expert divers and can stay underwater for up to 20 seconds, during which time they forage along the bottom, sometimes turning over stones while hunting for animal prey, especially mollusks and crustaceans. They prefer water that is 12 feet (4 m) deep or less. Other diving ducks, including the mergansers, hunt for bigger prey, especially fish, in the shallow waters. The red-breasted merganser (*Mergus serrator*) uses its wings for swimming as well as flying, thus improving its speed and maneuverability when chasing fish. It concen-

The pied-billed grebe is a common diving bird of the wetlands of North America.   *(Lee Karney, U.S. Fish and Wildlife Service)*

trates on small fish about 3–4 inches (8–10 cm) in length. Mergansers often hunt in pairs or in small groups and often stay underwater for over half a minute, and occasionally as long as two minutes.

Grebes (family Podicipedidae), such as the pied-billed grebe shown in the photograph, are also highly specialized predators of fish, and they have legs set well back on their bodies, close to the tail. As a consequence, they are very efficient swimmers, especially underwater, but are extremely ungainly on land. In fact, they spend very little time on land except when they are breeding. Their unusual build has also permitted the development of some extraordinary courtship behavior in certain species, such as the western grebe (*Aechmophorus occidentalis*) and the eared grebe (*Podiceps nigricollis*), during which the courting birds elevate the front ends of their bodies, curve their necks in unison, and appear to skate in synchrony over the surface of the water. Many of the grebe species have elaborate courtship techniques but none as distinctive as the western grebe.

Grebes and coots, unlike ducks and geese, have only partially webbed feet, the webbing forming lobes along the sides of the toes. Closely related to the coots are the rails and gallinules, which include the moorhen (*Gallinula chloropus*), but these birds have no webbing at all on their large feet. Although the moorhen swims readily, the rails and gallinules avoid open water wherever possible, preferring to skulk among emergent plants along the edges of marshes. Some, like the American purple gallinule (*Porphyrula martinica*) and the moorhen, are good climbers and may climb up the branches of willow trees or even, in the case of the gallinule, up the stalks of reeds. The Old World purple gallinule (*P. porphyrio*) is bigger than the American species and has a much stouter bill. Like its American relation, it is omnivorous, but it is quite capable of climbing wetland shrubs and trees to prey upon nestlings. The large feet of the rails and gallinules enable them to spread their weight over the surface of floating aquatics and reed bed rafts, allowing them to walk where others have to swim.

Even more skilled at walking on unstable surfaces are the jacanas. These tropical and subtropical wetland birds have enormous feet in relation to their size but are perfectly adapted to the fringes of shallow water habitats. Their tread is so light that they can easily walk over the surfaces of water lilies. Like many of the birds of these habitats, they nest at water level, building a small platform to support their eggs. In the case of the jacanas, the platform is considerably less substantial than the bulky structures of moorhens and coots, but some species of jacana have the ability to move their eggs if they are threatened by rising water or any other danger. The pheasant-tailed jacana (*Hydrophasianus chirurgus*) of Sri Lanka has been observed tucking its eggs one at a time beneath its chin and carrying them to a new location some distance away.

All of the birds mentioned here, apart from the grebes, have one other aspect of their breeding behavior in common—they all lay large clutches of eggs. Moorhens and coots usually lay 5–10 eggs per brood and often produce a second brood in one breeding season. Sometimes even more eggs than this are found in the nests, but this is usually because two females are sharing a single nest (as well as a common male). Among the coots and moorhen species, the juveniles from the first brood are often observed to help feed their siblings in the second brood. This may well contribute to better survival in the large family groups. Mallards and most other ducks also have large clutches, generally 9–13 eggs, and large groups of ducklings are often present with the mothers after hatching. The prolific egg-laying of these birds suggests a very high mortality rate, which is indeed the case. Many predators, including fish, birds, mammals, and reptiles, prey upon the vulnerable young birds, so the high productivity is a means of compensating for the anticipated high level of losses.

◼ AQUATIC BIRDS: FISHERS IN THE REEDS

Dipping and diving from the water surface are not the only techniques appropriate for birds that feed on fish around the margins of aquatic regions. Diving from the air is a more spectacular approach and can prove most successful. It has been adopted by a range of different bird groups, including kingfishers, pelicans, terns, and some raptors (birds of prey).

The kingfishers are mainly stocky birds with very short legs and sturdy bills set in heavy heads. Perhaps the most

top-heavy species is the stork-billed kingfisher (*Pelargopsis capensis*) of Sri Lanka, whose bill takes up about a third of its full length. These birds observe their fish prey from above the water, either by hovering, as in the case of the belted kingfisher (*Ceryle alcyon*) of North America, or by perching on a branch as does, for example, the Eurasian kingfisher (*Alcedo atthis*). Even small kingfishers, such as the brilliantly colored blue Eurasian kingfisher, can dive from great heights (more than 30 feet [10 m]) and, using the force achieved by such a drop, they can penetrate to depths of at least 3 feet (1 m) into the water. Generally, small fish are taken and are swallowed headfirst to avoid fins and scales becoming stuck in their gullets. A kingfisher with a fish held the other way round, with the head emerging from the beak, is undoubtedly on its way to feed a mate or young at the nest.

Brown pelicans (*Pelecanus occidentalis*) are found in same coastal wetlands, and they are expert divers from great heights. White pelicans (*P. erythrorhynchos*), on the other hand, do not dive but fish from the surface. Some pelicans, such as the spotted-billed pelican (*P. phillippensis*) of India, fish in groups; an arc of the birds thrashes the water with their wings, driving panic-stricken fish into tight groups. The birds then collect them in their capacious beaks.

Many species of terns breed in inland waters, especially in North America and in Asia. Some terns, such as the common tern (*Sterna hirundo*), dive on their prey from well above the water surface, but others, such as the black tern (*Chlidonias niger*), fly close to the water and dip into it to catch small fish, insects, crustaceans, and amphibians.

Among the raptors, few are better adapted to a fish-eating life than the osprey (*Pandion haliaetus*). This large predator, with a wingspan of about 6 feet (almost 2 m), soars and hovers above the water, sometimes at heights of 250 feet (70 m), and then plunges feet first to catch large fish close to the surface. The grip of the osprey's claws is so efficient that it can emerge from the water holding even a very large catch with just one foot. There are instances on record, however, of the osprey's talons becoming caught in such a powerful fish that the bird is dragged underwater and drowned. Being at the top of the food chain, these impressive birds have declined very severely in many parts of the world as a result of accumulating pesticides, particularly DDT, in their bodies, with fatal results. Careful conservation and protection is now leading to recovery in both North America and Europe.

Other large predatory birds of marshes include the African fish eagle (*Haliaetus vocifer*), a noisy bird that is relatively frequent in African wetlands, such as those around Lake Victoria and the Okavango swamps. Like the osprey, it will plummet from a great height to catch fish in the upper layers of water, but it will also take other prey, such as snakes and young birds. Its American equivalent is the bald eagle (*H. leucocephalus*), another species gradually recover-

Bitterns

Bitterns belong to the same family as herons and egrets, the Ardeidae. They share with these birds a spearlike bill, long neck, and long legs that enable them to wade through shallow water as they hunt for fish, as shown in the photograph. They are more closely associated with reedbeds and marshes than are the other herons, however, and spend much of their time skulking among tall reeds or fishing along the edges of enclosed pools deep within the marsh. They are less inclined to fly than herons and egrets but will take to the wing when disturbed or when they need to move from one area to another. Their plumage enables them to blend with their background of reeds, being brown with streaks and spots on their breasts that make them difficult to detect when they are standing still among dead reeds. They also have a distinctive behavior that assists in their camouflage. When alarmed they raise their beaks toward the sky so that their exposed throats and breasts look even more like the rushes and reeds that surround them. But they are still able to observe anyone approaching because their eyes are situated in widely spaced positions on either side of the bill so that they are able to see around their erect beak and keep watch on the surrounding area. In fact, they can see better when in this position than when looking over the top of the beak because their field of vision when looking over the bill does not overlap as effectively so that they have a central blind spot and do not have full binocular vision.

Both the American bittern (*Botaurus lentiginosus*) and the Eurasian bittern (*B. stellaris*) have deep thumping notes that they emit during the mating season and that can carry for more than a mile. Some have likened the call to the noise made by an old wooden pump, a kind of "pump erlunk, pump erlunk." It is also somewhat

American bittern wading through shallow water   *(Gary Kramer, U.S. Fish and Wildlife Service)*

reminiscent of a ship's foghorn or someone blowing short bursts of air over a large jug. It is a sound that once heard can never be forgotten and is very distinctively the sound of a wild and lonely wetland.

---

ing from population decline. This is also mainly a fish-eating species but is sadly much less common. The northern harrier (*Circus cyaneus*), previously known erroneously as "marsh hawk," and called "hen harrier" in Europe, takes small mammals and frogs, while some specialist predators of the reed beds, such as the snail kite of Florida, feed on gastropod mollusks.

An alternative "technique" for birds hunting in the wetlands is to possess long legs that enable them to wade in shallow water and a long sharp bill with which to snatch underwater prey. The herons, ibises, egrets, bitterns, and storks have evolved this type of strategy. A great range of size is found within these groups and reflects the range of

prey species that can be eaten. Small herons of 18 inches (50 cm) or less, such as the Indian pond heron (*Ardeola grayii*) and the green heron (*Butorides virescens*) of Asia, Africa, and America, spend their time around shallow pools, taking small fish and crustaceans, while at the other end of the scale are the goliath heron (*Ardea goliath*) of Africa, which is five feet [150 cm] in height, and the great blue heron (*A. herodias*), which is four feet [120 cm] tall and can swallow whole very large fish or even mammals the size of a rabbit.

The bitterns comprise a group of herons that is strongly associated with reed beds and marshes rather than open waters (see sidebar). Some bitterns are very small, such as the least bittern (*Ixobrychus exilis*) in America and the little bittern (*I. minutus*) in Europe. These tiny herons (only about a foot [32 cm] high) are even more secretive than their larger relations, but they are able to climb reed and cattail stems to observe their surroundings and to escape aquatic predators.

The bittern is widely distributed and relatively abundant in North America, summering in the northern states and Canada and heading south for the winter. The Eurasian species has declined very severely, however, mainly because of habitat loss as a result of the drainage of wetlands. It became extinct in England during the 19th century but returned to breed in 1911 and has subsequently managed to maintain a small but threatened breeding presence in the country. The experience of Europe should act as a warning to the conservationists of North America that this is a vulnerable bird because it is so dependent on a habitat that is itself endangered.

Storks are similar to herons in build, and many are strictly wetland species, but they have generally adopted a wider range of lifestyles, and their diet is less specifically aquatic in nature. They are mainly tropical and subtropical in distribution and are a distinctive feature of many Indian wetlands, such as Bharatpur, where painted storks (*Mycteria leucocephala*) form elegant flocks in shallow pools. Spoonbills (*Platalea* species), on the other hand, have a much more specialized feeding technique. Their bills are wider at the tips, and the birds feed in shallow water by sweeping their heads from side to side, mouths open, and when nerve endings in the linings of their bills make contact with crustaceans, slugs, mollusks, or small fish, their bills snap shut—capturing their prey.

Finally, there are many types of small bird that spend most of their lives within the dense forest of reeds and cattails, only occasionally flying above the cover that the vegetation provides. One of the most distinctive of these is the bearded tit (*Panurus biarmicus*), or reedling, which lives mainly in central Asia but has populations in isolated wetland areas of Europe. Its long tail provides it with the capacity to move with acrobatic precision among the vertical stems of reeds, and, although infrequently seen above the vegetation, it maintains social contact with other members of the flock by a piercing, bell-like call. The cupped nest is built low down among the reeds and is often furnished with a protective roof. In North America, the common yellowthroat (*Geothylipis trichas*) breeds in similar marshy locations, but, unlike the bearded tit, it is migratory over much of its range.

ANIMALS OF THE FORESTED WETLANDS

Forested wetlands differ from marsh and reed bed habitats both in the stature and in the permanence of their architectural structure. Trees, rising above the waterlogged soils, provide a whole new range of habitats in which both plants and animals can live. Their permanence from year to year lends stability to the structure so that this three-dimensional environment can always be relied on. Perennial plant epiphytes (plants growing on the trunks and branches of other plants), from bromeliads to hanging strings of lichens, are assured of long-term anchorage. Nesting birds can return every year to the same sites, and many animals, from reptiles to primates, can spend their entire life in the canopy without any need to venture into the waters or onto the land below.

Even the animals of the forest floor benefit from the new resources available to them in the form of timber. Perhaps the best known of these is the beaver (*Castor canadensis*). This wetland rodent has webbed hind feet, a rudderlike tail, and an ability to remain underwater for up to 15 minutes, although five minutes is more typical. It is essentially an aquatic animal, but its way of life requires the presence of trees. Beavers are vegetarian in diet, eating a wide range of plants through the summer and concentrating on the bark of willows and poplars through the winter. Their strong incisor teeth are capable not only of bark stripping but also of felling small trees (up to about 20 inches [50 cm] in diameter). The beaver's upper lip closes behind the incisors and protect the mucous membranes and gums from splinters. By felling trees and building dams (up to 3,000 feet [914 m] in length and 10 feet [3 m] high), beavers can control their local environment quite closely, particularly with respect to water level, as shown in the photograph. The beaver pond remains at constant level, unless drought or damage to the dam causes a temporary lowering. This means that the beaver lodges built within the ponds (see the diagram on page 146) can have underwater entrances but do not become flooded inside.

The beaver lodge may be up to 20 feet (7 m) in diameter and may have several rooms. Up to 10 beavers can live within a family lodge in which bedding of shredded wood and bark makes them comfortable. Within the lodge, beavers are safe from the predation of lynx, coyote, and bear. Extra branches are kept underwater through the winter months, where the bark stays fresh beneath the ice. The construction of dams and lodges makes the beaver an extremely influential environmental engineer that actually creates its own wetlands. Its action can change the course of succession. By building dams and raising water tables, the beaver recreates an early, aquatic stage of succession from which vegetation development begins anew.

The raising of the water table by beavers can lead to the flooding and death of some trees, but this is not necessarily harmful to wildlife. Standing deadwood soon becomes infested with fungi, followed by wood-boring insects and then, inevitably, woodpeckers. Temperate forested wetlands, in particular, are rich in woodpecker species, most of which,

Beaver pond in the boreal forest zone *(U.S. Fish and Wildlife Service)*

however, are not confined to wetland forest in their distributions. One species that is distinctively a wetland woodpecker is the ivory-billed woodpecker (*Campepephilus principalis*)

of the southeastern United States. This very large (reaching up to 20 inches [50 cm]) woodpecker, however, has been severely reduced in numbers and has quite possibly passed over the brink into extinction. Sightings have been reported in recent years, but confirmation of its survival is lacking and doubts remain strong.

Mammals also inhabit the canopy layers of wooded wetlands, squirrels in the temperate areas, and monkeys in the tropics and subtropics. The red uakari (*Cacajao* species),

Cross section of a beaver dam and lodge. The lodge is a domed structure built in a shallow site in the center of a lake created by the beavers' dam. The lodge is constructed in such a way that it can be entered only by an underwater route.

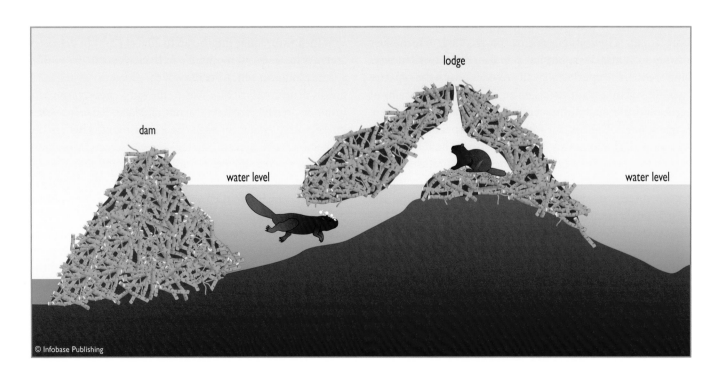

© Infobase Publishing

for example, is a monkey of the Amazon forests and has a remarkable, hairless red face, which gives it a somewhat human appearance. Groups forage in the tree canopy as high as 130 feet (40 m) above the floodwaters of the Amazon basin forest, seeking the seeds and flowers on which they feed. These, and many other monkey species, spend their entire lives in the canopy, so it makes little difference to them whether they are living in a wetland or a dryland forest.

The same is true of many bird species, but one bird of the Amazon flooded forest is very typical of wetland sites, namely the hoatzin (*Opisthocomus hoazin*). Having the general appearance of a scruffy chicken with a bristly crest, this bird is about 2 feet (0.6 m) long, has large wings and tail, but is not fond of flying. It is herbivorous, feeding on flowers, fruits, and foliage, and often stuffs its crop until it becomes unbalanced by the excess weight at the front of its body. Like all leaf-eating animals, it needs to eat a great deal of herbage to supply its nutritional demands. It climbs around in the canopy using its feet, beak, and even wings, which in the juvenile have two claws present on the main joint. It often descends to the water, where it can swim quite well and then drag itself out on to low-hanging branches using its various climbing appendages. Although it has a very primitive appearance, somewhat resembling the fossil *Archaeopterix,* the hoatzin is actually related to game birds and is not particularly primitive in an evolutionary sense. It really is a wetland chicken.

Forest wetland trees also provide a safe, stable location where many of the wading, fish-eating birds of the marshes can nest and roost. Ibises, storks, herons, egrets, and night herons can often be found in large numbers in the trees of these wetlands. Sometimes, favored trees are used so frequently that the droppings of the birds kill the foliage and, ultimately, even the tree itself. The wetland ecosystem around these roosts is highly eutrophic as a result of both the guano (droppings) and the regurgitated bones, fish scales, mollusk shells, and so on produced by the roosting and nesting birds. Despite the noise, chemical pollution, and the smell, these sites can be visually impressive, especially in the case of brightly colored wetland birds, such as the painted storks of India or the scarlet ibises (*Eudocimus ruber*) of Central and South America. Roosting and nesting in these large colonies helps to deter predators and certainly provides an efficient early warning system in times of danger.

Wetland forests harbor many predators. Mink hunt on the ground level of temperate forests in both the Old World and the New World. The American species of mink (*Mustela vison*) is larger than the Asian one (*M. lutreola*) and has been introduced accidentally in Europe where it is now a major pest because of its intense predation on wetland birds and mammals, such as the water vole. Snakes also occur in the forested wetlands. For example, the Indian python (*Python molurus*), which can grow up to a length of 20 feet (6 m), can climb trees and so presents a threat to the nests and young of the wetland birds. The green python (*Chondropython viridis*) of New Guinea regularly hunts its prey in the treetops. The pythons and the boas of South America kill their food by constriction and then swallow it whole.

PLANT LIFE IN BOGS: BOG MOSSES AND LICHENS

True bogs are defined by their total dependence on rainfall for their water supply (see "Temperate Raised Bogs," pages 103–105). They are also totally dependent on rainfall for their supply of mineral elements, apart from some nitrogen that may be fixed by a few plants with associated symbiotic bacteria (such as the bog myrtle [*Myrica gale*] and some blue-green bacteria either free-living or associated with fungi in lichens; see the sidebar on page 148). It is inevitable, therefore, that bog surfaces will generally be poor in nutrient elements (an exception being the bogs of extremely oceanic climates where seawater may arrive suspended in the air). The plant life of bogs, therefore, is undemanding with respect to most nutrients required for growth (nitrogen, phosphorus, potassium, silicon, calcium, magnesium, and so on). Indeed, if these elements are added artificially to a bog, the vegetation changes as more demanding and faster growing plants gain dominance and the true bog species are shaded out. In terms of competition, therefore, the bog plant should not only be able to survive in nutrient poor conditions but should also, if possible, maintain the nutrient poverty of the ecosystem by taking nutrients out of circulation. This is precisely where the bog mosses excel. The environmental changes these mosses produce may not be as obvious as that of the beaver (see "Animals of the Forested Wetlands," pages 145–147), but overall it is even more effective.

The bog mosses belong to just one genus, *Sphagnum*. The group is very different from all other mosses and liverworts and is usually placed in its own class, Sphagnopsida, within the division (equivalent to the animal phylum) Bryophyta. It is a class with only one family, the Sphagnaceae, which in turn has only one genus. Exactly how many species of *Sphagnum* there are in the world is still unclear, but there may be about 200. The confusion arises partly because many geographical areas have not yet been fully surveyed for these mosses and partly because it is very difficult to distinguish between some of the species, so botanists disagree about how many species there really are.

Geographically, *Sphagnum* is a very widespread group, being found in all but the very arid, the ice-covered, and the most continental parts of the world. They are found from the Northwest Territories of Canada to Tierra del Fuego in

Lichens

A lichen consists of an association of a fungus with an alga or with a cyanobacterium. The closeness of that association is quite variable and can simply consist of the two organisms living alongside one another. Such a loose association would not normally be called a lichen. One definition of a lichen is a stable, persistent, and self-supporting association of a mycobiont (the fungal component) with a photobiont (the photosynthetic partner) in which the mycobiont forms a protective cover for the photobiont. Delicate photosynthetic cells of a micro-alga, or cyanobacterium (see sidebar "Cyanobacteria," page 134), are enclosed within a sheath of fungal mycelium. These cells are protected by the surrounding mycelium, and although they desiccate quickly under dry conditions, they are able recover remarkably well. Light penetrates the mycelium, allowing the photobiont to photosynthesize, and the fungus derives its nutrition from this source. Both partners thus benefit from the association, and these benefits allow the combined "organism" to occupy positions that neither component could achieve on their own. They are able to grow on rock surfaces, the surface of stony soils, or, as epiphytes, on the branches of trees and shrubs, coping with desiccation even in dry and exposed conditions. In wetlands they may grow as epiphytes on trees or shrubs or on the peat surface of bogs, usually on relatively dry hummocks. They are particularly abundant and important in Arctic mires, as shown in the photograph.

The structure of the combined thallus is quite complicated and often takes on a distinctive and recognizable form, which is why lichens can be classified and identified visually. The "lichenized" thallus sometimes produces distinctive chemicals, secondary products of metabolism that protect them from grazing animals by making them unpalatable. Reproduction in the lichens can take the form of small particles, containing both fungal and algal components that become detached from the main thallus and are carried away by wind or

Lichens of the genus *Cladonia* are important members of the surface communities of many wetlands, particularly bogs. They fix nitrogen from the atmosphere and enhance wetland nitrogen cycling.   *(South Florida Wetlands Management Department)*

water. Alternatively, the fungus component may produce fruiting bodies and release airborne fungal spores that become dispersed and land on soil or bark surfaces, where they may encounter the appropriate alga and form a new lichen thallus.

The algal or cyanobacterial component of lichens are well known as independent species that can grow without the aid of the fungus, though usually in a more restricted range of habitats. It was once believed that the fungal component survived only in combination with its symbiont, but recent studies using molecular techniques have shown that the fungi of lichens do grow independently of the photobiont but take on such a very different form that their involvement in the lichen thallus has never been suspected. This poses a real problem for taxonomists involved in classifying organisms and raises the question once more as to whether lichens should be regarded as true species.

South America and from Siberia to South Africa and New Zealand. So, although so specialized in habitat, the genus is an exceptionally successful one.

The basic structure of all *Sphagnum* moss species is very uniform, which is the main reason they can be very difficult to identify. The long main stems are capable of virtually indefinite rapid upward growth, and this property, com-

bined with their capacity to hold water and reduce microbial decay, allows them to build up considerable depths of peat. Occasionally the stems divide in two, so that the stem density within the growing cushions of moss increases and seals any gaps in the surface. At their tips, the stems bear a dense mass of short branches, as shown in the diagram (opposite), arranged as a tight head. As the stem elongates, the branches

Ⓐ Single strand of *Sphagnum*

Ⓑ Profile of *Sphagnum* carpet

head or capitulum

spreading branches

pendulous, hanging branches

dense layer of heads

subcanopy zone

water table

© Infobase Publishing

Architecture of *Sphagnum* moss. When the moss grows in a carpet, the heads form an intact platform, below which lies a shaded, moist microhabitat for a wide variety of invertebrate animals.

can be seen to be attached in dense clusters both spreading from the stem and extending back down it. Both branches and stems are covered by approximately triangular leaves without stalks. These are densely packed and the spaces between them become filled with water, giving the whole plant an immense water-holding capacity. *Sphagnum* moss can hold up to about 20 times its own dry weight in water.

The growth form of the *Sphagnum* moss cushion, with tall stems, loose lower branches, and dense upper heads bears a close resemblance in cross section to the profile of a miniature forest. The complexity of the layers within the profile creates a very varied series of microclimates, where the upper surface is occasionally hot in the light of the sun and sometimes even quite dry, while the lower layers are increasingly cool, wet, and poorly lit as one descends into the waterlogged peats. This provides many different opportunities for smaller plants, such as the straggling, shade-tolerant but drought-intolerant liverworts, which act as minute "lianes" or climbers in the lower canopy. The complex architecture also offers a range of microhabitats for invertebrate animals, some of which move up and down the profile

according to the time of day (see "Bog Invertebrates," pages 156–157).

The microscopic structure of *Sphagnum* leaves and stems, shown in the diagram above, is also extremely unusual and very well adapted to its habitat and ecological niche. In the leaves, there are two types of cell: living green photosynthetic cells, which are small and narrow, and dead colorless cells, which are large, barrel-shaped, and inflated. The green cells form a network surrounding the colorless cells and communicate with one another at their junctions. The dead cells' swollen structure is maintained by rounded strips of thickening in their cell walls that prevent their collapse. The walls of these colorless cells also have large pores in them so that they are usually filled with water, which further adds to the water-holding capacity of the moss. The outer layers of the stems have unusual cells, called *retort cells* because of their similarity to the piece of glassware of that name used by chemists. They are fat, inflated cells that narrow to a neck at their upper end, which then opens in a pore.

In addition to absorbing water from its surroundings, the *Sphagnum* plant takes in the dissolved elements present in the weak solution of acidic bog water. The cell walls of *Sphagnum* are rich in chemical compounds called polyuronic acids which have the capacity to form bonds with the positively charged ions in the water (see "Element Cycling in Wetlands," pages 56–60), exchanging them for hydrogen ions. The result is that the mineral nutrients become locked up in the bodies of the *Sphagnum* moss, eventually

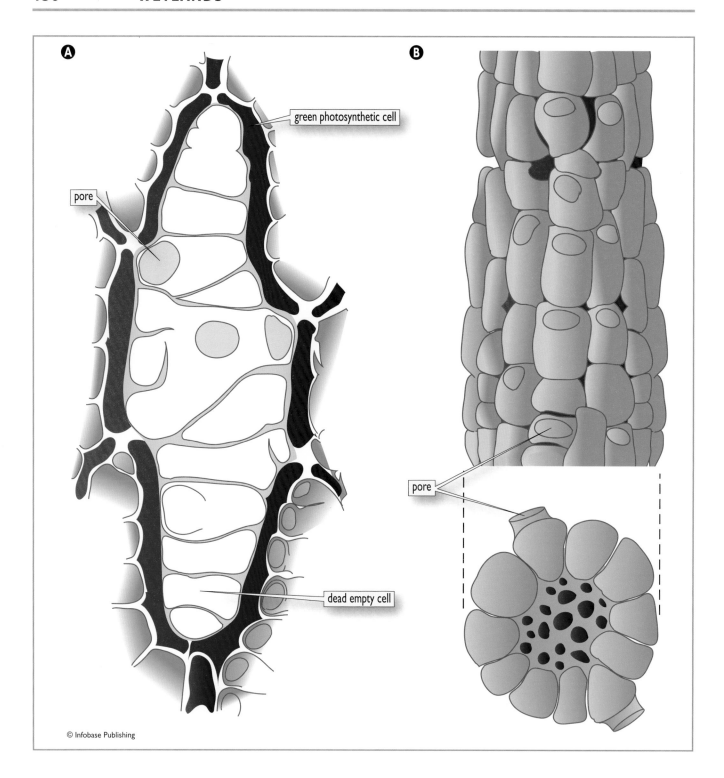

green photosynthetic cell

pore

dead empty cell

pore

© Infobase Publishing

become incorporated into the peat, and the surrounding water becomes richer in hydrogen ions; in other words, it becomes more acid. So the growth of this moss makes its environment wetter (by water retention), poorer in nutrient elements (by chemical exchange), and more acidic (by the release of hydrogen ions). By this combination of processes, *Sphagnum* makes it more difficult for many of its potential competitors, including many flowering plants and even trees, to enter the ecosystem and compete for space.

Sphagnum mosses may even succeed in replacing trees, such as alder, by preventing further seedling establishment.

Sphagnum also has the capacity to reduce the decomposition activity of bacteria and fungi. This antimicrobial property, combined with its absorptive capacity, has led to the moss being used as a wound dressing (as in World War I) and as a primitive diaper for babies. *Sphagnum* has also proved very effective as an absorbent for oil spills, readily taking up organic liquids.

(opposite page) Cell structure of *Sphagnum* moss. (A) Leaf cells consist of two types: Narrow chlorophyll-containing cells surround dead, colorless cells with pores that permit them to become filled with water. Bands of wall thickening act as supports to these large dead cells. (B) Stems of *Sphagnum* have outer layers of large cells, some of which have pores and act as water-retaining vessels.

Different species of *Sphagnum* have various requirements for the conditions in which they live. Many prefer open, well-lit habitats, but some grow in shade, under the canopies of small shrubs or trees. Some grow permanently submerged in bog pools, while others form dense cushions that may become elevated up to 2 feet (0.6 m) or so above the bog surface. Some species will grow only under extremely acid conditions, while others can tolerate higher levels of nutrient (particularly calcium) input. Since several species often grow together in a bog, their different growth mechanisms and environmental preferences can add much to the diversity of the bog's surface topography and habitats.

■ PLANTS IN BOGS: CARNIVOROUS PLANTS

The nutrient poverty of the acid waters of bogs presents difficulties to many plants. Nitrogen (in the form of nitrates or ammonium ions) is needed to build proteins, and phosphorus (as phosphates) is needed for cell membrane construction, nucleic acids, and other vital biochemicals. Animals, including invertebrates, are generally richer in both of these elements than are plants, so it is not surprising that some plants have evolved organs that enable them to trap, digest, and absorb these elements from animal life on the bog surface.

Unlike animal predators, carnivorous plants are unable to stalk and hunt their prey. They are forced to adopt a passive approach, and their predation depends on trapping the mobile but unwary animal. Leaves are the most appropriate organs to use as traps, but many different adaptations have arisen to convert them into lethal weapons. In all cases where leaves are used for insect trapping, however, the chlorophyll pigment is still present, and the plant continues its photosynthesis to gain carbon and construct carbohydrates. The plant's diet of animal tissues is an addition to, rather than a replacement of, standard plant nutrition.

The simplest mechanism for insect trapping is the development of a sticky surface on the leaf. The sundews (*Drosera* species), such as the round-leaved sundew (*D. rotundifolia*) shown in the photograph below, and the butterworts (*Pinguicula* species), such as the common butterwort (*P. vulgaris*) shown in the photograph on page 152, are the most common examples of this technique on bogs. In both plants, all leaves have adopted the dual function of photosynthesis and insect trapping, although the sundews often have an additional red (anthocyanin) pigment in the leaf which probably serves to attract the attention of insects in the generally green world of the bog. The sundews have large, long-stalked tentacles with glistening, sticky, glandu-

The round-leaved sundew (*Drosera rotundifolia*) is an insectivorous plant that catches its prey using sticky tentacles. A struggling fly rapidly becomes entangled with the adhesive secretions of the plant's glandular hairs. *(Gary M. Stolz, U.S. Fish and Wildlife Service)*

Butterwort (*Pinguicula vulgaris*) traps insects by having small sticky glands covering its leaves. The insects are digested and absorbed by the leaf surface. Ants sometimes steal dead insects from these plants, but it is a dangerous activity because they often become stuck themselves.   *(Peter D. Moore)*

Sphagnum moss, for in these locations they are more likely to encounter their sun-basking insect prey. In some Spanish sites, however, the summer sun is so hot that insects tend to avoid open, sunny conditions, and, if the plant is to be successful in trapping its prey, it must occupy the shade. Sometimes this means that the plant has to trade some of its photosynthetic activity for the advantages of a trapper's life in the shade.

An alternative trapping method that does not leave the prey open to theft is the pitfall trap. In the pitcher plants of bogs (e.g., *Sarracenia* species) and of the tropical swamp forests (e.g., *Nepenthes* species), either the whole leaf or part of it has rolled inward to form a tube in which the digestive juices are contained, as shown in the photograph below and the diagram opposite. The lips and upper part of the tube are attractive to insects (by bright coloration and the secretion of sugary nectars) and are also slippery (as a result of wax secretion in which the wax takes the form of overlapping, detachable flakes), so that visitors easily slip over the edge of the lip and find themselves in deep water. Drowning, decomposition, and digestion follow, though not necessar-

lar tips, which adhere very strongly to any surface they contact. The struggles of a trapped insect soon cause it to come into contact with other glands, but there is also a remarkable transmission of signals from one tentacle to another that causes neighboring stalks to bend their sticky tips in the direction of the trapped prey. Gradually, the whole leaf begins to curl so that the insect is brought into contact with as many glands as possible, guided by a primitive electrical signal, similar to that of nerves in animals.

In the butterworts, the glands are much smaller and less evident than in the sundews and are of two types, stalked and stalkless. Again, the function of the stalked glands is to trap the prey, but the stalkless glands in the butterwort are simply involved in enzyme secretion for digestion, followed by the absorption of the digested materials. In both butterworts and sundews, the prey animal is digested while still alive, but once all the soft tissues have been assimilated into the plant, the leaf unrolls to reveal the shriveled exoskeleton, the sole remains of the unfortunate insect.

A problem with this sticky trap technique is the possibility of the theft of the prey by insectivorous (or detritivorous) animals. Ants are often attracted to the array of insects stuck on the leaves of sundews and butterworts and will steal the bodies around the edges of leaves where they can remove them without becoming entangled themselves. This so-called *kleptoparasitism* is a very dangerous way of gaining a meal, and many ants end up becoming prey items themselves.

The sticky trap plants of the high latitude bogs usually occupy open sunny positions, often on hummocks of

The leaf tip of a carnivorous pitcher plant, showing the waxy rim of the trap, where insects slip and fall into the pool of digestive juices below.   *(Amanda Rhode)*

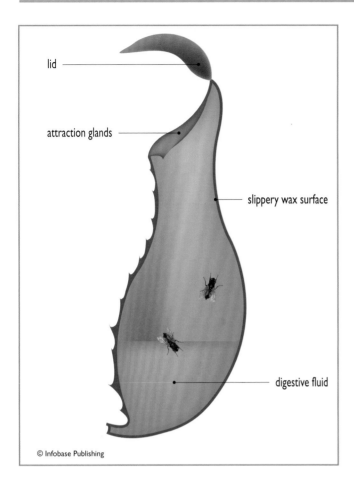

lid

attraction glands

slippery wax surface

digestive fluid

© Infobase Publishing

Cross section of the insect-catching pitcher of a pitcher plant. The lip of the pitcher is attractively colored and scented, but the surface of wax plates is extremely slippery and causes any visiting insect to fall into the pot of digestive fluid. Here it drowns and is digested and absorbed by the leaf surface.

ily in that order. Water originally accumulates in the pitchers from the rain, but enzyme-secreting glands lie inside the pitcher, and these are stimulated into action by the touch of the struggling insect. There are some insects, including species of mosquito, that can live within the enzyme lakes of the pitcher plant. The mosquito *Wyeomyia smithii*, for example, breeds in the pitchers of the northern pitcher plant, *Sarracenia purpurea,* which is found on bogs in eastern North America from Labrador right down to Florida.

An even more elegant method of trapping prey and avoiding kleptoparasitism has been evolved in the Venus flytrap (*Dionaea muscipula*), a rare plant found wild only in North and South Carolina. Its leaves have two halves joined by a flexible hinge, and the leaf surfaces have sensitive hairs. These hairs, when stimulated by movement, send an electrical signal to the hinge that causes water to move out of hinge cells, and the two halves of the leaf snap shut very quickly—quickly enough to catch any visiting fly (see sidebar). The leaf is fringed by long, stiff hairs that eventually form the

Plant Movements

Plants may appear static, but they actually move their various organs throughout their lives. Flowers open and close during the day and night; some flowers track the Sun through the course of the day, and many climbing plants wave their tendrils around to seek an appropriate support. These movements are relatively slow and cannot be detected by eye, but they are clearly visible when time-lapse photography is used to study them. Other movements of plants are sufficiently rapid for the human eye to observe them, such as the collapse of a leaf of the sensitive plant *Mimosa pudica* when its leaves are touched. The closure of the leaf of the Venus flytrap (*Dionaea muscipula*) is even more rapid; its hinged leaf needs to shut sufficiently rapidly to trap a fly upon its surface. Its movement, therefore, needs to be faster than that of an alert fly, ready to depart on the slightest pretext. The leaf of the flytrap has long bristles, so these help because the fly is imprisoned as soon as the two sets of bristles intersect and the leaf does not need to be completely closed in order to achieve this. The closure time for a leaf is about 0.5 seconds, which is fast for a plant.

The signal for action comes from sensitive hairs on the leaf surface, but the leaf is not easily fooled. Mechanical stimulation of a single hair does not release the closing mechanism, but when two hairs are touched the leaf closes. The transfer of the signal requires a rapid system of transmission, and several possible ideas to account for this have been put forward. Chemical signaling is widespread in plants, but this could not be achieved in the time needed because chemicals are slow to diffuse from one cell to another. Hydraulic techniques, using the water content of cells, is clearly involved because the cells below the epidermis lose water as the leaf closes, but this is more likely to be an effect of the reaction rather than a signal for a response. The speed of transmission resembles that of nerve action in animals; therefore, an electrical stimulus must be involved, and the movement of charged ions on cell membranes is undoubtedly implicated, but the precise mechanism is still unclear. One important observation, however, is that the level of ATP in the surface leaf cells declines rapidly, showing that a great deal of energy is invested in the rapid closure of the leaf.

bars of the trapped fly's prison. Chemicals from the struggling fly then cause the secretion of digestive enzymes from the surface glands on the leaf; mechanical stimulation by an inanimate object does not cause enzyme release.

Another active trap of this kind is found in the bladderworts (*Utricularia* species) of bog pools. These are submerged aquatic plants with their stems and leaves under water and only their flowers raised above the surface. Some of the submerged leaves are modified into cylindrical bladders (about 0.1 inch [2–3 mm] long) with trapdoors at their tips and with sensitive trigger hairs at the entrance. Water and air are extruded from the bladder by the compression of its sides, so that the trap is sprung. When the prey item (usually a small crustacean, such as a water flea) swims by and touches the hairs, an electrical message is sent to the trapdoor, which springs open, resulting in a rush of water that carries the prey animal into the bladder because of the reduced pressure inside. Digestion takes place within the bladder.

Carnivorous plants have long fascinated ecologists, including the great English naturalist Charles Darwin (1809–82) himself, both because of their reversal of normal feeding relationships and also because of the nervelike processes involved in the plant responses. One further ironic feature of these remarkable plants, it must be noted, is that they all depend on insects for pollination.

PLANTS IN BOGS: DWARF SHRUBS AND DROUGHT

One aspect of the biology of mires that has long puzzled ecologists is why so many of the plants, particularly the small woody shrub species, appear to be so well adapted to drought. Although some peatlands become dry on their surfaces during particularly hot and dry summers, drought is not usually long-lasting, nor is it usually severe. Yet many of the dwarf shrubs of peatlands, especially those in the temperate zone, have small, leathery, rolled, evergreen leaves, which are more often associated with the vegetation of dry conditions, such as chapparal and Mediterranean scrub. These structural characters associated with drought in plants are termed *xeromorphic*.

In the arctic mires and many northern temperate peatlands, the black crowberry (*Empetrum nigrum*) is a common member of the surface vegetation. This low-lying shrub is evergreen and has shiny, almost cylindrical leaves. On close inspection it can be seen that the leaves are actually tightly rolled in upon themselves, leaving only a very narrow, white-colored strip on the lower side where a thin gap leads to the delicate undersurface of the leaf. The upper part of the leaf is glossy with the presence of wax, which protects its surface cells from desiccation. Heather (*Calluna*

vulgaris) also has short, blunt, needlelike, evergreen leaves that are rolled and hairy. In fact, many species in the family Ericaceae, including the cross-leaved heath (*Erica tetralix*) and the cranberry (*Vaccinium macrocarpon*), are similarly equipped with drought-resistant leaves.

Many of the blueberries have small, leathery leaves, including the velvetleaf blueberry (*Vaccinium myrtilloides*), late low blueberry (*V. angustifolium*), and Eurasian bilberry (*V. myrtillus*). Many of the blueberries lose their leaves in winter, leaving green, stiff stems, but the leaves are tough and waxy and many blueberry species are able to survive on dry soils. The Labrador tea (*Ledum groenlandicum*) has large, leathery, evergreen leaves, which are not tightly rolled (although they do curl inward at their edges) but have dense masses of hair on the leaf undersides that prevent air movements and thus minimize the plant's evaporation and water loss. The leatherleaf (*Chamaedaphne calyculata*), another woody shrub of peatlands, has quite narrow evergreen leaves whose undersides are covered with small scales that serve a similar purpose of reducing evaporation.

In fact, almost all of the small, woody plants of temperate peatlands exhibit some degree of adaptation to drought, and this type of structure appears, at first sight, most inappropriate for such wet habitats. Plant ecologists and physiologists have developed several theories to account for the phenomenon, and it is entirely probable that a number of different causes lie behind this surprising aspect of mire vegetation.

Perhaps the most obvious explanation is that the plants of the high latitude peatlands experience physiological drought. This term means that although water may be present in abundance in the environment, it remains unavailable to the plant, which therefore suffers drought. The water is unavailable because of its very low temperature (or even its frozen state) in winter. Some experiments have shown that plants reduce their water uptake when the soil they are grown in is cooled, but this is not universally the case. It could be argued that under cold conditions the transpiration rate would be low anyway, so reducing evaporation in winter is not a problem. The idea that physiological drought can account for the xeromorphic characters of wetland plants is no longer strongly favored.

An alternative explanation, currently receiving considerable support from experimental work, is that reduced transpiration from leaves can cut down the quantity of toxins absorbed from the wetland soils. The main toxins proposed are iron and manganese. These elements are needed in small quantities by plants but in excess can prove harmful, and both elements are present in mire waters, especially under acid conditions when they become more soluble. It is interesting to note that xeromorphic plants are mainly associated with acidic wetlands. The toxicity problem is overcome in part by the diffusion of oxygen from wetland plant roots (see "Living in the Wet," pages 132–133). Oxygen passes from the atmosphere, through the plant, and leaks from the roots, where it intercepts iron and manganese, forming oxides that precipitate in the soil and therefore do not enter the plant. A fast rate of transpiration, however, can result in a greater inward movement of these toxins, leading to their accumulation in the leaves and causing death in sensitive species. In one very informative experiment, a sensitive species, bell heather (*Erica cinerea*), was sprayed with silicone to prevent transpiration, following which there was less uptake of iron, and the plant survived. So the xeromorphic adaptations of these wetland species may simply be a means of protecting themselves from accumulating unwanted metals in their tissues.

This explanation may account for the adaptations that reduce transpiration rates in wetland plants, but it does not entirely answer the question of the preponderance of evergreen leaves among mire shrubs. If excessive transpiration can be harmful to a wetland plant, then is it not more appropriate for such plants to be equipped with deciduous leaves? Leaves lost in winter will not transpire, so no toxins can be accumulated.

The explanation for the evergreen leaf may lie in a very different area of wetland ecology, namely that of nutrient conservation (see sidebar).

The occurrence of fire over wetland surfaces further complicates the issue. Most evergreen shrubs have chemicals within their tissues that protect them from the grazing activities of animals, particularly invertebrates. The presence of resins, terpenoids, and polyphenols makes them distasteful to many potential grazers, but it also renders them highly flammable, and these habitats are consequently fire prone. Burning plant tissues release much of their nitrogen content

Why Be an Evergreen?

Many plants in wetlands, particularly the northern bogs, have evergreen leaves. This does not mean that the leaves last forever. The leaves may survive only two or three years, but they are not discarded at the end of each summer in the manner of deciduous leaves, so the plant has a green appearance even in winter. For most plants the main function of the leaf is to photosynthesize, trapping solar energy and using it to fix carbon dioxide from the atmosphere and produce sugars. Why is the evergreen habit advantageous in bogs? One possible reason that would apply to high latitude bogs is that the plant is ready to begin photosynthesis as soon as conditions become suitable in the spring. As days become longer, the temperature rises, the cover of snow melts, the evergreen leaf is fully constructed and ready to begin its activity. A deciduous plant has to build new leaves, which means that it has to commence a high level of metabolic activity very early in the season when the temperature may still be quite low. Stored food reserves must be mobilized, new elements must be taken up from the soil, and the energetically expensive business of leaf building must commence at a time when the plant has just survived the most stressful time of its year. An early start without the necessity for new leaf construction may be worth all the energy losses associated with additional respiration through the long winter nights.

One must also consider the total energy balance of the leaf over its entire lifetime. Building a leaf costs energy, so if it can be made to last longer then it may prove to have better value. Renewing leaves each year is expensive in terms of energy investment, especially for plants that grow in the tundra, where growing seasons are short and all energy must be used with parsimony and efficiency. Not only do leaves cost energy, they also require nutrient elements for their construction. Nitrogen is needed for protein building, and phosphorus for membranes and nucleic acids, and these must be obtained from bog peat that is generally poor in these elements. If a leaf lasts longer, then less effort needs to be expended in extracting these elements from the peat, so the investment in an evergreen leaf may prove worthwhile. Evergreen leaves, in other words, are an economically sound option. Nutrient conservation is likely to be an especially important consideration on acid, nutrient-poor bogs, which receive their only nutrient input by rainfall. Evergreen leaves may be the most efficient way of conserving these scarce resources.

Evergreen leaves are also frequently tough and leathery and are well protected against water loss by transpiration. This may seem irrelevant to wetland plants, where water is abundant, but in winter in the high latitude mires this water may be very cold or even frozen, in which case it is not available to the plant. Even bog plants can desiccate in winter. So, the evergreen leaf may also be advantageous in conserving water during the winter cold.

to the atmosphere as oxides of nitrogen, so the entire habitat can become even more impoverished in nitrogen as a result of such fire. Despite such risks, the evergreen habit is clearly a successful one in many mire ecosystems.

BOG INVERTEBRATES

The surface of a bog in summer is heaving with life. This may not always be apparent, for most of the living creatures are very small, and, relative to their size, they live in a complex forest of tussocks, dwarf shrubs, and bog mosses, so that they can easily remain hidden from a casual human observer. Some are too small to be seen by the human eye and require microscopy for their study. Others are visible but so small that a magnifying lens is needed to check their identity, but some, such as flies, spiders, ants, bees, and wasps, are quite large enough to be apparent on the surface of the mire.

Each film of water on the surface of a *Sphagnum* leaf is a tiny microcosm, pulsating with microscopic life. Single-celled, photosynthetic algae are present in all well-lit situations and are readily consumed by a wide range of protozoans and rotifers and also more complex organisms such as crustaceans. These include the tiny water fleas, *Daphnia,* which actively, and jerkily, forage for their food in the water. There are also detritivores and decomposers, feeding on the dead fragments of plant and animal material that abound in bog waters. These include bacteria, protozoan ciliates, which swim by means of small hairs over their surfaces and use the same hairs to direct food into their gullets, and amoebas that flow like gelatin and engulf their food in liquid arms, or pseudopodia.

Among the amoebas, a particular group, called testate rhizopods, are very characteristic of *Sphagnum* bogs. These are distinct in having shells, usually made of protein or silica, which protect the delicate protoplasm of the amoeba. The shells, or tests, come in a range of different shapes. Often they are spherical or flask-shaped; sometimes they are formed like discs or hemispheres. They are usually sculptured in distinctive patterns (webs, or overlapping plates like shingles, or covered in tubercles) and are also resistant to decay so that their remains are abundant in the fossil records of peats. The microscopic examination of *Sphagnum* leaves can reveal a whole community of these minute scavengers.

At a higher level of organization and size, the plants of bogs are usually richly endowed with bugs (Hemiptera), which make a living by sucking the sap from plants with long, syringelike mouthparts. Flies (Diptera) are also well represented on bogs, as in most habitats, for this is a diverse group of insects that has evolved forms to fit almost all habitats. The flies of bogs mainly have larvae that eat plant detritus, as in the case of the crane flies or daddy longlegs (Tipulidae). Many different species of tipulid flies are found, and the species vary from one bog to another, depending on climate, relative wetness, and the different vegetation types. All have fat, tough-coated larvae ("leather-jackets") from which the slender, long-legged, clumsily flying adults eventually develop.

Invertebrates that feed on dead plant material are called detritivores, and among the most common detritivores of bog surfaces are the springtails (Collembola), the mites (Acari), and the nematodes (Nematoda). The springtail name is appropriate, as this invertebrate can usually be observed on the surface of mosses when disturbed, for they flick their tails and leap into the air to a height many times their own body length. Mites are less active and crawl slowly among the plant litter. Nematodes are among the most abundant and diverse of all animal groups but are inconspicuous, unsegmented worms that spend their lives burrowing among plant detritus.

The ants (Formicidae) are usually well represented on bog surfaces, especially where there are relatively dry hummocks. Here they build their communal nests and construct chambers for their eggs and pupae that are above the normal flood levels of the mire. They forage widely over the mire surface and are catholic in their feeding tastes. The more adventurous individuals will even risk robbing insectivorous plants of their prey (see "Plants in Bogs: Carnivorous Plants," pages 151–154).

Ants, as long as they keep out of the water, have no problems with breathing in a bog, for they are always in contact with the atmosphere. Submerged animals, however, may experience the shortage of oxygen that is so often a feature of wetland life. Many of the insect larvae of bog pools, such as mosquitoes, flies, and some beetles, swim to the surface and take in air through specially adapted organs, often with telescopic, or snorkel-like structures to tap the air above. A few mosquito and chrysomelid beetle larvae, however, have an even more sophisticated approach to the problem of lack of air. They use their siphon organs to penetrate the air spaces of semiaquatic plants—they literally breathe through plant roots.

The invertebrates that most effectively attract the attention of humans on bogs are undoubtedly the bloodsucking insects. Mosquitoes (Culicidae) and tabanid flies (*Tabanus* species) are the main pests in southern mires, with blackflies (Simuliidae) and biting midges (Ceratopogonidae) making life almost intolerable for vertebrates on the tundra bogs in summer. In all of these insects, it is usually the female that bites, removing only a small quantity of blood as a sort of aperitif prior to reproduction.

Invertebrate carnivores are also found in abundance in bogs. Dragonflies (Odonata) hunt on the wing as adults but are aquatic hunters in the bog pool environment during their larval stages. Other carnivores of the pools include the dytiscid water beetles (*Dytiscus* species), while ground beetles (Carabidae) are found over the drier surfaces of the mire. Spiders are also much in evidence in bogs and exploit the

Bog Spiders

The spiders of bog surfaces generally carry their egg masses around with them, both for safety and to ensure that they can be taken to warm locations that will encourage embryo development. In a study of a Danish bog, ecologists found that the density of *Lycosa pullata,* a surface-dwelling spider, could be as high as 12 per square foot (120 m⁻²). These spiders spend their entire life running around on the heads of *Sphagnum* moss. Another species of spider, *Pirata piraticus,* lives underneath *Sphagnum* heads in the cool, moist forest of moss stems. When the females are brooding their egg masses, they build a tube up to the surface, through the moss canopy, and they sit near the top of the tube with their tails and egg cocoons protuding into the warm sun above. When danger threatens, they retreat down the tube into the safety of the moss layers.

Ecologists conducted experiments on these spiders in the laboratory, and, when they were offered a choice between a cool environment (64–75°F [18–24°C]) or a warm one (79–90°F [26–32°C]), *Lycosa* always picked the warm option. *Pirata* males, together with *Pirata* females who had no egg cocoons, chose the cool option, but the females with eggs preferred the warmth. Here one can see behavioral differences that ensure that the two hunter spiders forage and hunt in different micro-environments, thus reducing competition. There is also a shift in behavior when the female is brooding eggs, which ensures that the egg masses are taken to an appropriately warm location. The surface of a bog may look simple, but organization of life upon it and within it is quite complex.

range of different microhabitats and microclimates found on bog surfaces. The orb-web spiders (Araneidae) weave complex webs in the superstructure of the dwarf shrubs above the bog surface, where they are able to catch flying insects. Others hunt over the surface of the bog mosses or even below their canopy in the cool, most layers (see sidebar).

■ VERTEBRATES OF THE BOG

The raised bogs and blanket bogs of the temperate regions and the tundra mires of the high latitudes have in common an open vegetation, often lacking in trees. The structural complexity of these habitats is therefore poor, consisting of tussock sedges, pools, open areas of bog, and occasional eroding peats. It is a habitat of poor productivity that can support only limited food webs. Invertebrates (see "Bog Invertebrates," pages 156–157) may be abundant for the short growing season, and plants provide a resource that attracts herbivorous vertebrates while the spring and summer flush is taking place. Some of the more southerly raised bogs, especially the relatively frostfree oceanic ones, even continue to provide a source of grazing for migratory herbivores, both mammals and birds, through the winter period.

One of the best-known mammals of the arctic mires is the lemming (*Lemmus* species). There are several different lemming species found in the arctic wetlands, but they are certainly not restricted to wet habitats. Indeed, they are reluctant to swim, though very capable of doing so when required. Their presence on mire surfaces is usually apparent because of their elaborate systems of tunnels running through the tussocks of sedge and cotton-grass vegetation, which are maintained even under a cover of snow. Lemmings are, of course, renowned for their occasional population explosions and mass migrations, which are probably related to food availability. There is often a four-year cycle of abundance, but not every peak in this cycle results in mass movements. Lemmings are preyed on by arctic foxes (*Alopex lagopus*) and by snowy owls (*Nyctea scandiaca*), and the two predators often interact aggressively in their competition for prey. Arctic foxes often kill far more lemmings than they need and bury the remainder for future use. The vegetation of the tundra mires can be significantly modified by the grazing activities of these small herbivores. During peak population years, the biomass of vegetation is quite clearly reduced by their consumption, sometimes by as much as 50 percent.

Another small mammalian grazer of the arctic mires is the arctic hare (*Lepus arcticus*). In winter (when the mean temperature of the coldest month may approach -40°F (-40°C) they flock together in groups of about 100, which probably helps them to avoid predators by sheer confusion. Caribou (*Rangifer tarandus*), called reindeer in Eurasia, also graze the summer mires and range over the entire arctic region, but they exist as a number of subspecies which remain distinct as a result of separate migratory and breeding habits within the herds. Like lemmings, they exhibit population fluctuations, although with much longer intervals between peaks; in the case of the Greenland population, the cycle takes about 60 years. The predatory impact of wolf packs may influence these cycles, but so also may food supply and climatic shifts. The caribou often calve in the cotton-grass (*Eriophorum vaginatum*) tussock mires. During the high summer, much of their feeding activity also takes place within the polygon mires (see "Arctic Polygon Mires," pages 114–116).

Many waterfowl migrate into the tundra mires to breed, including geese, ducks, and swans. Snow goose (*Chen caerulescens*), barnacle goose (*Branta leucopsis*), eider duck (*Somateria mollissima*), and tundra swan (*Cygnus columbarius*) are among the many wetland species that rely on the arctic summer to breed. These birds begin to arrive in April or May, when the pack ice begins to break up, but a sudden drop in temperature at this time of year can cause great mortality. In 1964, for example, the Canadian Arctic experienced one of its coldest summers on record, and it was estimated that 100,000 king eiders (*Somateria spectabilis*) died as a consequence. The onset of an early fall cold spell can similarly create havoc in the gathering flocks of birds, especially the season's young ones.

Wading birds also use the high-latitude mires as their breeding location, exploiting the long hours of daylight. Pectoral sandpiper (*Calidris melanotos*), Baird's sandpiper (*C. bairdii*), dunlin (*C. alpina*), sanderling (*C. alba*), black-bellied plover (*Pluvialis squatorola*), and golden plover (*P. dominica*) all breed in the northern mires. The golden plover is also associated with the aapa mires (see "Aapa Mires and String Bogs," pages 89–91) of the boreal regions, where breeding pairs occupy territories along the ridges of the peatland. Some of these breeding waders run short of calcium in their bodies as a result of the heavy demands of egg laying, together with the general scarcity of this element in the acid environment. A study of wader diets in northern Alaska revealed that many of the waders' stomachs contained fragments of the bones and teeth of lemmings, suggesting that these birds forage for any spare calcium they can find to make good their deficits. When they hatch, young birds have the advantage of an abundance of invertebrate food, which allows them to grow rapidly and provides them with a rich source of protein.

Although Arctic foxes will prey on young birds and eggs, the main predators of the birds of these tundra mires are the peregrine falcon (*Falco peregrinus*), the gyrfalcon (*F. rusticolus*), and the rough-legged hawk (*Buteo lagopus*). It is often noticeable, however, that groups of geese breed close to the nests of these predators, for though they rarely hunt in close proximity to their nests, yet these raptors offer some protection from Arctic foxes. The peregrine is essentially migratory in its habits, moving into the Arctic from its southerly winter quarters simply to exploit the supply of food and the lack of disturbance available through the summer. The gyrfalcon, on the other hand, is a permanent resident of the tundra, moving only a little way south during the winter.

The common crane (*Grus grus*) is a bird most particularly associated with open bog habitats in Europe and Asia. It migrates each spring from the Sudd swamps of Africa, from India, and from Southeast Asia to breed in the bogs and tundra of the north. The common crane breeds on the ground but always chooses inaccessible locations surrounded by pools or quaking *Sphagnum* lawns. It builds a low mound for its nest, just sufficient to elevate the eggs above the influence of water, and uses the same site in successive years. Perhaps surprisingly, for such a large and long-billed bird, it is largely vegetarian, although it may take some animal food in summer.

North America has two native cranes, the whooping crane (*G. americana*) and the sandhill crane (*G. canadensis*). The sandhill crane, shown in the photograph below, is fortunately still locally common, breeding in the tundra regions of Canada and also in marshes and grasslands in the United States as far south as southern California. Although they breed in the wetlands, these birds migrate, often at very great altitude, to agricultural fields, where they spend the winter, often in large flocks (see photograph, opposite). Whooping cranes are also migratory, but they are now very scarce. They breed mainly in the marshes of Wood Buffalo National Park, in Alberta, Canada, and then migrate to the Gulf coast of Texas in winter. The breeding population is

The sandhill crane (*Grus canadensis*) breeds in the tundra and northern marshes of North America and migrates south to spend the winter in the southern part of the United States and in northern Mexico. *(Kim Hart, Oklahoma Department of Wildlife Conservation)*

Sandhill cranes migrate in flocks. The red on the face distinguishes adult birds from juveniles. *(John and Karen Hollingsworth, U.S. Fish and Wildlife Service)*

down to only about 100 birds, but they do show some signs of recovery as a result of careful protection.

In the Russian tundra, another crane, the Siberian white crane (*G. leucogeranus*), is now restricted to just two isolated areas 2,000 miles apart. Like all cranes, these spectacular birds have an elaborate display dance, which they perform on arrival at their breeding grounds. They make their nests on the damp peaty vegetation and lay just two eggs. Their low egg number, the fact that they do not reach sexual maturity for six years, and their long and hazardous migration to Iran, India, and China each winter puts these bogland birds in a very vulnerable position. There are probably less than 2,000 Siberian white cranes left in the world, and concerted efforts are now being made to conserve them, both in their breeding grounds, on their migrations and in their wintering locations.

WETLANDS AND BIRD MIGRATION

Many wetland birds migrate. In response to unfavorable weather conditions, some migrate quite locally, including some species of heron and coot, which may move from inland wetlands to the coast during periods of particularly cold weather when frost makes it difficult for them to obtain food. Salt water at the coast has a lower freezing point, so the seas are less likely to become frozen, and invertebrate and fish life in coastal regions continue to be active and available as prey.

Other wetland birds, including many ducks, geese, and swans, undertake long journeys as part of a regular seasonal pattern, sometimes traveling between the Northern and Southern Hemispheres in their annual movements. Unlike "hard weather movements," these migrations are deeply implanted into the genetic makeup of the birds and are instigated by seasonal cues, such as shortening or lengthening days, only slightly modified by local weather conditions. The fact that such patterns have developed and persisted in certain wetland species indicates that there is survival advantage to be gained by migration, despite the obvious stress and risk involved with long journeys.

Migration offers a number of positive advantages to a bird species. A summer move into northern latitudes can offer a rich reward of food reserves for the short season, resulting from the flush of summer productivity and the consequent supply of plant and invertebrate food. Since there are few or no resident birds to exploit these resources, there is little competition, except from other immigrant populations. The long midsummer days (24 hours of daylight north of the Arctic Circle) provide unlimited feeding opportunities, which may be critical in supplying the demands of hungry young. The low levels of human disturbance in the far north may also encourage breeding, and communal nesting (a practice found in many geese) provides defense against the depredations of carnivores, such as the Arctic fox. When added up, these advantages weigh heavily against the negative effects of the risks encountered on the journey.

The continued success of migratory wetland birds depends on the persistence of their required habitat in the breeding grounds, the survival of suitable winter habitats at the other end of the journey, and the reliability of stopover sites along the migration route. Conservation of wetland migrant species demands that all three of these factors are given due attention. National policies for protection and conservation of these long-range migrants are often inadequate for ensuring their survival because their migration often takes them over international boundaries.

Some examples of wetland migration movements will illustrate this. The whistling swan of North America and the Bewick's swan of Eurasia are now believed to be races or subspecies of the same species, *Cygnus columbianus*, which is now often called the tundra swan to emphasize its genetic unity. It is a vegetarian bird, feeding on the leaves, shoots, roots, rhizomes, tubers, and seeds of a wide range of aquatic and wetland plants. It breeds in the high tundra mires on elevated banks among the wetlands during June, July, and August. In September the adults and young leave the tundra regions together and fly south to their wintering grounds. Different populations move in different directions, eastern Siberian birds moving to the western Pacific rim, Alaskan birds moving to the west coast of the United States or even across the entire continent to join Canadian tundra birds on the Atlantic coast. Birds from the western Russian Arctic move into western Europe for their winter quarters, apart from a small population that heads for the inland Caspian Sea in northern Iran. This splitting of populations has undoubtedly assisted in the development of distinctive races (told apart visually by the amount of yellow on the base of their black bills).

The swans are large birds and are strong flyers, but they cannot complete such long journeys in a single flight. They need to rest at points along their routes and to replenish their fat reserves, so the reliable presence of stopover sites is vital for their survival. The unexpected loss of such a location could prove fatal for birds traveling at the limit of their physical ability. In North America these journeys involve traversing only two countries, Canada and the United States, so conservation measures to ensure the survival of stopover wetlands are relatively easy to plan and develop, whereas in

The major flyways of waterfowl in their fall migration. Wetland and other birds use regular routes on their migration and depend upon the presence of locations where they can rest and refuel.

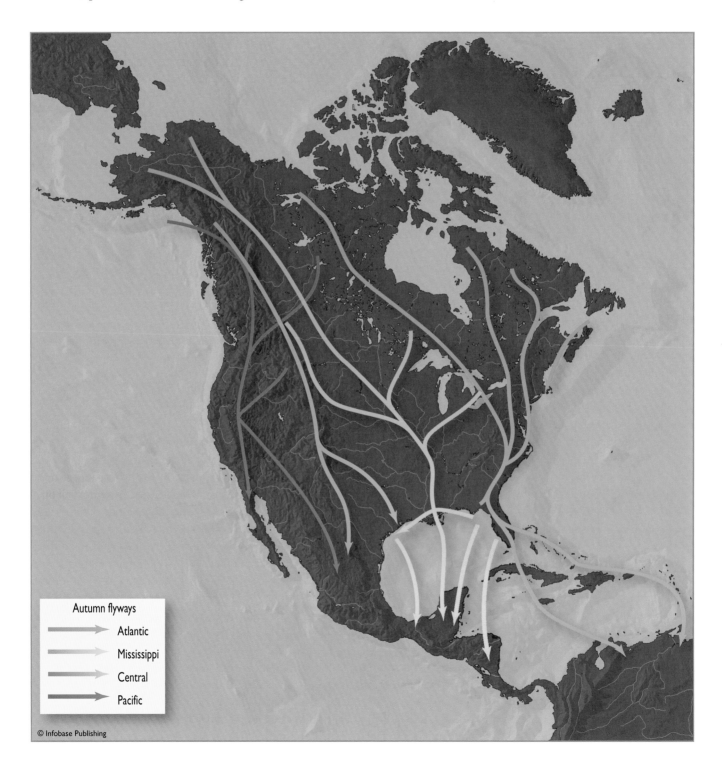

Autumn flyways

Atlantic
Mississippi
Central
Pacific

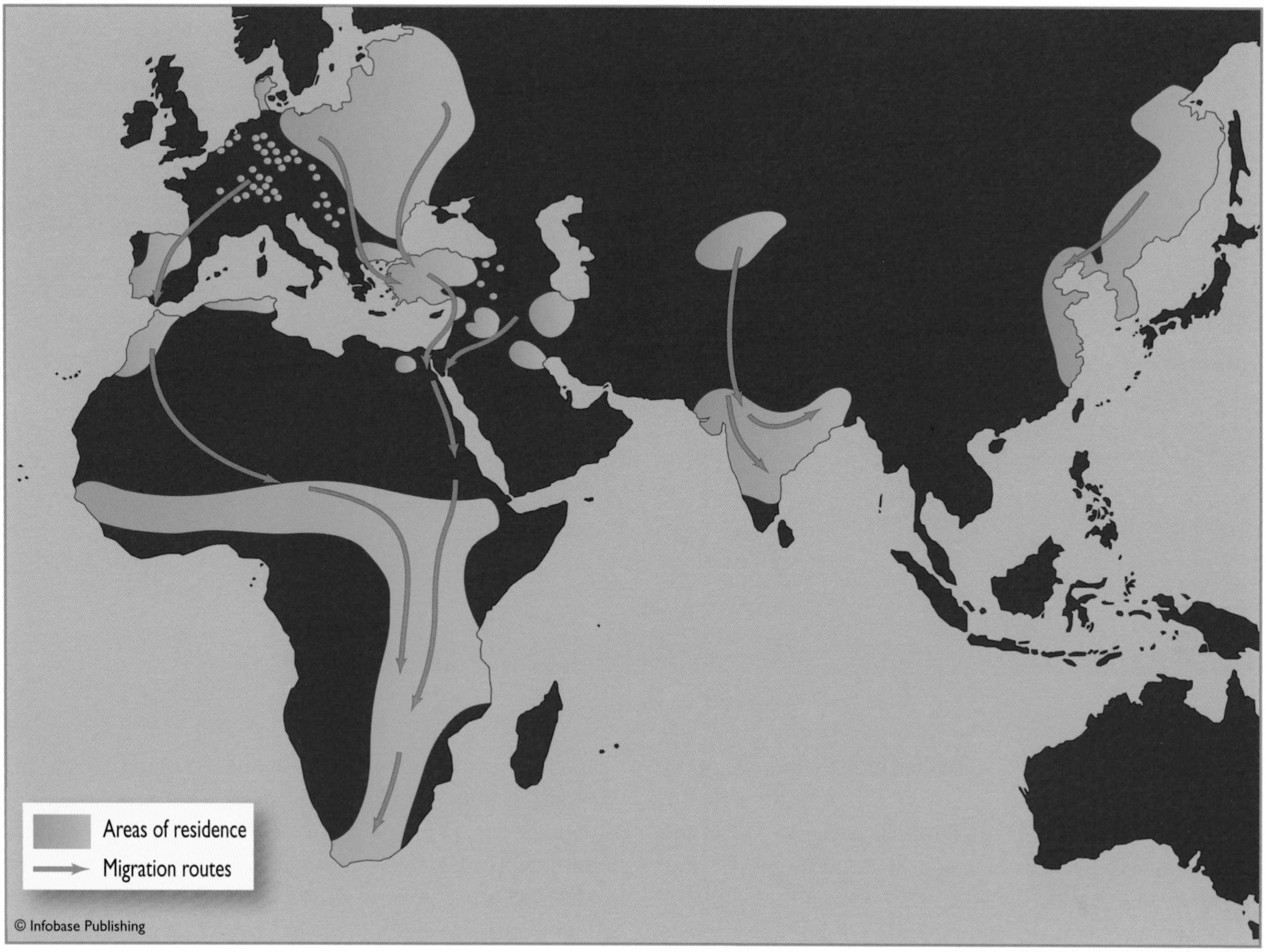

Areas of residence

Migration routes

© Infobase Publishing

Pattern of fall migration of the white stork (*Ciconia ciconia*) in Europe, Asia, and Africa. This wetland bird must cross desert barriers during each migration.

Europe and Asia, many national boundaries are crossed by the migrating swans, and the protection of these and other wildfowl species demands greater international collaboration, which is not always easy.

Extensive studies of migration patterns among the wildfowl of North America have revealed four major flyways used by these wetland birds, as well as by many other types of migrant birds. These are known respectively as the Pacific, Central, Mississippi, and Atlantic flyways and are shown on the map above. The existence of these clearly defined and preferred routes for migrant birds makes the task of conservation a little simpler because efforts can be concentrated on maintaining stopover sites along them. In the case of the Mississippi flyway, for example, the prairie pothole mires are important as a staging area for migrants as well as being breeding ground for ducks in their own right (see "Pothole Mires," pages 100–102). The reduction in wetland area from 18 percent of the land in 1850 to about 8 percent today is a trend that must be halted if this flyway is to be maintained.

Disturbance of wildfowl during their "refueling" stops can also endanger these birds as they replenish their energy reserves. The black brant goose (*Branta bernicla*), for exam-

ple, has a critical staging post at Izembek Lagoon near the tip of the Alaska Peninsula, which is used by almost all birds from the high Arctic of Siberia and Alaska on their way south. The area is disturbed by aircraft approximately every two hours, and this interrupts the feeding patterns of the geese and can reduce their gain in weight. It has been calculated that if interruptions to feeding exceed about 50 times a day, the geese would fail to gain weight at this stopover.

Migration between Northern and Southern Hemispheres presents additional geographical difficulties for wetland birds, including movements over alien and dangerous habitats, such as the open sea or the arid deserts. An example of such movements is afforded by the white stork (*Ciconia ciconia*) of Eurasia. European populations heading for their wintering grounds in Africa have to fly over or around the Mediterranean Sea, as shown in the map above. The shortest

Freshwater wetlands can occur even in the desert. At Aswan in southern Egypt the Nile River provides an important migration route for many wetland bird species, including white storks. *(Peter D. Moore)*

crossing is at the Straits of Gibraltar in the west, and many birds gather to move over into West Africa at this location. The alternative is to migrate down the eastern coast of the Mediterranean, through Lebanon, Israel, and Egypt. They may then continue up the Nile valley, shown in the photograph, into the Sudd wetlands of Sudan or into the swamps of East Africa. The Gibraltar migrants, however, have no option but to cross the western Sahara to reach the moister regions of the Sahel to the south. In the case of such long-range migrants, the conservation of the wintering grounds, as well as that of the stopover sites, is of importance. The high incidence of drought in the sub-Saharan region during the 1970s had an impact on the winter wetlands and, hence, on these migrant wetland birds, just as it had upon the resident human populations.

PLANT DISPERSAL IN WETLANDS

Many animals have no problems dispersing from one wetland to another. Birds and flying insects, for example, can take to the wing and move widely in the hope of encountering further sites where they can continue their lives. Many mammals, from caribou to lemmings, can move considerable distances on foot. Microscopic organisms probably travel as passengers on the feet of migratory birds or even blown in the air, as is the case with many juvenile spiders. Other animals, however, have limited powers of dispersal,

and such animals are often rare and confined to certain wetland locations (endemics).

In the case of plants, passive dispersal is the only option, but this still provides a wide range of opportunities for movement, and most wetland plants are well adapted to move from one place to another. Movement, in fact, is often essential for wetland plants and animals because of the temporary nature of the environment. Wetlands are constantly changing in their hydrology and environmental conditions as they become silted and undergo succession (see "Succession: Wetland Changes in Time," pages 14–15). This means that the habitat becomes unsuitable for one set of plants and animals and more appropriate for others, so some become locally extinct while others invade. Wetland plants, therefore, are always on the move, constantly being carried to new locations by a range of different transport options.

An important feature of many wetlands is the fact that water flows through them (rheotrophic systems), and in such circumstances there is an obvious means of plant movement—they can float with the flow. Movement upstream, of course, is not possible in this manner, but downstream is no problem. For many wetland plants, the simple solution to the challenge of moving to new regions is the fragmentation of the whole plant and vegetative dispersal. Many of the large, floating aquatic plants use this method most effectively, including the bulrushes, reedmaces, reeds, papyrus, water hyacinth, and many more. Some of the most serious aquatic pest plants, including the water fern (*Salvinia molesta*), the water lettuce (*Pistia stratiotes*), and the estuarine cord grass (*Spartina anglica*), use this method for rapid

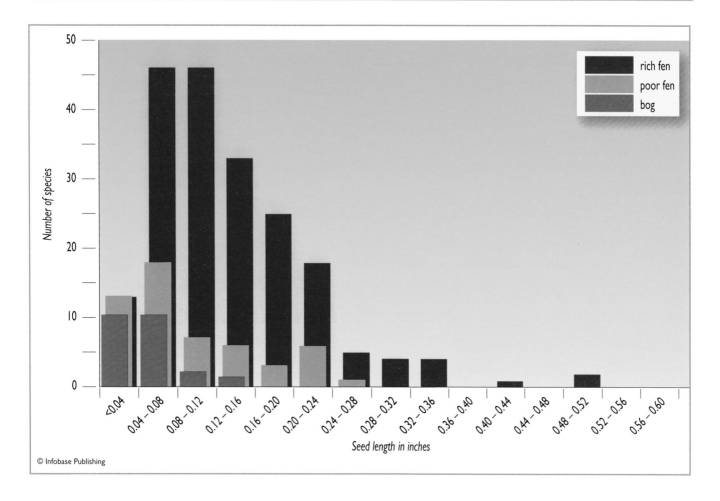

© Infobase Publishing

The number of plant species with particular seed sizes found in three types of mire within the British Isles. Rich fen (rheotrophic) mires have high species diversity with many large-seeded species. Poor fens (also rheotrophic, but with lower concentrations of minerals) have fewer species, but relatively large-seeded species are still found. Bogs (ombrotrophic) have even fewer species and almost all are small-seeded.

and efficient spread. The one disadvantage of the method is that the new plants are genetic clones of their parents, so there is no opportunity for genetic variation and adaptation to new conditions. If, on the other hand, the old formula works well, then such variability may not be particularly important.

Fruits and seeds can also be transported by water, and this method combines the efficiency of wide dispersal with the variability inherent in sexual reproduction. The first requirement of such propagules (organs used in the dispersal of plants) is that they should float, and many have woody coats with air pockets to give them buoyancy. Among the most remarkable of such floating seeds are the sea beans derived from a member of the pea family, *Entada gigas,* that grow in the coastal tropical wetlands of the Caribbean.

If these seeds are washed out to sea they can even cross the Atlantic Ocean without becoming waterlogged. Many other wetland plants travel less ambitious distances using such methods, including many sedges (*Carex* species and *Cladium* species), flags and irises (*Iris* species), bog beans (*Menyanthes trifoliata*), and alders (*Alnus* species). All of these examples have large fruits or seeds, and there is a clear advantage in size. Flotation requires airy, spongy tissues, which demand space. Also, establishment in a new site is made easier if there are substantial supplies of food reserves in the seed. Size is, therefore, an advantage to a water-dispersed fruit, an extreme example being the coconut (*Cocos nucifera*).

Transport by animals is an alternative mechanism. This has the advantage that animals (particularly migratory wetland birds, such as waterfowl) are likely to move to an appropriate new site for the spreading plant. The simplest means of exploiting such movement is to become lodged on part of the animal's body in mud. The feet of geese and ducks are probably responsible for the spread of many plants from one wetland to another, especially those which grow around muddy edges, such as the starworts (*Callitriche* species) and duckweeds (*Lemna* species).

An alternative strategy is to be transported inside an animal's gut. This has the advantage that the animal will actually seek out the fruits as a food resource rather than collecting them accidentally, but it has the disadvantage that the seeds may be damaged in the digestive processes or may be discharged in feces before the animal has traveled very far. There is also the additional "cost" to be met of making the fruit attractive to the animal. This will involve the provision of a food resource, together with some "packaging" to attract the animal's attention, such as bright colors, and the encasing of the seed in a protective coat. These costs have evidently not deterred many plants from using this mechanism, including many ericaceous plants—those related to heather (such as the blueberries, *Vaccinium* species, and bog rosemary, *Andromeda polifolia*) and water arum (*Calla palustris*). There are records of the bog rosemary being carried in the gut of snow buntings from Europe to the island of Surtsey, off Iceland.

Another means of movement is by air. Light, wind-dispersed fruits are found in many wetland plants such as the willow herbs (*Epilobium* species), some of the heaths and heathers (*Erica* species and *Calluna vulgaris*), the cotton grasses (*Eriophorum* species), and the rushes (*Juncus* species). Some of these, such as the rushes, simply have a very tiny seed of just .001 inch (0.3 mm), weighing only 0.000001 ounce (0.000026g), which behaves as dust in the atmosphere and is easily suspended in moving air. Wetland ferns and mosses (including *Sphagnum*) spread by tiny spores, which also disperse very widely in the air. Other plants, such as the willow herbs and cotton grasses, have fruits with hairy appendages that catch the wind and keep them in the air. As a dispersal mechanism, air travel can lead to wide movements, but the direction is determined only by the wind. The relatively random nature of such dispersal means that large numbers of seeds have to be produced to ensure that some, at least, will arrive at a suitable place for germination.

Examining all the options open to plants in moving from one wetland to another, one finds a distinct pattern emerging in relation to the water regimes of the various wetland types. Water transport is available only to rheotrophic mire species. Ombrotrophic mires are, by their very nature, isolated units of elevated peatland, and dispersal by water is unlikely to take a fruit or a seed to a suitable new location. Ombrotrophic mires are like islands in a sea of alien environmental conditions, and the most appropriate means of dispersal to other "islands" is by air. The effect of this is seen in the sizes of fruits and seeds from different types of wetlands. The diagram on page 163 shows an analysis of European wetland plants by their seed or fruit sizes, and it can be seen that rheotrophic mires have a wide range of seed sizes (generally smaller in the mesotrophic, poor fen locations), whereas ombrotrophic mires have mainly species with small seed sizes (for air dispersal). Bogs are, indeed, islands, and their plants (and animals) must be capable of air transport if they are to move to other islands in their vicinity.

The fact that wetland plants are so well adapted to dispersal may account for the fact that so many of them have wide distributions throughout the world. The bog mosses, for example, are found in acid mires almost throughout the world. Where major barriers to dispersal exist, such as the larger oceans, human beings have often provided the vital link in dispersal processes by either deliberately or accidentally transferring wetland plants from one continent to another. Often these introduced alien plants become wetland pests, as in the case of the water fern (*Salvinia molesta*) in Australia and the water hyacinth (*Eichornia crassipes*) in Asia.

■ THE BIOLOGY OF INLAND SALINE WETLANDS

Wetlands in enclosed basins, which have no outlet streams or rivers, lose water only by evaporation, and this leads to the accumulation of salts in their waters and sediments, left behind as the water turns to vapor. This situation is most frequently seen in hot, dry areas, where evaporation rates are high. These are the inland saline wetlands (see "Inland Saline Wetlands," pages 116–118), whose chemistry makes them very different from all other wetland types. Life in saline conditions presents many problems, and the organisms that survive in such sites are distinctive and highly adapted to these extreme conditions.

Around the fringes of saline wetlands, some higher plants are able to grow. These are mainly halophytes (salt-tolerant plants), including many members of the Chenopodiaceae family, including saltworts (*Salsola* species) and sea blites (*Sueda* species), together with the Plumbaginaceae, including the sea lavenders (*Limonium* species), which survive by pumping excess salt out of special glands on their leaf surfaces. The cell saps of these plants are rich in salt, despite the mechanisms they possess to exclude this unwanted chemical, and this makes them less palatable to many grazing animals. However, some vertebrate herbivores of these regions, including the Asian wild ass (*Equus hemionus*) and Asian wild sheep (*Ovis ammon*), are prepared to consume this salty diet. This places a strain on their kidneys, the organs responsible for the removal of excess salt from the blood. Another shrub typical of these salty wetland edges is the tamarisk (*Tamarix* species). This tree occupies dried-up river beds (wadis) and saline regions in dry lands, and it, like the sea lavender, is also able to excrete excess salt through leaf glands.

Within the salty waters of the wetlands, relatively few organisms can survive. Some bacteria can cope with high salinity, especially those which oxidize iron. In the soda lakes of the East African Rift Valley, the activities of these iron bacteria can sometimes color the water red due to the excessive formation of iron oxide. Some blue-green bacteria are also able to survive in high salinity. These bacteria, once called blue-green algae (see sidebar, "Cyanobacteria," page 134), are photosynthetic organisms and so are able to fix the energy of sunlight, and they form the basis of many food webs within the soda lakes. They are also able to fix nitrogen, so they are not dependent on the weathering of rocks for a supply of this element, and their rich protein component (built in part from this nitrogen) makes them a valuable food source for animal life.

Very few fish are tolerant of high salinity conditions. In East Africa, for example, there are very many species of the genus *Tilapia*, but only one race of one species (*T. alcalica grahami*) is adept at survival in extremely saline waters. This fish evolved in the highly alkaline Lake Magadi in Kenya, but it has now been introduced into Lake Nakuru also, and this has resulted in a great increase in bird biodiversity at that lake because of the new food resources available.

The most characteristic and best-known vertebrate of the saline lakes is the flamingo, particularly the lesser flamingo (*Phoeniconaias minor*) of the East African Rift Valley. The greater flamingo (*Phoenicopterus antiquorum*) is a much larger bird than the lesser flamingo and demands less extremely saline conditions because of its preference for bulkier particulate food than that preferred by the lesser flamingo. All flamingos are adapted for feeding on microscopic food, which they obtain by hanging their heads upside down in the water as they wade (or swim) through it and filter out the material contained. The beak of the flamingo is remarkable among birds, for it operates in an inverted manner, as shown in the diagram opposite and in the photograph on page 118. Whereas most birds have beaks that are relatively fixed in the lower mandible and flexible in the upper mandible, the flamingo has the reverse. The upper mandible remains fixed when feeding, and the lower mandible moves. In the greater flamingo, the upper mandible is shallow and water is pumped over the grooves in the bill by a constant opening and shutting of the beak (by raising and lowering the lower mandible). In the lesser flamingo, the upper mandible is triangular in cross section and has much finer grooves (about 50 per 0.39 inch [1 cm], compared with 15 for the greater flamingo). The flamingo pumps water through the bill by using its tongue as a piston, pushing it in and out, so the beak does not constantly open and shut. So the lesser flamingo is able to trap the tiny particulate matter derived from blue-green bacteria in the highly saline lakes, whereas the greater flamingo catches larger materials, such as shrimps and other crustaceans, from the less saline sites. In both species, the pumping action allows the bird to filter out food without swallowing too much of the salty water.

The lesser flamingo's fine trapping mechanism is so efficient that it has been calculated that a million flamingos

(the kind of population found in some East African saline lakes) could remove 65,000 tons of blue-green bacteria in the course of a year. So the lesser flamingo taps the energy resource of the lake right at the initial level of fixation with the photosynthetic bacteria, which is the most efficient way of obtaining a resource that is in short supply. Feeding on planktonic bacteria means that the lesser flamingo prefers shallow water (2–6 inches [5–15 cm]), whereas the greater flamingo wades in deeper water (up to 3 feet [1 m]) and may even feed while swimming.

Most crustaceans, including shrimps, are unable to tolerate the very high salt concentrations of the East African soda lakes, hence the predominance of lesser flamingos there. One species of shrimp, the brine shrimp (*Artemia salina*), is able to cope with salinites up to 10 times that of seawater. This shrimp is found in the saline lagoons of California, Tunisia, the Ran of Kutch in Pakistan, and the French Camargue. As the shrimp embryo develops, it produces glycerol that counteracts the osmotic impact of the high salinity in its surroundings and is thus able to avoid dehydration.

Beaks and feeding methods of two flamingo species. The greater flamingo is better adapted for filtering large particles, including crustaceans, from mud. The lesser flamingo filters water and extracts tiny particles, such as plankton, including blue-green bacteria.

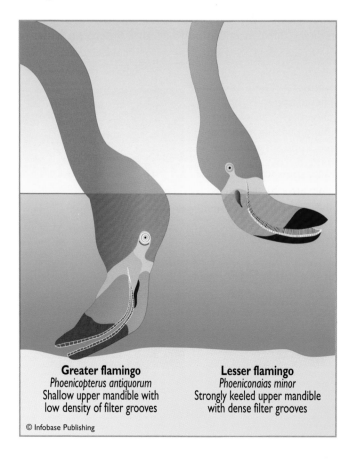

Greater flamingo
Phoenicopterus antiquorum
Shallow upper mandible with
low density of filter grooves

Lesser flamingo
Phoeniconaias minor
Strongly keeled upper mandible
with dense filter grooves

© Infobase Publishing

Many inland saline wetlands are temporary in nature. They are flooded during the wet season and then become dry for the remainder of the year. This is the case with the Lake Eyre basin in South Australia, for example, and also with the open salty flats of the Iranian desert south of the Caspian Sea, the Great Kavir. In that region, the winter snows on the Alborz Mountains melt in the spring, and their waters accumulate in the flat, alluvial plains of the desert. The pools formed are salt-rich as a result of a long history of repeated evaporation, but they provide important stopover sites for migrating wetland birds, including black-winged stilts (*Himantopus himantopus*) and marsh sandpipers (*Tringa stagnatilis*), in this arid region. Local desert species, such as sandgrouse and bustards, also use these temporary wetlands as a source of water. In such sites, however, the relatively immobile species such as fish and amphibians have to face the problem of survival through the subsequent drought (see "Wetlands in Drought," pages 168–170).

■ THE BIOLOGY OF COASTAL WETLANDS

Coastal wetlands do not generally experience the high salinities found in the inland saline sites, where evaporation leads to increasing salt concentration. Only in isolated lagoons and pools can salinity rise to high levels, and then it is usually temporary, for the next incursion of the tide reduces salt concentration to about 3.5 percent, which is tolerable to very many organisms. Coastal wetlands are, therefore, productive and diverse, unlike the biological deserts found in the highly saline sites.

The tropical mangrove swamps (see "Mangrove Swamps," pages 123–126) are particularly diverse and productive. The movements of water through this ecosystem are relatively gentle, and sediment accumulates steadily around the roots of the trees, which grow on sheltered shores. Organic material from leaves and the bodies of planktonic organisms is added to the imported mineral sediments and builds up in layers around the roots. Microbial decay in these stable deposits deprives them of oxygen, and they soon become black with iron sulfide and suitable only for the survival of animals that can cope with lack of oxygen. Even the roots of the plants find it difficult to obtain sufficient oxygen for respiration, and many of the mangrove trees (such as *Sonneratia*) have developed "breathing roots," or *pneumorrhizae,* which defy the normal inclination of roots to grow downward and extend against the force of gravity into the air above the water. Here they take up oxygen from the air to supply their underground parts. Other types of mangrove tree, such as *Bruguiera,* have roots that arch in the air above the water and return to the mud, leaving loops of root exposed to the air, through which they respire (see photograph below). Different types of mangrove roots are shown in the diagram on page 167.

The architectural complexity of the submerged root systems of mangroves provides ideal shelter for fish, and this habitat is extremely rich in these aquatic vertebrates. The archer fish (*Toxotes* species) of Southeast Asia is able to prey

Pneumatophores, or pneumorrhizae, of black mangroves are roots that grow upward out of the mud to assist in respiration. Stilt roots, shown here, give the trees stability in shifting tidal mud. *(South Florida Wetlands Management Department)*

 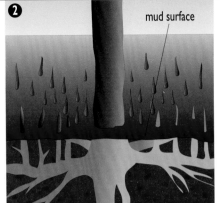 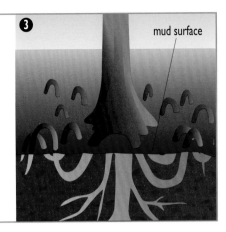

The various root systems of mangrove trees. (1) Stilt roots of *Rhizophora* species. (2) Pneumorrhizae ("breathing roots") of *Avicennia* and *Sonneratia*. (3) Knee roots of *Bruguiera*.

on the insects that land on emergent plants above the water surface by squirting water from its mouth and knocking them into the water. Other fish, such as the mudskippers (family Gobiidae), are able to crawl out of the water using their pectoral fins and lie on the surface of exposed mud banks, breathing air through their mouth cavities rather than taking in dissolved oxygen through the gills. There are many species of mudskipper, all leading varying lives. One species is capable of climbing trees. Another prefers to live on the edge of the water and harvests algae by sweeping its head from side to side. Yet another species, *Periophthalmodon schlosseri*, maintains complex systems of burrows in the mud which it fills with air that it carries down by taking gulps into its mouth cavity at the surface and conveying it into the burrow where it is released once again. These burrows can be as much as 4 feet (125 cm) deep and can contain up to a gallon (4.5 liters) of air.

The mangrove fishes are preyed on by a range of birds, from kingfishers to storks. Crabs are also frequent on the surface of the mangrove mud banks and are generally omnivorous. The fiddler crabs are most noticeable because of their enlarged right claw, which is used for attracting a mate and also for threatening rival males. These have very many enemies, including storks, water snakes that may pursue them into their burrows, a crab-eating frog (*Rana cancrivora*), and even a crab-eating macaque monkey (*Macaca fasciculoaris*) in Southeast Asia, which hunts fiddler crabs while wading in the tidal pools and shallows.

Several monkeys inhabit the canopy of the mangrove swamps of Asia, including dusky langurs and silvered langurs (*Presbytis* species), together with the very distinctive proboscis monkey (*Nasalis larvatus*). All three of these monkeys feed on the leaves of the mangrove trees.

The temperate saline wetlands include the salt marshes (see "Salt Marshes," pages 118–123). These have a simpler aboveground architecture than the tropical mangroves, being dominated by herbaceous plants, most of which die down to ground level during the winter season. They support a range of animal life at all seasons, however, and are particularly important for their bird populations in winter, when geese feed on the remaining plant organs, such as rhizomes and shoot bases, which remain unfrozen in most temperate regions because of the low freezing point of salt water. Wading birds also use these wetlands for feeding, both in summer and in winter, and probe the mud for animal prey living below its surface.

Living beneath the mud surface, however, is not easy for an oxygen-demanding animal. The oxygen concentration in water contained within muds and fine sands drops from about 100 percent at the surface to less than 5 percent only 2 inches (5 cm) below the surface. The marine lugworm (*Arenicola marina*) builds a U-shaped tunnel in the mud that is open at both ends and maintains a flow of water through the shaft with the aid of fanlike appendages. This additional oxygen supply in the tunnel may even result in the formation of concretions of iron oxides on the tunnel walls. One small crustacean, *Corophium volutator*, actually seeks out anaerobic muds in which to burrow because it has a high tolerance of low oxygen tensions and can avoid competition in this way. Small animals (about 0.04 inches [1 mm]) of this species have shallow burrows, about one-fifth of an inch (0.5 cm) in depth. Older, larger animals (up to about one-third of an inch [0.8 cm]) dig burrows about an inch (2.4 cm) in depth. The wading birds, with their different lengths of beak, specialize in different depths of feeding and, therefore, on different age groups of *Corophium*. The redshank (*Tringa totanus*), a wading bird with a long bill, eats as many as 40,000 of the larger, deeper-dwelling *Corophium* adults in a single day.

Bivalve mollusks, including clams and cockles, are also found within the muds of coastal wetlands. Although these mollusks inhabit layers well below the surface of the mud, they use the moving waters above for respiration by extending their siphons upward through their burrows to emerge on the mud surface. The wading birds that feed on these mollusks need more robust bills to break or lever open the shells. A typical mollusk-feeding bird of the coastal wetlands is the American oystercatcher (*Haematopus palliatus*). There is evidence to suggest that its Eurasian equivalent, *Haematopus ostralegus,* contains two races within the species, one with a thick bill, used to hammer open prey, and the other with a thinner, more pointed bill, used to pry open the valves of cockles. This could be an example of speciation taking place as different groups of birds specialize in their hunting strategies.

Above the surface, salt-marsh vegetation produces masses of seeds and fruits that are mainly dispersed by tidal waters. The organic debris of the tide line is often rich in seeds that may have traveled considerable distances. These resources are attractive to seed-eating birds, and some, such as the southern Californian subspecies of the savanna sparrow (*Passerculus sandwichensis,* subspecies *beldingi*), are largely confined to this type of habitat. Many other seed-eating birds exploit the salt-marsh environment, especially in winter when food may be scarce or snow-covered elsewhere.

◾ WETLANDS IN DROUGHT

Most wetlands, apart from those washed daily by the oceans, run the risk of occasional drought. This may be a rare event, occurring perhaps only once a century, in which case it may have catastrophic effects on plant and animal life, or it may be a regular feature of the wetland's history, perhaps an annual event, when the climate of the area displays distinct wet and dry seasons. In this case, the impact may still appear catastrophic, but the plants and animals have usually developed life cycles or behavior patterns that permit their survival during what is essentially a predictable part of their lives.

One of the largest wetlands in the world, the Pantanal of western Brazil, passes through an annual cycle of wet and dry seasons. The region consists of an extensive floodplain fed by rivers draining from the Andes. Parts of the wetland are permanent, occupying about 54,000 square miles (140,000 sq km), but rains in the mountains bring floodwaters, and the wetlands regularly overflow to cover a region of 96,500 square miles (250,000 km²) with water. This annual growth and shrinkage of the wetland leads to the spread of fish populations over surrounding lands, together with their bird and reptilian predators, so the wetland animals regularly expand and contract both their populations and their ranges with the floods. The plants, including many swamp trees in the flooded regions, time their growth, flowering,

and fruiting seasons to correspond with the arrival of the waters. Human populations, on the other hand, who mainly use the periodically flooded land for cattle ranching, have to withdraw their animals to higher, drier ground during the wet season. They benefit, however, from the increased herbage production resulting from the annual flooding.

In southern Africa, to the north of the Kalahari Desert in Botswana, lies a similarly periodic wetland, the Okavango delta. This inland swamp is also fed by rivers draining from the mountains, in this case the highlands of Angola, from which the waters drain south in a narrow valley, eventually opening out into the largest inland delta in the world, stretching for some 11,000 square miles (28,000 km²). About half of this area is permanent swamp, and the remainder is flooded during the wet season. The arrival of the floodwaters is often sudden, leading to a rise in water level of 3 feet (1 m) or more overnight. The flooded areas have previously been baked by the tropical sun, but the buried stems and seeds of some plants, such as the broad-leaved water lily (*Nymphaea capensis*), remain alive yet dormant until the floodwaters appear. Growth is then very rapid as the plants must produce new floating leaves, flowers, fruits, and a store of food reserves before the waters recede once more.

As in the case of the Pantanal, the fish population of the Okavango increases and spreads out during the floods. Among the most characteristic fish groups of the seasonal swamps and marshes are the mormyrids (family: Mormyridae), or "elephant-snouts." As their name suggests, they have mouthparts that extend into a snout, resembling the trunk of an elephant. More than a hundred different species are found in Africa. They survive the dry season in small pools and channels and then spread out with the floodwaters, preying on worms, insects, and mollusks. Living in the opaque, muddy waters of the floods, they have very poor eyesight, but they navigate by a remarkable system resembling radar, in which they generate an electric field around themselves and use this for the detection of objects in their path. They form an important food resource for local human populations, although the pregnant women of some tribes refuse to eat them because of a superstitious belief that these elephant-headed fish can have a harmful effect on the development of their fetus.

Mormyrids avoid the rigors of the dry season simply by moving with the waters, receding when the time of drought arrives. Those that do not move to more permanent waters die. Many other animals, including crocodiles and hippopotamuses, also migrate with the water, but water lilies remain in the same location in a dormant and drought-resistant form, waiting for the water to return. This strategy is adopted by many plants, with either vegetative parts or seeds remaining in the drying mud. Seeds are particularly efficient in surviving drought, often having tough outer coats and a very low rate of respiration so that their food reserves can last for long periods of time. There are records of wetland seeds, such as those of the Indian lotus (*Nelumbo nucifera*), surviving several hundred years in a desiccated state within mud.

Smaller wetland organisms, from algae to protozoa, are also adept at drought survival by the formation of resistant cysts. Many of the filamentous green algae, including *Spirogyra*, produce a hard-coated spore after the fusion of their male and female gametes that is drought-resistant and can lie dormant in mud. The amoebas are also able to encyst in times of stress and to emerge when conditions improve. Larger, mud-dwelling mollusks are also well suited to this escape strategy, being buried in the drying muds and encased in protective materials.

One type of vertebrate animal that also prefers to remain on-site when the waters fall is the lungfish (superorder: Ceratodontimorpha). These fish are found in South America, Africa, and Australia. In normal, wet conditions they respire through gills, as do other fish, but they also have one or two "lungs" which are probably adapted swim bladders but which serve a respiratory function when the animal is living in oxygen-poor conditions or is aestivating, that is, surviving the summer drought conditions in a mud cocoon or deep in a mud tunnel. The lungfish's ability to live under foul conditions has served some species well in the modern era, when human pollution has depleted oxygen in many wetlands. In South America some lungfish have even been found living in sewage systems.

Among the most spectacular seasonal wetlands are the inland basins of Australia, such as the Lake Eyre Basin, which is located in the arid zone but is fed by four rivers. Unlike the other seasonal wetlands described, the input of water here is extremely erratic and unpredictable, so the survival stress on wetland plants and animals resident in the area is high. The filling of the basin with floodwaters results in massive invasions of mobile creatures, including wildfowl, from outside.

In Europe, the wetlands of the Coto Doñana National Park in southern Spain are an extremely important habitat for large flocks of wildfowl and flamingos during the winter. This region is close to North Africa and has a similar climate, which sometimes results in the failure of winter rains. When this happens the extensive marshes are reduced to desert, as shown in the photograph, and many birds fail to survive the winter.

The wetlands of the Coto Doñana National Park in southern Spain. In summer drought periods these extensive wetlands can become very arid.   *(Peter D. Moore)*

Not all temporary wetlands are as extensive as these famous examples, however. Small, short-lived wetlands are abundant all around the world but especially where there is a contrast between wet winters and dry summers, as in the Mediterranean climate region. In California (which experiences a Mediterranean-type climate), for example, small temporary wetlands are relatively frequent but are also endangered simply because they are not spectacular. Within them are found many threatened species of wildlife, including small crustaceans such as the fairy shrimp (order: Anostraca). Each spring, populations of shrimps abound and breed to produce the eggs with tough coats that survive through the summer drought period. In warm, wet conditions, these eggs can hatch and the young mature within a week. In the San Diego area, it has been estimated that 90 percent of the small ephemeral pools have been lost in the last 70 years. Since many of the waterfowl that spend the winter in the region depend on the crustaceans in these pools for survival, the loss of these seemingly insignificant temporary wetlands could be serious for migratory ducks.

■ DROUGHT AND FIRE IN PEATLANDS

A warmer world, especially if it involves reduced precipitation in some areas, will have an impact on many wetlands. The effect of drought will be felt strongly by ombrotrophic peatlands, where reduced water input or additional evaporation will lower the water table for at least part of the year. The impact on rheotrophic wetlands will be felt indirectly, as less water is received by catchments and there is a slower flow of water into these sites. Drier surface peat and vegetation will lead to death or even local extinction of some plants and animals. It will also lead to the expansion of those plants that become more competitive under less waterlogged conditions. Drought can also lead to peatland ecosystems becoming fire-prone—a process that has many serious ecological consequences.

Most peatlands are not permanently saturated with water. The upper layers of peat are alternately saturated and then drained of their water content as the excess moisture moves laterally and drains out of the mire. This alternately wet and dry layer (the *acrotelm*—see "Decomposition and Peat Growth," pages 54–56), which is the site of most of the microbial decomposition of litter, is underlaid by the permanently saturated peats of the catotelm, where very little further decomposition occurs. The effect of reducing the incoming water (from precipitation or drainage) or increasing the evaporation rate, would be to lower the height of the boundary between acrotelm and catotelm, which would, in turn, reduce the rate of peat accumulation.

So the first effect of drying a mire surface is to increase the rate of peat deposition, which can be detected in the peat stratigraphy by the formation of a dark, well-decomposed layer of peat.

Continued drought can lead to an increase in microbial activity deeper in the profile, leading to the breakdown of peat that has already formed. This peat "wastage" comes about as the organic components of the peat are broken down to carbon dioxide by bacteria and fungi and are released into the atmosphere. The surface level of the peat then begins to fall in relation to the surrounding land surface, as has been observed in the fens of eastern England, which have been artificially drained and whose surface has subsequently fallen by as much as 20 feet (6 m) in places. The carbon contained in these peats is returned to the atmosphere and contributes further to the greenhouse effect; so, drought in peatlands could have a positive feedback effect on the process of climatic warming (see "Wetlands and the Carbon Cycle," pages 73–77).

The penetration of air into the upper layers of peat permits the establishment of some species of plants that have a limited tolerance of waterlogging. Various birch tree species, for example, may invade bogs following a period of drought. The cotton grass (*Eriophorum vaginatum*) is also drought-tolerant and can come to dominate mires affected in that way. Its tussocky growth actually produces local drainage patterns around its own roots. The purple moor grass (*Molinia caerulea*), a native of Europe that has also been introduced into North America, is another tussock species that rapidly invades and spreads over the surface of mires suffering from drought. This species seems to thrive in conditions of fluctuating water tables.

Species that demand wetter conditions, on the other hand, will suffer during drought. Most of the *Sphagnum* species, the bog mosses, require an abundant supply of water for their survival and are among the first evident victims of drought. Their drought tolerance, however, varies between species, and many are capable of recovery even after years of desiccation. Spores also survive, as do buds, and these can provide the first colonists in the process of revegetation of degraded peatlands (see "Bog Rehabilitation," pages 232–234). One of the most drought-sensitive bog moss species is *Sphagnum imbricatum,* and this has fallen from its erstwhile position of dominance to one of rarity in many raised bogs, partly as a consequence of recurrent droughts and fires.

Fire, of course, is a frequent consequence of drought because the drying peat of the surface layers is extremely combustible. Peatland fires are likely to become increasingly frequent if the pattern of increasing global temperature continues. Fire is often used in peatland "management" by farmers who have a superstitious feeling that in some way the treatment cleanses areas of wilderness and controls them. While the peat water table is high, this process does limited damage, but in times of drought it can be extremely harmful to the peatland ecosystem.

Fire in a peatland dominated by mosses and dwarf ericaceous shrubs may reach temperatures of 900–1100°F (500 to 600°C) in the canopy for a period of 2–4 minutes (by which time the upper vegetation has been consumed), and this does not normally kill the bases of the shrubs. Below the peat surface, the temperature may rise by only a few degrees. In drought conditions, however, when the peat itself ignites, the temperature is higher, the period of exposure to high temperature is longer, and the depth of heat penetration is also greater. The possibilities of mosses or invertebrate animals surviving in such an intense fire are low, and even the seed bank in the peat will be lost. The situation is worst where the fire advances against the wind, as this leads to higher temperatures that last longer.

Recurrent drought will, of course, lead to the frequent incidence of fire. The study of fire frequency during the course of history, especially in the boreal forests and peatlands of North America and northern Europe, has greatly improved our knowledge of changing patterns of fire. Examination of fire scars in trees and of charcoal bands in lake and peat sediments has enabled the local history of fire to be reconstructed in many areas of the world. In the boreal forests and wetlands of the Algonquin region of Ontario, Canada, for example, the fire frequency before the arrival of European settlers was approximately once every 80 years. Under this regime, the general vegetation of the area remained relatively stable. This kind of frequency seems to apply equally well to the northern regions of Europe (such as Finland) prior to intensive human management. Dense settlement and agricultural developments have often resulted in increased fire frequency, although the last century has seen effective fire-control policies in North America. Such control, however, has not always proved to be the best management policy, either for forests or for certain wetlands.

Fire can at times be beneficial to certain mires. In reed beds, for example, the reed *Phragmites australis* grows faster following a spring burn. The release of nutrients from dead stems and litter by burning can enhance the regrowth of new shoots from the old stem bases. Bands of charcoal in the reed bed sediments of Europe have been found that date back about 9,000 years to the early part of the current interglacial episode and appear to have been caused by human management of reed beds by fire.

The limitation and control of fires has occasionally been detrimental to the management of some wetlands. Among the nutrient-poor, pine-dominated regions of the Louisiana, Mississippi, and Alabama coastal plain are found many small wetland areas rich in pitcher plants. Several species of these insectivorous plants (including *Sarracenia alata, S. leucophylla,* and *S. rubra*) are found in these coastal plain wetlands where they have competitive advantages over other plants

by their ability to obtain nutrients from captured insects (see "Plants in Bogs: Carnivorous Plants," pages 151–154). The region is frequently subjected to fire, but fire frequency has recently been diminished by the development of roads, which act as natural firebreaks. In the absence of frequent fire, other, more aggressive plants are assuming dominance in these mires, and the pitcher plants are suffering. Here, it is necessary to maintain a relatively frequent fire regime in order to conserve these particular plant species.

■ CONCLUSIONS

Biodiversity is a concept that includes the full range of species in a habitat, sometimes called species richness, but it also covers much more than this. Biodiversity also takes into account the relative evenness of a community by examining whether the various species are represented in an even, or equitable manner, or whether there is dominance by some species. A biodiverse system is one that is relatively even. Biodiversity also includes the genetic variability of the different species present because a genetically diverse species is better equipped to cope with environmental change. Finally, biodiversity also takes into account the range of different habitats and microhabitats in a landscape because these elements provide more opportunities for species diversification. Wetlands are extremely variable because they are found in all parts of the world and thus develop within a great range of climates, geology, topography, and hydrology. Wetlands, therefore, rank among the most diverse of all biomes.

Wetlands change during the course of time as pioneer species are replaced by later colonists in the process of succession. This development also adds to the biodiversity of wetlands. Changing conditions provide opportunities for species with different techniques for survival and different environmental requirements. These are called ecological niches. The full potential of any species, called its fundamental niche, is rarely achieved because other species usually prove more efficient within part of the organism's potential range, so what is observed is the realized niche, within which the organism is more efficient than any of its competitors.

The niches of wetland species include the capacity to survive in aquatic or semiaquatic habitats. One of the major problems they face is obtaining sufficient oxygen because dissolved oxygen diffuses much more slowly through water than gaseous oxygen moves through air. Both plants and animals are in danger of accumulating toxic materials as a result of incomplete respiration.

Microbes occur in abundance within wetland habitats, even though a lack of oxygen may limit their diversity and activity. Some decomposer microbes are able to operate by using sulfate ions, reducing them to sulfide ions. Other microbes, including the cyanobacteria, are photosynthetic, and in the presence of light they can act as autotrophs, fixing their own carbon from dissolved carbon dioxide.

Submerged aquatic plants have the advantage over terrestrial plants in that the density of water supports their tissues, giving them buoyancy, and there is no need for woody support structures, but they need to supply their roots with oxygen, and those with floating leaves often achieve this by means of hollow leafstalks allowing the diffusion of gases to the roots. Dissolved carbon dioxide can become scarce in shallow ponds on a warm summer day, and some aquatic plants have developed a photosynthetic mechanism more often associated with desert plants in which carbon is temporarily fixed in the nighttime when it is more freely available, stored as organic acids, and released for normal photosynthesis in the day. The mechanism is known as Crassulacean Acid Metabolism.

Among the most successful group of aquatic vertebrates are fish, which often use the shelter provided by submerged and floating plants as a location for breeding. In the Amazon basin the wetlands regularly flood and invade the surrounding forest, giving fish an opportunity to feed upon the abundant life of the forest floor. Many of these fish of the flooded forest feed upon the fruits and seeds of the trees. Some large vertebrate herbivores also lead a largely aquatic life, such as the hippopotamus, and vertebrate carnivores include caimans and crocodiles. Invertebrate life abounds in wetlands, including mosquitoes and mayflies, and these are fed upon by water beetles, amphibians, and birds, such as the dipper.

Above the water level, many vertebrate herbivores take advantage of the high primary productivity of some wetland ecosystems, including moose, water buffalo, capybara, waterbuck, and lechwe. The different marshes of the world each have their range of wetland vertebrate grazers. Large carnivores feed upon the vertebrate grazers, including lions in the Old World and anacondas and jaguars in the New World. Many insects live below water in their larval stages and above the water when they mature, including dragonflies. Butterflies, on the other hand, spend their entire life cycle above water level but use wetlands for a variety of purposes, sometimes seeking mud as a source of sodium that is needed for their reproduction.

Birds, including ducks and grebes, use wetlands extensively for feeding and breeding, spending most of their time above water, but some have the ability to exploit the underwater environment by diving for their food. Others dabble, by upending in the water, or graze on the emergent vegetation. Gallinules and jacanas have large unwebbed feet, which enable them to spread their weight upon the leaves of water lilies and walk over the floating surface. Some birds dive from the air to capture fish, such as the kingfishers and some pelicans; others wade through the shallows and spear fish with long bills, including the herons and egrets. Bitterns are a type of heron with camouflage plumage that spend their lives in the cover of dense reed beds and marshes. Some small perching birds, such as the yellowthroat, also make their home among reeds and cattails and erect nests that are suspended among the stalks of the plants.

Forested wetlands and swamps provide a more complex environment in which a canopy life can be combined with a wetland habitat. Many birds prefer wet forest, such as the South American hoatzin, and many snakes, including the Indian python, can both swim and climb trees. In the boreal forested mires of the northern latitudes, beavers are a typical wetland mammal, constructing dams and creating their own wetlands in which they build their lodges protected from predators by moats.

Ombrotrophic rain-fed bogs are relatively low in their plant diversity, but the plants able to survive have complex adaptations to cope with the low levels of nutrients and high degree of acidity in their environment. The bog mosses, *Sphagnum* species, have a spongelike structure with many empty cells that take up water and hold it against the force of gravity. They also take up plant nutrients very efficiently and acidify the surrounding water, making life difficult for most other competitor species. Carnivorous plants occupy an unusual position for a plant in the food web, taking their nutrients from invertebrate animals by trapping them in specialized leaves, digesting them, and absorbing the scarce nitrogen-containing molecules that are needed for life. Some animals, such as ants, act as kleptoparasites, stealing trapped insects from carnivorous plants, but this is a very risky way of obtaining food.

Many of the dwarf shrubs of boreal and Arctic mires have leaves that appear to be adapted to conditions of drought, being small, leathery, thick cuticled, and hairy beneath. One possible reason is that the water present in the wetland environment is not available to the plant when conditions are very cold, but a more likely explanation is that controlling water loss cuts down the accumulation of toxins, including iron and manganese, from their surroundings. The evergreen leaf is probably a means of conserving nutrients such as nitrogen by constructing leaves that last longer.

Sphagnum bogs have a complex microstructure in their surface layers, the moss forming a miniature forest canopy above the water table. Within this canopy, there are many invertebrate animals, including detritivores such as amoebas, mites, and nematode worms, and small predators, including ants and spiders. The spiders vary in their behavior patterns, some seeking the warmth of the bog surface and others preferring the cool, humid conditions below the moss canopy. When carrying egg cocoons on their tails, female spiders may build tunnels to the surface and hold their eggs in the warm conditions above the mosses.

Bog vertebrates may be grazers, including lemmings and caribou, or predators, including snowy owls, cranes, and wading birds. Many wildfowl and waders migrate north into the Arctic wetlands for the breeding season, taking advantage of the short burst of productivity and the long days in order to feed their demanding young. Many of these birds migrate long distances along well established flyways in order to move between their summer breeding grounds and their winter quarters. Wildfowl in North America use flyways that usually correspond to river systems, where wetland stopover sites will be available for them to replenish their food supplies. In Europe, however, where many migrants need to cross into southern Africa for the winter, there is the barrier of the Sahara to cross. The white stork, for example, may either choose a western route and cross the desert or an eastern route and make its way south up the valley of the Nile River.

Plants do not migrate, but they must disperse into new areas to keep their populations viable. Wetland plants, especially those of rheotrophic, flowing wetlands, can use water movements to disperse their seeds or vegetative fragments downstream. These plants often have large seeds that float well, but bog plants are elevated above water flow and could not reach new bogs by means of moving water, so they depend on air transport. Bog plants, therefore, tend to have small seeds, apart from the few that use bird transport.

Inland saline wetlands are challenging habitats for wildlife. The salinity of the waters in hot conditions can rise considerably above that of seawater, and there are few aquatic organisms apart from specialized bacteria that can survive in them, but these provide food for flamingos, which are the most characteristic bird of this habitat. The lesser flamingo filters bacteria from the saline water by holding its head upside down and pumping water through its beak using its tongue. The greater flamingo eats larger food, including shrimps, which it extracts from the water by pumping the lower mandible of its bill. By filtering their food in these ways, both species avoid taking up too much salt. Coastal wetlands are also saline and include the tropical mangrove swamps where tree roots avoid suffocation in the anaerobic mud by growing upward and producing breathing roots. Animals in the mud, such as mudskipper fish, build tunnels in which they store air or, in the case of some annelid worms, keep aerated by generating water currents that circulate through the burrow. Both mangroves and the temperate salt marshes are rich in bird species, which feed upon the wealth of mud invertebrates.

Some wetlands occur in regions of seasonal climate and may be subject to regular periods of drought. Most plants survive this as buried fragments or seeds that can regenerate when the water returns in the wet season. Some specialized fish, such as the mormyrids and the lungfishes, either contract their range into surviving pools in the drought period or bury themselves in mud and aestivate, which is the summer drought equivalent to hibernation in temperate animals. Invertebrates, including fairy shrimps, also resort to this survival mechanism. Peatlands are not subjected to regular drought, or they would not survive, but they may experience occasional, irregular drought episodes. When this happens, the upper layers of peat dry out as the water table falls, and they become very subject to fire. Fast-moving fires may not cause much damage to a peatland, but prolonged intense fire generates high temperatures and ignites the peat itself, which is very damaging to the ecosystem because it destroys the seeds and fragments of moss that would otherwise survive and regenerate when the fire and drought have passed. Fire in marshes and reed beds can be beneficial, removing dead biomass and invasive trees and releasing nutrients into the water that encourage new growth of emergent plants. Fire can thus be used as a management tool in this kind of wetland. Fire may be a necessary requirement for some plant species, such as the pitcher plants of the southern United States, which have suffered as a result of more efficient fire prevention in recent years.

Wetlands not only occur in many different parts of the world with varied flora and fauna, but they also experience a very wide range of ecological pressures—climatic, physical, and chemical. Their plants and animals, therefore, have developed a varied series of adaptations to cope with the many trials they must face. This is why wetlands have such a high biodiversity.

6

Wetlands in History

Wetlands have a history almost as old as life itself. Some of the earliest records of living things on Earth are found in rocks that were formed in shallow coastal wetlands. Layers of very primitive bacterial and algal cells bound together by silica and lime developed in shallow, saline coastal waters and were eventually fossilized. They are still preserved today, despite their great age of more than three billion years, and the concretions of rock that they formed are called *stromatolites*. Modern equivalents to the fossil stromatolites can be observed in a living state around the west coast of Australia.

It was during Devonian times, 416 million to 360 million years ago, that plants and animals began to colonize land surfaces and also the freshwater environments of the developing Earth, and many of the earliest fossils of land plants are preserved in sediments that developed in shallow wetlands. In part, this reflects the fact that most of the plants of that time (primitive ferns, mosses, and liverworts) were dependent on wet environments for at least part of their life cycle but also result from the better conditions for preservation and fossilization that wetlands provide. The fossil record, in other words, is actually biased toward wetland organisms because of the poor decomposition rates found in these ecosystems.

■ COAL-FORMING MIRES

Peatlands are first evident in the Carboniferous period, which occurred 360 million to 299 million years ago. By this time, vegetation was abundant over land surfaces and included forests that were as tall and as complex in their structure as modern tropical forests, though they differed greatly in their precise composition. It was the litter and detritus from such tropical wet forests that eventually accumulated to form peat and, ultimately, coal. It is an interesting point that our modern industrial society is based on and developed as a result of the accumulation of peat in the tropical wetlands of more than 300 million years ago.

The Carboniferous coal deposits were formed in tropical and subtropical conditions, but it is important to remember that both the shapes and the positions of the continents were different at that time. What is now North America, together with China and Southeast Asia, lay on the equator at that time, and these are regions rich in Carboniferous coal deposits. The coal-forming mires are known to have been tropical because the fossil logs contained within them have no seasonal growth rings, so the climate must have been uniformly hot and moist throughout the year.

The vegetation of the Carboniferous coal swamps was very different from anything currently present on Earth, but the same principles operated as far as the accumulation of peat is concerned, so it is possible to speculate on the types of mires that were involved in coal formation. The following are certain characteristics of coal deposits that help in trying to reconstruct the conditions under which they were formed:

1. Many coal deposits contain little inorganic mineral material. The best coals, from the human point of view, are those with very small contents of silica but high levels of organic compounds, which combust easily and leave little ash. In modern peats, deposits of this type are formed only under ombrotrophic conditions, where water arrives in a relatively pure form by rainfall, and no eroded sediments are brought by runoff waters. In other words, many of the Carboniferous mires must have been bogs, in the true sense of the word (see "Hydrology of Wetland Catchments," pages 23–24). Their peat surfaces must have been raised above the surrounding groundwater tables.

2. Coal deposits are often extensive, covering many square miles. This implies that the mires that formed them were also extensive, covering wide areas of the contemporaneous landscape. Either their domes of peat stretched over many miles, or they formed a complex series of smaller domes, packed together and interspered with drainage systems.

3. Coal deposits are often deep. There are limits to the depth that can be achieved by a modern peat deposit, depending on the rate of accumulation of litter at the top of the sediment and the decomposition rates within the different layers (see "Decomposition and Peat Growth," pages 54–56). It is difficult to know exactly how deep were the peat deposits that have given rise to coal because of their subsequent compression, perhaps by about a factor of six, meaning that they are now only one-sixth their original thickness. The great depths of some coal deposits can only be explained by a process of continual subsidence of the land surface, burying older peat deposits and allowing new mires to form above them. Coal deposits, therefore, tend to be stacked, as shown in the diagram, with seams of rich, organic sediments being interspersed with inorganic materials separating these rich seams.

The Carboniferous coal-forming mire was therefore an extensive raised bog (or collection of raised bogs), with a vegetation dominated by primitive plants with tree forms, growing in tropical conditions and in sites where the land surface was constantly subsiding. Although there are no modern-day equivalents, the nearest set of conditions to these are found in the raised bog forests of the coastal regions around some of the southeast Asian islands, such as Papua New Guinea and Sumatra (see "Tropical Raised Mires," pages 106–107).

As these peats became older, their chemical composition changed. Decomposition resulted first in the production of carbon dioxide (while oxygen was still available) and then methane (when conditions became increasingly anaerobic). Sulfur in the sediments became concentrated as sulfides, and this component of coal is currently a problem because its combustion leads to the emission of the pollutant gas sulfur dioxide. High pressures, as the peats became buried more deeply, and high temperatures, as the rocks were subjected to even deeper burial and contact with volcanic activity, resulted in compaction and the chemical and physical alteration of the peats, ultimately forming coal.

Coal deposits, which are essentially fossil peats, thus were formed by ancient photosynthesis. The carbon dioxide that plants absorbed and retained from the atmosphere became deeply buried and was taken out of general circulation for millions of years. As we burn these fossil fuels in our modern industrial processes, we return carbon to the atmosphere that has been trapped for 300 million years, and we are now beginning to reconstruct an atmosphere that is richer in carbon dioxide, one which more closely resembles the "greenhouse" atmosphere that preceded the Carboniferous period.

Coal-forming systems are by no means restricted to the Carboniferous period. Coals were formed through the subsequent Permian and Triassic Periods and on into the age of the dinosaurs, the Jurassic (200 million to 145 million years ago). Younger coals, and even less mature *lignites,* or brown coals, have also been formed within the last 65 million years (the Tertiary period), and it is within these deposits that the first remains of fossil *Sphagnum* moss have been discovered. These fossil wetlands may well be the true progenitors of modern raised bogs.

■ VEGETATION OF THE COAL SWAMPS

Many ancient wetlands have left a record of their past existence in the form of peats, lignites, and coal, from which geologists can reconstruct something of their structure and vegetation. The chemical composition and high organic content of coals suggests that many of these ancient peatlands were forested bogs with hydrological systems resembling those of modern raised bogs or elevated bog forests. The vegetation that formed the coal deposits, however, was quite different from anything currently in existence on Earth, so paleontologists are faced with great difficulties when they try to reconstruct the types of peat-forming communities and ecosystems that must have existed in the past.

One of the main difficulties to overcome is assembling lists of plant species from their various scattered component parts. The geological record may contain fragments of root tissue, isolated leaves, wood from trunks, and reproductive structures such as cones, as well as the microfossils—the spores that are dispersed from fruiting bodies. Only rarely

Stacked coal deposits formed in ancient estuarine environments in which peat-forming ecosystems were subjected to tectonic sinkage of land surfaces. This resulted in a relative rise in sea level, the marine flooding of the swamps, and the formation of silt and clay layers in the peat profile. New domed mires then began to form, and the stratigraphy consequently shows alternating organic and mineral layers.

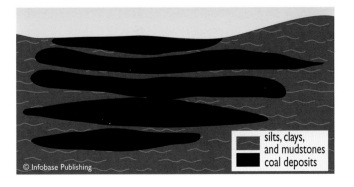

silts, clays, and mudstones

coal deposits

are the different parts found together (such as spores within a fruiting cone), and this means that the different parts may be regarded as derived from different plant species until they can eventually be reassembled as particular species. So the process of simply identifying how many plant species are involved in a fossil peat-forming community is a slow and painstaking one.

The next problem is determining which species are associated together into true communities. Just as modern wetlands have different groups of plants that characterize certain sets of conditions (high water table, nutrient-rich conditions, elevated hummocks, etc.), so these ancient mires would have been occupied by mosaics of different types of vegetation. It is also probable that the vegetation changed in the course of successional development as the sediments and peats accumulated and raised the mire surface above the influence of groundwater. At present, knowledge of the detailed ecology of the ancient wetlands is still sketchy, but new information is rapidly being gathered.

The most ancient of freshwater wetlands that bore peat-forming vegetation date back to Devonian times (416 million–360 million years ago) and have been investigated in some detail in the deposits of West Virginia, Belgium, and Siberia. At that time the vegetation seems to have been dominated by a primitive ancestor of the true ferns, called *Rhacophyton*. This plant resembled modern ferns in its divided, leafy fronds that arise from a base. It was relatively small in stature, with some fronds achieving a length of about 14 inches (35 cm), but it appears to have formed a dominant cover in some of these early mires.

During the Carboniferous period, when many of the world's major coal deposits were formed, the peat-forming vegetation became increasingly dominated by massive plants called *lycopsids,* closely related to our modern club mosses, such as *Lycopodium* species. One of the best known genera among this group is *Lepidodendron.* Some of these plants were treelike in form with a height of about 120 feet (40 m) and a trunk diameter of about six feet (2 m). They were erect, with branched upper parts bearing leafy twigs. As they grew, the lower branches fell, leaving distinctive scars on the trunk, and it is these branches that formed the main detritus from which the developing peat was derived. They appear to have had shallow, spreading systems of roots, a fact that correlates well with the arrangements found in modern trees growing in waterlogged conditions, but it is difficult to be certain that this was so because the compression of coal deposits tends to exaggerate this laterally spreading effect.

It is probable that there was a complexity of canopy structure, as in modern tropical peatland forests, with younger lycopsids, together with tree ferns and herbaceous lycopsids forming subsidiary layers. The overall impression from the fossil record is that the diversity of the Carboniferous coal swamps was relatively low.

Sequential analyses of layers in coals have shown that vegetation changed over time, just as modern peat deposits record past successional sequences. Compression of such sequences, however, makes it particularly difficult to follow the successions of the past, and the record of fossil spores in the coal layers provides a greater degree of time resolution than can be achieved using the megafossils (wood, leaves, and so on). The trouble is that the spores cannot always be related to the particular lycopsid species involved, so it is difficult to infer ecological changes with accuracy. Some of the most detailed studies, however, have shown that the Carboniferous swamps began with a high water table and high nutrient-status ecosystems and were dominated by *Lepidodendron.* The upper layers have microspores that are believed to belong to such lycopsids as *Sporangiostrobus,* typical of the drier conditions of the elevated peat surface. Once the succession is complete, there is often evidence of a flooding event, presumably due to a change in geology of the basin, involving subsidence, and leading to the recommencement of the succession from a wet swamp once again, often dominated by the lycopsid *Sigillaria.*

By the later part of the Carboniferous period (called the Pennsylvanian), the coal deposits of Europe and North America became increasingly dominated by tree ferns, while the lycopsids continued to thrive in what is now China. Little is known of the vegetation of coal-forming wetlands in the early Mesozoic (between 250 million and 145 million years ago), except that conifers became increasingly important, especially in the Jurassic coals of Australia. By the Cretaceous period (250 million–65 million years ago), the coal swamps were dominated by conifers of the Taxodiaceae, a family of plants that includes the modern swamp cypress (*Taxodium distichum*) of the Florida Everglades, with an understory of tree ferns and an increasing number of flowering plants, including both palms and herbaceous plants. These mires were much more diverse than their predecessors.

The taxodiaceous trees differed from the lycopsids in having rooting systems that penetrated down into the underlying sediments and in having the ability to produce new roots from trunks when they were subjected to flooding. These roots may well have increased sediment trapping in the swamp, leading to peat buildup. Whereas the lycopsid swamps experienced an irregular input of organic matter from the shedding of branches in the canopy, the conifer and angiosperm swamps were characterized by large quantities of shed leaves, the remains of which are a major component of the coal-fossil assemblage. The branching of the conifers was more extensive than that of the lycopsids, so the vegetation probably formed a denser canopy, casting heavier shade. The more complex light environment and the general diversity of the microhabitats in these later swamp forests probably increased the opportunity for the development of a greater degree of biodiversity.

The coal-forming swamps of the geological past have, therefore, been occupied by very different vegetation types from those of all modern wetlands, and there is no real equivalent to them in present-day vegetation. Some parts of the Everglades are still dominated by taxodiaceous conifers, and some of the peat-forming swamps of Sumatra are similar to the ancient mires in their hydrology, but a full understanding of the ecology of the coal swamps is unlikely to be achieved because of the lack of precise modern analogues.

It is worthwhile to reflect on the importance of these fossil peat deposits in the social development of the human species. Since the early stages in human cultural development, the exploitation of trapped energy in plant biomass has been part of people's way of life. Burning wood and other organic fuels has been a means of increasing the temperature of our immediate environment, enabling people to survive in cold environments, and has supplied a means of cooking food, rendering it more easily digestible. The campfire may well also have been an important factor in maintaining the cohesive social structure of early human groups. In more recent times, we have used sources of biomass energy, especially forests, together with the energy of streams and wind (for driving mills), in the development of primitive industry. The Industrial Revolution of the past few centuries, however, has been driven by the discovery of the vast energy resources contained within fossil wetland deposits. With the aid of this resource, metals could be extracted from ores, machines built, and industrial plants fueled. Many of the most important industrial regions of the world have become established in association with coal deposits, the global distribution of which is shown in the accompanying map. The Sun's energy, locked away in ancient peat, has supplied the means by which humanity has entered its new state of development. It is difficult to overemphasize our debt to the wetlands of the past. Throughout human history and prehistory people have developed a strong association with wetlands.

Distribution of the major coal deposits of the world

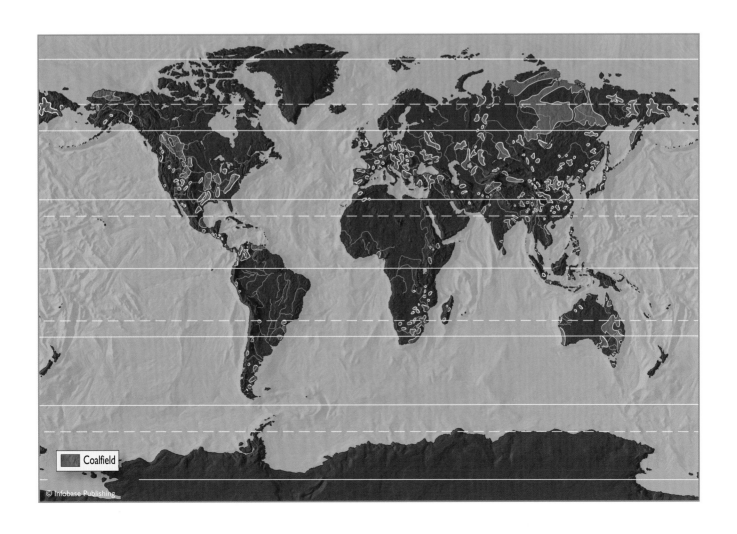

Coalfield

WETLANDS IN THE PLEISTOCENE

The present climate of the world is considerably cooler than it was before the formation of the Carboniferous coal swamps, and one of the reasons for this is that the carbon dioxide gas in the atmosphere has been reduced by the fossilization of such large masses of organic matter in the ground. When plants first invaded the land, about 400 million years ago, the concentration of CO_2 in the atmosphere was approximately 12 times greater than the present level, but this had fallen to similar levels to the present by the late Carboniferous period, 300 million years ago. The "greenhouse Earth" of early times was converted into a more temperate habitat for living things. There have been considerable fluctuations in CO_2 since then, but never rising to more than four times the current concentration.

In the last two million years, which geologists divide into late Pliocene, Pleistocene, and Holocene epochs, the Earth has experienced periodic episodes of cold that have plunged the polar regions into ice ages. Ice sheets have expanded and retreated at intervals of approximately 100,000 years. At their maximum extent, they reached beyond the Great Lakes in North America and covered much of Britain and northern Germany in Europe. Global mean temperatures varied by as much as 18°F (10°C) between the times of maximum warmth, the interglacials, and those of maximum cold, the glacials.

The distribution of the biomes of the Earth is largely controlled by climate, so these climatic fluctuations have involved shifts in the positions of the biome boundaries. Tundra has expanded in the glacial episodes and moved toward the equator, crushing the temperate forest biome in the process. Wetlands are not strictly zoned like most terrestrial biomes, but different types of wetland are associated with different climates (see "Climate and the Global Distribution of Wetlands," pages 4–9). Consequently, the shifts in the patterns of the other biomes were associated with changing types of wetlands in different parts of the world. The areas with a cover of ice sheets were devoid of wetlands, but the tundra biome on the ice sheet fringes contained wetlands now associated with Arctic conditions (see "Palsa Mires," pages 112–114, and "Arctic Polygon Mires," pages 114–116). Geologists have detected evidence for the past occurrence of polygon mires in the northern United States and in northern Europe by examining modern soil profiles. The deep, ice-filled crevices, or ice wedges, formed around the perimeter of polygons leave permanent scars on the landscape. When the ice melts, the gaps become filled with eroded soils, and if these are viewed in cross section, as when a ditch or a road excavation cuts through them, the V-shaped notches in the ancient soils are visible. Polygon mires have thus left clear signs of their former existence in areas far removed from their present locations.

Some open-water wetlands survived intact through the glacial episodes if they lay beyond the reach of the ice sheets. In southern France, for example, there are lake sediments in the craters of long-extinct volcanoes that extend back more than 100,000 years, covering the last glacial and interglacial episodes. Similar lake deposits have been found in the Tropics, in the Andes foothills, and in northern Australia. So the history of open-water sites is locally documented in these sediments.

The history of peatlands is less well known because peat-forming wetlands are most abundant in the cool temperate zone, and this is precisely where the destructive scouring of the glaciers was most active, so the majority of Pleistocene interglacial peat deposits have been destroyed by the subsequent expansion of the ice sheets. Geologists have found some fragments of peat, often occurring as peat lenses in glacial detritus, which may have been lifted and carried intact by glaciers or simply buried by glacial deposition, as shown in the photograph on page 179. These fragments provide evidence of tundra mires and are useful in tracing the past locations of different plant and insect species found fossil within them, but detailed information on the history of peatlands is restricted to the late Pleistocene and Holocene, the last 14,000 years when the ice sheets were finally in retreat.

Late Pleistocene mires in the temperate zone were almost entirely rheotrophic. They included shallow-water glacial lakes, marshes, and spring mires developing in the bare landscapes left by glacial retreat. The soils were fresh and unleached, with mineral particles recently scoured from rocks and ground into fine fragments by the action of ice. Chemical nutrients, such as calcium, potassium, and phosphorus, were therefore in relatively rich supply, and the marshes and fens were bathed in nutrient-rich waters. There is little evidence for oligotrophic, nutrient-poor mires, and fossils of *Sphagnum* bog mosses are scarce in the late stages of the Pleistocene and the early Holocene.

The early part of the Holocene, from 10,000 years ago, experienced a very rapid rise in temperature. The fossil remains of beetles in sediments from that time demonstrate that the temperature within the temperate zone rose by as much as 11°F (6°C) in a matter of decades. The climate of southern Britain, for example, was transformed from Arctic to Mediterranean conditions in only 40 or 50 years. Vegetation was slow to respond to these rapid climatic changes, but mobile creatures, such as beetles, could fly into new regions and colonize them rapidly, so their remains provide better evidence for the rapidity of local climate change at that time than do the plant fossils. Forest began to spread in response to the increasing warmth, and, although much slower than beetles, trees achieved quite rapid rates of spread. Most tree species appear to have moved northward into the retreating tundra at a pace of about 1,000 feet per year (300 m y^{-1}), which is quite fast for a tree. Evidence for

Cliff exposure in South Wales, revealing a peat band, which is a fossil wetland from a former interglacial period. *(Peter D. Moore)*

the rate of forest spread comes from the increasing abundance of tree pollen in lake sediments along their routes. The tundra habitat became increasingly scarce as the forest moved north, and between 8,000 and 6,000 years ago a forest of black spruce (*Picea mariana*), tamarack (*Larix laricina*), and paper birch (*Betula papyrifera*) invaded the far north of the North American mainland, eliminating the tundra and its associated tundra mires from all areas apart from the High Arctic islands.

Between 5,000 and 4,000 years ago, however, the climate took a sharp turn toward cooler conditions. The spruce trees of the far north became more stunted, and tundra mires began to spread back over the northern regions of the mainland of continental North America once more. The tundra in its present form began to establish itself at that time. It is likely that the climate had begun to cool even earlier, perhaps about 7,000 years ago or even before that, but it took some while for the climatic shift to affect the balance of the forest and tundra, and there was undoubtedly a significant downturn in temperature from 5,000 years ago onward. The wetlands developed within the boreal forest zone of those times resembled those of the present day (see "North American Wetlands," pages 126–128). Aapa mires ("string bogs") and acidic *Sphagnum*-dominated peatlands ("muskeg") were present within the coniferous forest.

The ombrotrophic raised bogs of the more oceanic regions of the temperate zone also began to develop approximately 7,000 years ago. They formed over the top of rheotrophic wetlands, usually marshes or swamps that had developed during the early Holocene. The invasion of

Sphagnum, and its rise to dominance in these ecosystems, is a major development in the Holocene wetlands, resulting in the creation of massive peat domes over much of the temperate zone. Ecologists and paleoclimatologists continue to argue about whether this major change in peatland types was a natural consequence of successional development or whether it was instigated by a climatic change. Undoubtedly climate was changing at that time, becoming wetter and possibly cooler. Local climate, however, was still influenced by the survival of the remnant of the great Laurentide Ice Sheet, which did not finally disappear from the region to the east of Hudson Bay until 6,500 years ago. Alaska may have achieved its warmest conditions as long ago as 9,000 or 10,000 years, while Labrador was probably at its warmest only 4,000 or 5,000 years ago. Patterns of climate history thus varied greatly with geography. Many of the temperate wetlands changed from rheotrophic to ombrotrophic hydrological systems during the period of climatic change between 7,000 and 5,000 years ago, however, and increasing climatic wetness undoubtedly contributed to these changes.

In the highly oceanic regions of western Europe, a new type of ombrotrophic wetland began its development after about 5,000 years ago—the blanket bog. Once again, the increasing precipitation and the declining temperature of that time made blanket bog development possible. In many regions that have been studied in detail in western Europe, the change from fen or wet woodland to ombrotrophic bog is often accompanied by layers of charcoal in the peat stratigraphy together with other evidence, such as indicators in the fossil pollen record, that prehistoric people were clearing forest, probably to improve grazing conditions for their domesticated stock. The arrival of pastoral agriculture was thus a final stage in the development of this new wetland type (see "Blanket Bogs," pages 107–111).

Just as the forest moved north in North America and invaded the tundra, so the prairie grasslands followed the forest and took over dominance as climatic changes brought summer heat and drought to the south of the forest and thus gave grasses a competitive advantage over trees. Many of these prairie regions still contained buried blocks of ice at that time, and these ice blocks slowly thawed to create depressions in the ground in which water accumulated and wetland succession began. The prairie pothole mires were thus created during the early Holocene, from 9,000 years ago (see "Pothole Mires," pages 100–102).

Many of the dry lands of the southwestern United States were wetter during the glacial period of the north. Death Valley in California, for example, has geomorphological features suggesting that there was an extensive freshwater lake present in glacial times, possibly 300 feet (100 m) deep. Several of the deserts of the southwest have fossil lake sediments containing the remains of freshwater mollusks, so wetlands were evidently much more extensive in these lands that are now so dry. The word *pluvial* is used to describe a period of climatic wetness in these currently dry areas. One piece of biological evidence for the existence of pluvial wetlands in the southwest of North America is provided by the distribution of a genus of fish called the desert pupfish (*Cyprinodon* species). These fish are found scattered in small waterholes and lakes in the generally dry landscape of western Nevada. In an area of 3,000 square miles (8,000 square km), there are 20 known populations of the pupfish that belong to four recognized species. There is great variation in the sizes and scale patterns of these fish, and biologists believe that they all have a common ancestor which was once widespread through the region at a time when wetland habitats were much more common and were linked with one another. The likely time is the pluvial period of the last ice age, lasting approximately from 25,000 years ago to 10,000 years ago. Since then the region has become drier and the wetlands have become fragmented and scarce. As the fish populations became isolated from one another they developed along different lines, resulting in the complex of variety of forms that currently inhabit the region.

The North American deserts were thus wet during the last glacial, and the same is true of the North Sahara and the Namib Desert of southern Africa, but not all of the world's dry lands were wet during the ice age. In much of Africa the wettest period was between 90,000 and 55,000 years ago, and the same is true of northern Australia. The dry scrublands of the eastern Mediterranean, such as the hills of Syria, were even drier than they are now during the last glacial period, as were the uplands of East Africa. Precipitation increased in these regions in the early Holocene, raising the levels of lakes and creating new wetlands in the arid areas. It is not possible to generalize for all of the arid regions of the world, therefore, but much of tropical Africa, the Middle East, and India was dry at the end of the glacial and became wet during the early Holocene. Wet tropical forests moved north into the Sahel region of the southern Sahara, pushing savanna vegetation before it even deeper into the dry wastes of the desert. In the Rajathan Desert of western India, extensive marshes of sedges developed where there had been dry valleys before. Rains over the eastern highlands of Africa brought a deluge of freshwater north along the Nile River, pouring into the eastern Mediterranean Sea and creating a stratified body of water in which low-density freshwater lay over the top of the higher-density salt water.

The early Holocene, from 11,500 years ago onward, therefore, was indeed a period when many deserts were transformed into wetlands and forests, but this was not to last. Conditions began to change once more about 5,000 years ago when the Sahara began its relentless spread south once more, and the wetlands of the Rajasthan Desert became more saline as evaporation increased. Thus the modern salty inland wetlands of the dry Tropics began to develop (see "Inland Saline Wetlands," pages 116–118). As the water disappeared from these regions and the wetlands became smaller and more saline, so the people who lived in these regions also began to withdraw. The cattle ranchers pulled out of the Sahara, and the entire Indus civilization of the Rajasthan Desert of India and Pakistan collapsed because of the impact of climate change. Wetlands and human cultures have thus responded in similar ways to the changing environment of the Earth. People and wetlands are inextricably linked.

■ WETLANDS IN PREHISTORY

People need water to sustain their bodies and also to provide for their domesticated plants and animals. It is not surprising, therefore, that early human cultures were often associated with wetland habitats. From the point of view of the archaeologist, wetlands have been a particularly fruitful resource and have yielded much information about the organization of human cultures long before the invention of writing.

The low level of microbial decay in wetland sediments means that archaeological materials are often very well preserved. Even wood artifacts, together with textiles and leather, survive in waterlogged conditions, providing a detailed record of human prehistory in the vicinity of wetlands, long before historical records become available. The abundance of wetland archaeological sites gives the impression that these habitats were particularly attractive to early human cultures, but this could in part be due to the over-representation and concentration of archaeological work on such sites. It is reasonable to suppose, however, that wetlands held many attractions for early settlers.

In the uplands of Wales, there are examples of shallow lake sediments dating from soon after the last ice age, about 8,000 to 10,000 years ago, in which are found successive and frequent narrow bands of charcoal. Evidently, the surrounding vegetation was subject to regular and repeated burning, but it is difficult in the face of such information to be sure whether this was the result of natural fires caused by lightning, or fires set deliberately by humans. The fact that the oak/elm/hazel forest of this oceanic, moist region of the world is very resistant to natural fire, coupled with the observation that such fires were restricted to certain basins and are absent from others, suggests that human activity is responsible. Why should the Middle Stone Age (Mesolithic) people of that time indulge in incendiary activities?

At that time, small, open-water wetlands in the uplands of Wales would have been particularly attractive to grazing animals, especially to the European red deer (*Cervus elaphus*), closely related to the North American elk or wapiti (*C. canadensis*). This animal was a major resource for the hunter-gatherer peoples of the period, who exploited the hunting opportunities provided by the gathering of deer around such wetlands and water holes. Burning the catchment was therefore a wise management practice, for it opened up the forest around the pool, giving clearer visibility for the hunt. It also led to the regrowth of appetizing young shoots of trees such as birch, elm, and hazel, which provided a further attraction for the deer. Thus, in Europe, long before the arrival of agricultural ideas, wetland sites were exploited as a hunting resource.

Open water also provided an opportunity for transport using primitive boats. Such a mode of travel would have been far swifter, and perhaps safer, than movement over land through dense forest and scrub. It is not surprising, therefore, that settlements on wetland fringes proved attractive to Mesolithic people. One of the best-known sites, dating back some 10,000 years, is found in northeast England at a lakeside settlement now called Star Carr. Here, the settlers seem to have taken advantage of an old beaver dam to collect ready-cut wood and build dwellings and jetties. They hunted deer, moose, wild pigs, and wild cattle, together with waterfowl, including ducks, storks, cranes, and grebes. Beaver remains have also been found from the site but, surprisingly, no fish.

Lakeside settlements dating from the Mesolithic period are recorded across Europe from Denmark through Germany and into Russia. Apart from a few pike skeletons in Estonia, however, the use of the wetlands as a food resource seems to be restricted to the hunting of terrestrial animals rather than fish. This absence of evidence for fish, however, could

be due to the fact that fish remains do not preserve as well as the bones and antlers of larger animals. Some nets and fishhooks have been found, together with a bone harpoon that was recovered from the seabed of the Dogger Bank, a shallow area of the North Sea, in the fishing gear of a modern-day trawler. The North Sea between eastern Britain and northern Germany was occupied by peatlands and brackish marshes at that time (see diagram), so Mesolithic people evidently hunted in the region.

As the glaciers and ice caps melted with the maintained warmth of the present interglacial period, so the sea level rose around the world. In the north of Europe, the North Sea became filled with water and consequently isolated Britain from the mainland. At that time, some 5,000 to 3,000 years ago, many preagricultural settlements are known to have developed around the new seashore and its attendant coastal wetlands. In eastern Denmark, for example, at a site called Tybrind Vig, a sophisticated semiaquatic settlement has been excavated that dates from this period. Cobbled slipways for boats and fish-traps made of hazel sticks have been recovered. Fishhooks were made out of bone (often the ribs of red deer) and ropes were constructed from willow fibers. Plant fibers were also woven into textiles, and the wooden paddles of boats were decorated with carvings. At this stage in the late Mesolithic era in Europe, wetland people certainly

As the ice sheets melted at the end of the last ice age, sea levels began to rise. In northwest Europe, Britain at that time was joined to the European continent, and the red line shows the coastline 10,000 years ago. As the sea rose, the low-lying lands were flooded, forming what are now the North Sea and the English Channel, thus separating Britain from the continental mainland.

© Infobase Publishing

depended heavily on fish, together with mollusks, seabirds, seals, whales, and dolphins.

In the northwest region of North America, settlements were present around the coastal wetlands 3,000 years ago. Villages were established in the region of Vancouver Island and Puget Sound, which clearly depended on fishing for their sustenance. The late-summer salmon run was evidently an attraction, but shellfish were also collected and stored in large baskets made from tree bark. Very well-preserved wooden carvings have also been recovered from the sediments. These have even included objects decorated with inlaid sea otter teeth.

In Switzerland and north Germany, prehistoric settlements around the edges of inland wetlands have been found to contain many houses with strong wooden supporting piles at their corners. Floors to the houses are no longer present, but it has been assumed that the remaining upright poles once supported the dwellings on stilts above the water surface. Current archaeological opinion, however, is that most of the houses were originally built with their floors at ground level, directly on the underlying peats, with deep piles at their corners to provide stability and support the roof. During the course of time, lake levels have fallen, leaving the dwellings perched above ground. They have consequently collapsed and been abandoned. The quaint image of "stilt villages" around the edges of Swiss lakes is therefore no longer regarded as correct. In Italy, however, some lakeside, pile-supported dwellings dating back 2,000 years have been described and their original stilt structure has been confirmed, so the controversy about these pile-supported dwellings is by no means over.

There is also the question of how long such wooden wetland villages were occupied. At Lake Annecy in southern France, archaeologists have calculated that a single oak/hornbeam/maple hut was occupied in total for about 18 years. Whole villages often seem to have been abandoned after only about 35 years, probably because local resources were exhausted. The organization of the buildings within the villages often gives the impression of being well planned, although the populations supported there were mainly quite small.

■ WETLAND BRONZE AND IRON AGE SETTLEMENTS

By the time the lake dwellings of Switzerland and Italy were being constructed, the concept of agriculture, both pastoral and arable, had spread across Europe, and this added a new dimension to the relationship between people and wetlands. A few crop plants, such as rice (*Oryza sativa*) in Asia and American wild rice (*Zizania aquatica*) in North America, together with the Eurasian water chestnut (*Trapa natans*), need aquatic conditions for their growth, but most of the early domesticated plants in Europe (wheat, oats, barley, rye) and America (maize, squash) were essentially dry land species. In time, a new, negative attitude to wetlands emerged as drainage for agriculture became a priority, but in prehistoric times the wetlands still had certain advantages to offer their human inhabitants. One of these advantages was physical security in times of intertribal violence.

The wetland village settlements of the Bronze and Iron Ages (between 3,500 and 2,000 years ago in northern Europe) were generally fitted with a high wooden stockade that completely encircled the village. No doubt this was partly intended to retain domestic animals within the limits of the village, but it also served as a means of defense when the community was under attack. Having one side, or even all sides, surrounded by water would also provide additional security since potential invaders had to arrive in open boats. Of these prehistoric settlements, the Bronze and Iron Age villages of northern Germany and Poland have, perhaps, received the greatest attention from archaeologists.

In the Friesland area of north Germany, within lowlying pasture lands now reclaimed from the sea, are found many slightly elevated hummocks of ground, rising just five–six feet (2 m) above the surrounding land and having diameters of about 500–1,000 feet (150–300 m). Archaeological excavations of these raised areas (known locally as *Wűrten*) have shown that they were artificially constructed by importing soil from the surrounding mainland, which was used to elevate village sites above the local influence of the sea. Often the soil was obtained by cutting turf from local heathlands, and as a result, it was acid, sandy, and rich in organic matter. When mixed with animal dung, the imported soil, besides giving added height to the settlement, also provided a resource for growing crops within and around the edges of the village. *Wűrten* were thus small habitable islands projecting from the flat wetland landscape of salt and brackish marshes.

Animals, including cattle, horses, sheep, pigs, and goats, were kept in these villages and grazed on surrounding salt marshes. At night they were housed in the same log houses as the people, occupying one end of the building and separated from the human family by wattle partitions. Heat from the animals' bodies may well have been welcome in winter on this bleak and windy coast. Among the cultivated plants of the *Wűrten*, cereals, beans, and flax (for its oily seed) were the most abundant. These isolated village communities must have been quite self-sufficient, although they would have needed to import such commodities as iron and timber from the mainland. They may have traded in food produce or skins, perhaps in later times with the Roman army of occupation. Many of the *Würten* were abandoned about 400 C.E., and it is thought that their inhabitants were among the Anglo-Saxon invaders of Britain following the fall of Rome. The original English were thus a wetland people from northern Germany.

In northern Poland, to the south of the present-day city of Gdansk, lies Biskupin, a remarkable wetland settlement dating from the Iron Age. It is situated on an artificial island in a freshwater lake and covers an area of about four acres (1.5 ha). Not only was it surrounded by a high palisade of wooden piles, but logs were also driven into the lake floor, projecting outward from the settlement at an angle of 45 degrees to deter approaching boats, rather in the manner of modern antitank defenses. A bridge provided access for carts from the mainland. Most remarkable were the rows of terraced houses (joined in rows end to end), resembling a Victorian urban city slum rather than a prehistoric village, and also the fortified two-story gatehouse controlling the entrance to the settlement. Much of the site has been painstakingly reconstructed, as shown in the photograph, and what is most striking is the immense quantity of prime timber that must have been used in the original construction. Archaeologists have calculated that 35,000 oak and pine trees would have been needed for the outer stockade alone. These trees must have been obtained from the surround-

ing catchment forests, and the cleared land would then have been used for agriculture to support the community, which, consisting of more than a hundred houses, could well have numbered between 500 and 1,000 people.

The very success of the Biskupin settlement may have ultimately led to its downfall. The site was once burned down but seems to have survived this catastrophe, only to be engulfed eventually by the rising waters of the lake. The extensive deforestation of the catchment that resulted from the building and maintenance of Biskupin could have been the major factor influencing hydrological changes in the basin. The removal of trees from a catchment results in greater quantities of runoff water, and this may well have caused the rising lake levels that destroyed the town. Wetland peoples evidently need to exercise care in the management of their environment.

In North America, fortified islands and peninsulas are known, but not on the scale or with the high degree of preservation found at the German and Polish sites. Settlements in the freshwater wetlands of Florida offer the

Biskupin in Poland was an Iron Age settlement on the edge of a lake. The village was constructed of wood, much of which has survived, and some buildings have been reconstructed to the original plan. *(Peter D. Moore)*

most promising opportunities to cast light on the lifestyles of the Native American wetland dwellers. In New Zealand, fortified settlements of Maori peoples that predate the arrival of Europeans have been found in marsh and swamp lands surrounding Lake Mangakaware in the North Island. These settlements consist of wooden palisades enclosing groups of about five houses that probably supported total populations of approximately 30 people.

The isolation of such wetland settlements was overcome to some extent by boat movements between them, but the construction of trackways to link settlements to the mainland and to one another is an important feature of many prehistoric wetland cultures. Some of these trackways were adequate for the transport of domestic animals and even carts. In Ireland, Iron Age tracks have been uncovered that consist of heavy planks, 10–16 feet (3–4 m) long, laid side by side for a distance of over half a mile (960 m). The changing water tables, combined with wear and tear on the wooden track, however, must have demanded frequent repair and replenishment.

In the setting of these apparently highly civilized wetland peoples, we must remember the human executions or sacrifices that led to the macabre interment of the bog bodies (see "Stratigraphy and Archaeology," pages 37–38).

WETLANDS—BARRIERS OR HIGHWAYS?

Wetlands have existed since long before our own species arrived on this planet. The first human societies probably originated among the tropical savannas of Africa, but the earliest hominids in East Africa settled along the lake edges in the Rift Valley. Eventually, people spread throughout the world, often encountering wetlands during their movements. For a bipedal, terrestrial omnivore, the wetlands were probably a source of food but also a difficult environment to exploit. An ability to swim, which enables many animals to subsist in and around wetlands, was not sufficient for technologically minded humans who used sophisticated weapons (harpoons, fishing hooks, nets, bows, and arrows) for hunting and who carried camping equipment and clothing with them on their travels. Many wetlands must have represented seemingly impenetrable barriers to the movement and spread of humankind.

In the Tropics, tall papyrus swamps with their many dangers acted as natural barriers to movement, as did the wet forested swamps of the coastal regions of the subtropics. In the more northerly temperate zone, the developing wetlands, culminating in raised bogs, were often hazardous and difficult to cross by travelers weighed down with equipment.

By the time domesticated animals were being moved from their original ranges into the geographically scattered settlements of developing civilizations, wetlands provided major barriers to movement and to trade. Despite the occurrence of such barriers, however, archaeologists are constantly discovering evidence that the urge to explore and to communicate and trade between peoples has led to the ultimate defeat of every barrier, from mountain chain to desert, and wetlands are no exception.

Some types of wetland, in fact, may have been very helpful in the spread of the human species and the development of exploration and trade among early peoples. Where open water exists, there is always the possibility that it can be exploited as a means of transport. The invention of the boat must have been a relatively obvious and simple technological innovation, providing a means not only of overcoming aquatic barriers but also a way to circumvent even greater problems represented by dense forests or difficult boglands. The wetlands, even through the course of recent human history, have often become the gateways to remote continental interiors.

Perhaps the simplest form of boat is the modified tree trunk, or log boat, in which the log is carved or burnt out to form a type of hollow wooden canoe. Powered by sails or wooden paddles, it provides a means of carrying humans, their equipment, and even small domestic animals from one place to another over the surface of water. It is difficult to determine when such boats first came into use, but it is likely that they have been used in temperate North America and Eurasia for at least 9,000 years, and it is possible that their history is much longer. The problem with reconstructing the story of their early use is that wood does not survive well unless it is preserved in the waterlogged sediments of the wetlands themselves. Fortunately, it is in precisely this type of location and habitat that many such boats ended their days, and many hundreds of log boats have been recovered in a fossil state from the wetlands of North America and Europe. Most of these date from later prehistoric times, from the last 4,000 years, but there is every reason to suppose that such craft existed also among earlier cultures.

No doubt these log boats came in a range of sizes, depending on the requirements of their makers, the availability of timber, and the time and effort required for their construction. Some log boats have been recorded that were almost 50 feet (15 m) in length, constructed from massive tree trunks (often oaks). A boat of this kind would require great effort in its construction because about 90 percent of its total wood volume would be removed in the hollowing process. Such a boat could carry 28 oarsmen together with 10 tons of cargo. Given such a heavy load, however, there would be only a few inches of freeboard above the waterline, so it would be safe only in calm waters. Log boats, therefore, were useful for moving people and equipment over relatively small open-water wetlands, along rivers and along the inshore regions of coasts.

The people of Polynesia were able to develop the idea of a dugout canoe further by the addition of outriggers to stabilize the hull even in the face of ocean swell. In this way they perfected long-distance travel by sea. Their craftsmanship also provides evidence of the quality of work possible with the aid of only stone and bone tools. In 1772, the French explorer, navigator, and botanist Louis-Antoine de Bougainville (1729–1811) expressed the view that the Samoans must have used steel tools to achieve such fine work, but he was wrong. Islands rich in tall trees, such as Kabara Island in Fiji, were specifically favored by these people. Their artifacts give some indication of the potential for boat making even among Stone Age cultures.

Alternatives to the labor-intensive process of hollowing out logs were available, however. The stretching of skins over timber frames is one of these techniques, and it is likely that such canoes, or *coracles*, also have a very long history, especially in areas such as the Arctic where trees are few. The survival of evidence for such boats is rare, however, and few remains have been found. In Europe, traces of what were probably Bronze Age coracles have been excavated. The use of bark instead of skin was extensive in both northern Europe and North America, particularly where trees (especially birch) were present, but canoes constructed this way did not achieve the size of log boats. Another alternative was to use planks for boat construction rather than whole tree trunks. This demanded very different tools and techniques, but it is an art that seems to have been mastered in Europe by about 1500 B.C.E. The planks were bound together by strips of willow, and the gaps between planks were plugged with mosses.

Propulsion of the boats was achieved by paddles or by poles, and fossils of both have been found in Finland that date back some 4,000 years. These objects were evidently treasured, being sculpted and decorated in an artistic manner that identifies them as more than merely functional.

In the case of high-latitude wetlands, covered by layers of ice in winter, there was the further opportunity to travel over the ice by foot, dragging sledges with equipment. The prehistoric remains of these sledges (dating back about 9,000 years), together with short broad skis (dating back around 4,500 years), provide irrefutable evidence that the frozen wetlands were a means of convenient and rapid transport in the untamed wilderness of the prehistoric world.

Evidence suggests, therefore, that the wetlands provided major transport systems for prehistoric people. Perhaps it is not surprising, therefore, that when Europeans began to explore the interiors of the dark continents of Africa, and North and South America from the 16th century onward, they too relied heavily on rivers and wetland systems as a means of access.

■ WETLAND EXPLORATION— AFRICA AND THE NILE

Of the many stories of endurance and perseverance in wetland exploration, there are few to match that of the White Nile in Africa. The lower reaches of this river in Egypt had been well known to the civilized world since the earliest of historical times. Indeed, the Nile River valley and delta were major focal points of early civilization, but where this great river began was still shrouded in mystery even in the middle of the 19th century. The source of the Nile, at that time, was regarded in geographical circles as the greatest unsolved secret remaining on Earth since the discovery of America.

This secret had survived millennia of exploration and speculation. The Greek explorer Herodotus (c. 484–c. 407 B.C.E.) traveled up the Nile about 460 B.C.E. and reached the first cataract at Aswan (see the map) but was able to go no farther. Two soldiers sent by the Roman emperor Nero had passed this point, but returned from the Sudan with stories of a great swamp, probably the Sudd of southern Sudan, which they regarded as impenetrable. Shipwrecked sailors and travelers from the east coast of Africa told of the existence of two great lakes some great distance inland, where the Nile was said to have its origin, and it was probably such tales that prompted the second century C.E. geographer Ptolemy to draw a speculative map showing these lakes, together with the source of the Nile in a mountain range that he dramatically called the Mountains of the Moon.

For more than 1,600 years no further progress was made in this affair. All of the great continents of the world had been discovered, and European colonialism and settlement had spread through most of them, but East Africa in the 1840s was still unknown to Europeans apart from the few intrepid Christian missionaries who had begun to settle in the so-called Dark Continent. It was they who sent home unlikely reports of high, snowcapped mountains in the equatorial regions (Mount Kilimanjaro and Mount Kenya), which roused new interest in Ptolemy's "Mountains of the Moon." It was also from these missionaries, together with the Arab slave traders of the coast, that news came of great inland lakes, including Lake Nyanza (later called Lake Victoria), thus giving further credence to Ptolemy's map.

In 1856 two British explorers, Richard Burton (1821–90) and John Speke (1827–64), set out to solve the mystery of the source of the Nile. They had the choice of two routes. The direct route followed the course of the river south from the Mediterranean past the Aswan cataracts, across the Sudan desert, past the confluence of the Blue and White Niles at Khartoum, and on into and beyond the impenetrable wetlands of the Sudd. The alternative was to start at the other end, to travel inland from the East African coast and explore the reported lakes. They chose the latter.

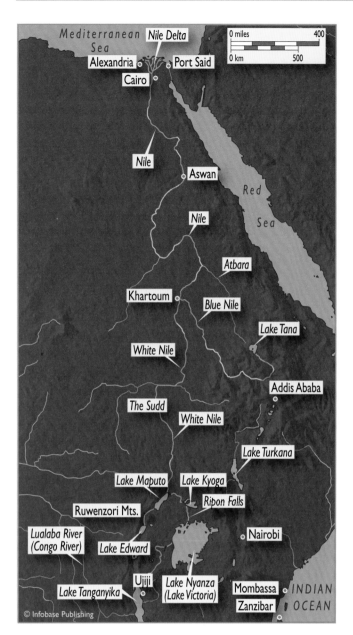

Map of the Nile River and surrounding lands. The Nile, its marshes, and its swamps were the focus of many explorations during Victorian times.

Heading east from Zanzibar, they eventually reached one of the lakes, Lake Tanganyika, but realized that the altitude of this wetland was too low to be the source of the Nile. Disappointed, and both seriously ill, they turned back toward the coast, pausing to recuperate at an Arab trading station at Kazeh. It was here that Speke decided to take a trip to the north to investigate rumors of another large lake, Lake Nyanza. He left Burton at Kazeh and headed for what was later known as Lake Victoria. After he had spent many days trekking over dry scrubland, the landscape began to take on a more humid appearance, with waterholes and swamps, eventually opening onto the vast expanse of the

lake itself. Far too wide to see the farther shore, its margins clothed in rich swamps of papyrus, the lake excited Speke greatly. In a sudden, intuitive reaction, he declared this lake to be the source of the Nile, even though he had no evidence for such a claim. After just three days at the lake, Speke returned to Kazeh, where Burton was awaiting him, and announced his discovery. Quite naturally, given the total lack of proof, Burton refused to accept that there was any scientific basis for claiming that this immense lake was the source of the Nile.

On Speke's return to London (Burton delayed his return while convalescing at Aden), the Royal Geographical Society listened to his claim with excitement. They immediately financed a further expedition for Speke to return to his lake, travel around to its north side, discover the presumed exit of the Nile, and, they hoped, follow it north to Egypt. For this second expedition Burton was not invited but was replaced by an army officer, James Grant. By 1860, the new team had returned to Lake Nyanza and skirted its western shore into the region that is now Uganda. They lived there for many months in the company of the local king, Mutesa, and it was not until July 1862 that Speke was able to make the final stage of the journey and discover the northern location of a great river's exit from the lake. Pouring over an impressive waterfall, 700 yards wide, the river descended through crocodile-infested rapids and made its way northward. There was no doubt in Speke's mind that this was the Nile, and Lake Nyanza was its source. What he was unable to do at this time was to follow the river on into Sudan and confirm that it was the same river that entered the great swamps of the Sudd. For 50 miles below the falls, the river could not provide safe transport to canoes, so trekking northward over land left the vestige of a doubt whether the river Nile that they later joined was indeed the same waterway they had left at the lake.

Farther downstream, however, they met European explorers heading up the Nile to meet them, and Speke returned to England in great triumph. He was greatly (and rightly) praised for his endurance and enterprise in solving the greatest of wetland mysteries, but Burton's voice was still raised in doubt. All that Speke had achieved, Burton contended, was the discovery of a great lake with a river that led out to the north. There was no absolute proof that this river was the Nile. The disagreement became heated and, at times, malicious. In September 1864, the debate was to be aired publicly at the meeting of the British Association for the Advancement of Science, held at the English country town of Bath. Here, both Burton and Speke were to deliver papers on their findings and their views. Each man arrived at Bath and encountered the other without exchanging a word. Then, on the day before the debate, Speke went out to relax with friends, hunting in the surrounding countryside. He never returned. He was discovered with a gunshot wound close to the heart and died almost immediately after he was found. No one witnessed the accident, and the possibility of suicide under such great emotional stress has never been fully ruled out. The irony of the whole story lies in the fact that Burton was wrong and Speke was right in his assertion that Lake Victoria is indeed the source of the Nile.

■ LIVINGSTONE, STANLEY, AND THE CONGO

Much of Central Africa is drained by the Congo River, a river of almost 3,000 miles (4,700 km) in length. Unlike the Nile River, this river remained almost completely uncharted, even over its lower reaches, until the 19th century. Its mangrove-choked mouth on the west coast of the continent epitomizes the mystical and impenetrable qualities of "darkest Africa." The mangrove wetlands of the Congo estuary were first discovered by Europeans in 1482, 10 years before Columbus sailed the Atlantic. The Portuguese explorer, Diogo Cão, first found the estuary but was unable to navigate far up the river because of the rapids encountered quite close to the river mouth, now known as Livingstone Falls. Consequently, the river and its interior remained unexplored by Europeans right into the mid-19th century.

The exploration of the Congo will always be associated with the American, Henry Morton Stanley (1841–1904) who, because of his Welsh birth, was eligible for the British knighthood that he eventually received in 1899 from Queen Victoria. Stanley's name, however, is also inseparable in the public mind from that of David Livingstone (1813–73), and it is with this Scottish missionary that an account of the penetration of the wetlands of interior Africa must begin.

Livingstone's river was the Zambesi. In 1855 he traveled down the river from the north in a canoe, becoming the first European to set eyes on what he called "Victoria Falls." He attempted to measure the height of this cascade, which takes the Zambesi from the central plateau of what is now Zambia into a gorge of basaltic rocks and on to the lowlands of southeastern Africa. He estimated the height of the Victoria Falls at 354 feet (105 m), twice the height of North America's Niagara Falls, by means of a strip of white cotton cloth weighted down with bullets, which he lowered over the edge of the precipice. Between 1858 and 1864 he also managed to make his way into central Africa by ascending the Zambesi from the coast and so reached Victoria Falls from the Zambesi estuary. David Livingstone was a medical doctor and a Christian missionary, one of whose greatest concerns was the defeat of the slave trade. He was also a scientist, concerned with gathering more information about the geography of central Africa and its habitats, including its wetlands.

Livingstone set off for Africa again in 1865, financed by the British Foreign Office and the Royal Geographical Society, with the intention of settling the question of the origin of the Nile (still disputed following the death of Speke). He decided to go alone, although he had invited one of his former colleagues, John Kirk, to accompany him, only to be declined. Livingstone was a solitary man and, by all accounts, a difficult companion on such expeditions. Three years after setting out from Zanzibar, he turned up at Ujiji, on the shores of Lake Tanganyika, in poor physical condition and sick with malaria. Communication with the outside world proved impossible because the Arab slave traders of the area refused to carry his letters, which, they correctly believed, contained information on their atrocities against the peoples of central Africa. From there he continued his journey eastward, despite his bad health, and discovered a river (the Lualaba) running north, which he believed to be the Nile, but which we now know to be the headwaters of the Congo. He finally returned to the village of Ujiji, where he remained until his famous meeting with Stanley on November 10, 1871.

Meanwhile, back home, Livingstone was believed by many to have died in interior Africa. London's Royal Geographical Society, however, launched an expedition to try to locate the lost explorer, but public interest was so great that the media were also concerned enough to join the search, for explorers at that time were always newsworthy. In 1869, James Gordon Bennett, Jr. (1841–1918), financial controller of the *New York Herald,* launched his own expedition to hunt for Livingstone and put in charge of it the journalist Henry Morton Stanley. Stanley (real name Rowlands, a Welshman who had developed his remarkable career as a reporter during the Civil War, adopting American citizenship) set off from Zanzibar and trekked east into Africa. The book he subsequently wrote concerning this journey (*How I Found Livingstone in Central Africa*) describes in vivid detail days of wading neck-deep in swamps, the perils of crocodiles, and the constant attacks of malaria. His final achievement of finding Livingstone alive, and his much-quoted greeting of "Dr. Livingstone, I presume," have ensured immortality for this pioneering journalist-explorer.

Stanley was a ruthless yet businesslike explorer. He undoubtedly sought personal fame and fortune rather than the advancement of science or the spread of the Christian religion or even the extension of the British Empire. His lack of high-minded idealism led to a degree of disfavor in Victorian society, but this did not in any way deter him from becoming one of the greatest of the African explorers. Despite the attitude of London society, in later life he again became a British citizen, perhaps to ensure that he would be eligible for the knighthood he eventually received. Although the meeting with Livingstone may be his most remembered achievement, his greatest triumph in terms of exploration was his travels on the River Congo.

In 1874, financed by the *New York Herald* and London's *Daily Telegraph,* he returned to Zanzibar to undertake one of the best-planned and best-equipped expeditions of the times. His first objective, which he achieved, was to circumnavigate Lake Nyanza (Lake Victoria) to check the exit streams. He found that Speke's discovery of the solitary river leaving the lake at the Ripon Falls was correct, although he hesitated to assert that this was indeed the Nile. There was yet the possibility that Livingstone's northward-flowing river might be the true Nile. On the west side of this lake, in modern Uganda, he also discovered the Ruwenzori Mountains, the "Mountains of the Moon" of Ptolemy's ancient map, and one of its peaks became known as Mount Stanley. Here, among the constant cover of cloud, were the as yet undiscovered equatorial blanket bogs.

Stanley then returned to the village of Ujiji, where he had met with Livingstone, and set out from there to retrace the old missionary's journeys. Finding Livingstone's Lualaba River, he followed it northward and determined to stay with it until it reached the sea, be it the Mediterranean or any other. In fact, the river turned out to be the Congo, and Stanley's journey through its tropical wetlands and forests makes a harrowing tale. He was attacked by the native peoples of the lands through which he passed, had all his supplies and medicines stolen, and was shipwrecked. Many of his colleagues were drowned or died of sickness during the journey; only 114 of the original 356 survived. Stanley was the only European to survive. Almost three years after leaving Zanzibar, he arrived at a settlement among the mangroves of the mouth of the Congo on the Atlantic coast of West Africa.

With the completion of this journey, the major features of African wetland geography were determined. The Ripon Falls at the north end of Lake Victoria must indeed be the source of the Nile, as Speke had intuitively asserted. The Lualaba River, on the other hand, which Livingstone had considered a possible contender for the headwaters of the Nile, was in fact a tributary of the Congo, the major African river draining into the Atlantic. The basic hydrological pattern of the great African wetlands was now established, and their detailed scientific study could begin.

■ WETLAND EXPLORATION— SOUTH AMERICA

Second only to the Nile River in length, at more than 4,000 miles (6,440 km), the Amazon River dominates the drainage system and the wetlands of South America. Its mouth, opening into the Atlantic Ocean, was discovered and explored at the beginning of the 16th century, when a Spanish navigator, Vicente Yáñez Pinzón (1460–1524),

who had previously sailed with Columbus, traveled in this region, beginning in the year 1500. So vast was the waterway he found that Pinzón considered it the entrance to an inland, freshwater sea.

The Spanish exploration of the South American interior soon led to the discovery of the Andes mountain chain, and further exploration of the Amazon River commenced from the western tributaries rather than proceeding upriver from its delta. It was a Spaniard, Francisco de Orellana (1511–46), who first managed to sail down the river (actually starting on the River Napo tributary) from Peru in 1541. Under the leadership of Gonzalo Pizarro (1506–48), with the intention of finding the fabled golden city of El Dorado, the expedition started out from Ecuador but became entangled in the Peruvian jungles. Running short of food, the explorers sought a way out by building a boat, and a group led by Orellana headed downriver, hoping to find a source of supplies. In their travels they encountered many difficulties, not least of which were the local tribes living in the wet forests of the region. Many were hostile to the strangers, and one armed and aggressive tribe of naked warriors made a particular impression on the Spaniards because it appeared to be entirely composed of women. When this story was recounted in later times, it naturally caused comparisons to be made with the Amazons of ancient Greek mythology, a race of warrior women of the Caucasus region, from whose queen Hercules had reputedly been commanded to steal a girdle as one of his superhuman tasks. It was this comparison that eventually led to the naming of the Amazon River, although Orellana had called it *El Río Mar,* the sea-river, because of its immense dimensions. Despite encounters with wetland tribes and wildlife, Orellana achieved an extraordinary feat and eventually emerged at the delta, subsequently making his way to the Caribbean. Pizarro and his remaining party, meanwhile, had given up all expectation of Orellana's return and made their way by foot back to Ecuador.

Unlike the Nile River, whose course is relatively straight and simple and whose tributaries are few (the Blue Nile being the principal one), the Amazon has numerous tributaries and a complex drainage system. Seven of its tributary rivers exceed 1,000 miles in length, so its drainage basin is enormous, and the volume of the river's flow is the greatest of all the rivers of the world (three times that of the Congo, 10 times that of the Mississippi, and 60 times that of the Nile). One-fifth of the freshwater of the world that is discharged into the oceans emerges from the Amazon. Such is the scale of this river and its associated forests and wetlands that much of it is still little known, and only a small fraction of its diversity of wildlife has yet been scientifically described. The exploration has continued with such remarkable people as Margaret Mee (1909–88), a botanist, artist, and conservationist, who spent the latter part of her life traveling and recording the flora of this region.

Among the other great wetlands of South America, the Pantanal has long been recognized as remarkable for its wildlife. It was a region that attracted the attention of European settlers because its annual flooding led to high productivity. The native peoples of the area lived by corn cultivation, together with hunting and fishing in the wetlands, and early European settlers (Spanish and Portuguese in the 17th century) followed these patterns. The high grass productivity immediately following clearance demonstrated that this region had a great potential for cattle ranching, and during the 18th century this development became extensive.

Charles Darwin (1809–82), during his famous voyage aboard the HMS *Beagle,* did not penetrate so far inland as the Pantanal, but in 1833 he visited the Parana wetlands and was able to row a small boat through the flooded swamps. "I rowed some distance up this creek. It was very narrow, winding and deep; on each side a wall 30 or 40 feet high, formed by trees intwined with creepers gave to the canal a singularly gloomy appearance." In these narrow channels, the local Indians told him stories of the jaguar and its attacks on capybara, horses, and people, and it was here that he first saw the remarkable "scissorbeaks" (skimmers), flying low over the surface of the water with their lower mandibles submerged.

Darwin was also among the first biologists to describe the blanket mires of Tierra del Fuego at the southern tip of South America: "The trees reach to an elevation of between 1,000 and 1,500 feet, and are succeeded by a band of peat, with minute alpine plants." The cold, windy, wet weather, however, precluded any lengthy exploration of these wetlands, and he seems to have been more fascinated by the local peoples, whose eyesight he found to be acute but whose reticence he mistook for lack of intelligence.

The high-altitude wetlands of the Andes, such as Lake Titicaca, on the borders of Peru and Bolivia, had been discovered and described much earlier, mainly as a consequence of the Spanish search for gold in the mountain spine of South America. By 1535, the conquistadores had subjugated the Inca of Peru. In that year, a party of more than 500 set out under the leadership of Diego de Almagro (1475–1538), heading south through the Andes and enduring the intense cold and the effects of the high altitude, including dizziness, sickness, and extreme fatigue in the limbs. When they eventually descended into the coastal lowlands of Chile, they had failed to find El Dorado or, indeed, any significant source of gold, but they had discovered the great lake of Titicaca high among the mountain peaks at 21,000 feet (6,400 m). Despite its great altitude, people (the Almara) living around its shores regarded the lake as sacred, and they grew what was to become an important crop in the temperate regions of the world—the potato. These local people constructed rafts and canoes of reeds and used the wetlands as a means of transport, which inspired later immigrants to carry up

to this high-level lake the parts of a steamship and to construct it for service on the lake. The first voyage took place in 1862.

The wetlands of South America are still greatly in need of exploration and detailed study. New discoveries are still being made. It was only in 1935, for example, that an American pilot, Jimmy Angel, flew a light biplane over southern Venezuela and discovered the world's highest waterfall (3,200 feet [980 m]), later called the Angel Falls in his honor. The little-known region around Mount Roraima, which was first visited by the English explorer Sir Walter Raleigh (1552–1618) in 1595 (also in search of El Dorado), was subjected to botanical survey by the British botanist Everard Thurn (1852–1932) nearly 300 years later in 1884. His report was the inspiration for Sir Arthur Conan Doyle's 1912 novel of prehistoric survival, *The Lost World.* The region can hardly be described as lost, but it certainly remains largely undiscovered.

■ WETLAND EXPLORATION— NORTH AMERICA

The idea of exploration and discovery may seem an essentially patronizing and colonialist one. The people of Uganda were perfectly familiar with the Ripon Falls and Lake Nyanza (Victoria) long before John Speke appeared on the scene. The native peoples of the Amazon delta would have been mystified at the concept of the Spanish sailors "discovering" the great river. Nevertheless, the knowledge that water passing over the Ripon Falls might eventually flow past the pyramids of Egypt and into the Mediterranean Sea, or that the Amazon River arose 4,000 miles away in the Andes, was not available to these peoples, and such discoveries did depend upon the technology, together with the bravery and intense motivation, of the explorers. The wetlands of North America were equally well known, of course, to the native Americans, but the development of a wider understanding of their relationships, both spatial and biological, began with the spread of European peoples on this continent.

European settlement of the eastern seaboard of America, beginning in the 16th century, brought the colonists into immediate contact with wetlands, such as those of Chesapeake Bay in the south and the St. Lawrence lowlands in the north. They found the wetland fringes inhabited by Native American peoples, who used the ecosystems' productivity to support their hunting, gathering, and cultivation of wild rice and other crops. European diseases decimated these native populations who had no immunity to them, and the growing immigration from Europe led to a relentless western movement of the new arrivals. By the

17th century, European explorers were wandering through the wetlands of the upper Mississippi valley. French Jesuit missionaries, soldiers, trappers, and traders led the way into the Midwest where they found very low population densities of the native peoples. Perhaps that population had already declined as part of the great population crash resulting from European contact.

Big game animals were abundant in the Midwest, especially in the vicinity of the wetland areas, and one young French soldier, Pierre Deliette, recorded in 1688 how he accompanied a band of Illinois on a hunting expedition. They tracked a herd of bison more than 60 miles between marshes and water holes, eventually bagging a total of more than 1,300 animals, drying their meat over fires after the kill. Animal furs were also a great attraction for Europeans seeking to make a living from the new lands. Even before the movements of white people to the west of the Great Lakes region, Hurons from the St. Lawrence region were bringing back pelts for trading, but by 1660 Europeans were making their own expeditions and reaping the rewards. The French hunters Groseilliers and Radisson, for example, explored Wisconsin and Minnesota and brought back canoes loaded to the gunwales with rich cargoes of skins. The use of the rivers and wetlands as a means of transport for hunting and exploration was a feature of the times. The rivers and lakes of the North American wetlands also permitted further penetration into uncharted territory. In this way Jean Nicolet arrived at Green Bay, Wisconsin, in 1634. Starting from the Great Lakes, trade routes developed along the major rivers, the Illinois, Missouri, and Mississippi to the south, and from Lake Superior to Lake Winnipeg and the Saskatchewan River into the interior of Canada to the north. From there, the Nelson River gave access to Hudson Bay and the great fur resources of the boreal and tundra regions.

There was, at that time, a great hope that an aquatic route through to the Pacific Ocean would be found, either by means of a great westward-draining river or by the much sought-after Northwest Passage sea route. In 1789, Alexander Mackenzie (1755–1850), a Scottish employee of the North West Company, was instructed to explore for a river route, setting out from Fort Chipewyan on the western shore of Lake Athabasca in Alberta. The first rivers he encountered flowed northwest into the Great Slave Lake, but from there he was delighted to find a major exit river heading to the west. He set off along it, together with his party of Native Americans and French Canadians, with every hope of arriving at the Pacific. This was not to be, however, for the Rocky Mountains arose before them and the river was deflected northward. The river became narrow and fast-flowing in places, such as the Rampart Rapids, where it flows through a seven-mile-long gorge, 200 feet in depth, before eventually opening into the Arctic Ocean. He called it the "River of Disappointment," but it has since

become known as the Mackenzie River. The vast tundra wetlands of the Mackenzie Delta, however, proved to be rich in wildlife that his company could plunder, from beavers to whales, so his river was no disappointment to his employers. The warmth of the summer season was short, but the Mackenzie River provided a transport route to the wetlands of the High Arctic.

Apart from hunting animal skins for profit, the exploration of the American wetlands as a wildlife and biodiversity resource was also under way by the early 19th century. Among the best known of the early American biologists was John James Audubon (1785–1851), who became known as the father of American ornithology. On Christmas Day in 1810, he was to be found with a party of Native Americans near the mouth of the Ohio River, hunting for trumpeter swans. In a single day, they managed to shoot more than 50 of these birds, whose skins were sold to supply the increasing demand for swansdown in Europe. It may seem an odd way for a naturalist to spend Christmas Day, or any other day for that matter, but Audubon used such opportunities to study and draw the birds, and the concept that human predation could have any lasting impact on wild animal populations would have been entirely strange to him.

In the southeastern corner of North America lay the last of the great American wetlands to be discovered and explored by European settlers. The Spanish explorers of the 16th century were familiar with what is now Florida but did not investigate the forested swamps of its interior. Spain nevertheless laid claim to this land until early in the 19th century.

The Everglade swamps became occupied by Seminole Indians in the 18th century as a result of the breaking up and migration of the Creek peoples of Georgia and Alabama. In the early 19th century these swamp-dwelling Native Americans befriended runaway slaves and thus found themselves at enmity with the U.S. Army. It was General (later President) Andrew Jackson (1767–1845) who eventually led an army into the swamps in a punitive expedition and thus started the First Seminole War of 1817–18. Once he became president of the United States (1829–37), Jackson ordered the relocation of the Seminoles and thus created the "Trail of Tears" as whole families were forced to march west to beyond the Mississippi River. Some militant Seminoles remained in Florida, led by Osceola, who for a while succeeded in resisting attempts to eliminate his residual band in a series of military encounters now known as the Second Seminole War of 1835–42. In 1837, Osceola was captured and died in captivity, but the war continued, and an estimated 1,500 white soldiers lost their lives trying to remove the Seminoles from the swamps. Even after the Third Seminole War of 1855–58, many Seminole families held on to their rights as residents of America's greatest wetland, the Everglades.

■ WETLAND EXPLORATION— SOUTHERN EUROPE

Since the terms *discovery* and *exploration* are often confined to the arrival of people of European origin into areas where they had not previously ventured, it follows that such activities could hardly occur within the continent of Europe itself (with the possible exception of high mountain peaks). As far as wetlands are concerned, their presence in Europe has long been well known, and most human effort was focused on getting rid of them. What has taken place in Europe, particularly in the 20th century, is the spread of awareness of the importance of wetlands for conservation of wildlife. The modern-day "explorers" have been those people who have traveled into the more remote wetlands, appreciated their great value, and sought to convince others of their worth.

A good example of this process is the Coto Doñana region on the southwestern tip of Europe, in southern Spain (see photograph). It lies on the northern side of the estu-

ary of the Río Guadalquivir, a river that drains westward through southern Spain, passing through the ancient cities of Córdoba and Seville. The northern side of the Guadalquivir as it approaches the sea is protected by a southward-running ridge of shingle (seaborne gravel) and sand dunes, and behind these are the extensive marshes, the *marismas,* that form one of the largest wetlands in western Europe. The area is flooded by winter rains but becomes dry through the hot summer months, when extensive mudflats are baked by the sun, turning the wetland into a desert. Other habitats abound in the area, including valley mires and small lakes on tributary rivers, moist fens between the dune ridges, woodlands of pine and of cork oak, both in the dunes and over the drier sandy flats, where heathland and scrub offer summer shelter to deer, wild pig, lynx, and semi-wild cattle, some of which end their days in the bullring at Seville.

The richness and diversity of the wildlife of this area, especially the wildfowl, deer, and boar, did not go unnoticed in the course of history. The region is mentioned in documents dating back to the 14th century and was then owned by the duke of Medina Sidonia, whose main residence was at Sanlucar, to the south of the river. Its potentialities as a hunting reserve, coupled with its infertile soils and its liability to flooding, which rendered it useless for agricultural reclamation, contributed to Doñana's survival in a reasonably natural state—an unusual occurrence for any ecosystem in Europe. The name *Doña Ana* (Lady Ana) was undoubtedly derived

The Coto Doñana wetlands in southern Spain are among the richest in Europe, but when drought strikes the effects on wildfowl populations can be catastrophic. *(Peter D. Moore)*

from one of the duchesses of Medina Sidonia, but precisely which one is uncertain; *Coto* means hunting reserve, so the region became known as the Coto de Doña Ana.

It is said that oaks from the drier regions of Coto Doñana were used in the construction of the Spanish Armada, which attacked Britain in 1588. This is by no means impossible, but fossil evidence from the pollen deposited in the sediments of these wetlands has not confirmed such an event. The present-day patches of oak woodland seem to be of recent origin, perhaps less than 200 years old.

The resident and migratory game of the reserve was hunted not only by the household of the duke but also by the king of Spain himself. There is a very full documentation of the visit in 1624 of King Felipe IV to the region, in whose honor the hunting lodge, or Palacio, in the center of the reserve was fully refurbished and rebuilt in the form in which it still survives. Sadly, the duke of Medina Sidonia of that time suffered a stroke during the hectic preparations for this visit and was unable to attend the arrival of the king, but 12,000 others were able to do so, at the village of El Rocio on the northern edge of the reserve. Women (with the exception of some traveling "entertainers") were forbidden to attend such a hunting party. The hunting seems to have consisted mainly of bullfights, duck-shooting, and the somewhat staged shooting of wild cattle, but the king declared himself thoroughly entertained.

Subsequent visitors to the Palacio of Doñana included the Austrian royal family and the great Spanish painter Francisco de Goya (1746–1828). Then, in 1901, the Medina Sidonia family sold the property, and the new owner began to recover his investment by selling pine timber from the reserve. Hunting continued, the rights being leased to a small group of Spanish and British sportsmen who, fortunately, were interested in the natural history of the area as well as in harvesting some of its larger mammals and birds. They recognized the significance of the geographical location of this reserve, its climate being essentially North African in character leading to a distinctly African element in its flora and fauna. They also speculated on its importance as a migration route for birds from Europe making their way to Africa for the winter. Subsequent management by a succession of Spanish owners in the 20th century has continued to concentrate on the hunting potential of Doñana. This has led to a healthy concern for habitat conservation and has meant that development for farming, eucalyptus plantations, or tourism, has not taken place.

The need for scientific exploration became apparent following an expedition to the area in 1952 by François Boulière, a Frenchman, Roger Tory Peterson, an American, and an Englishman, Guy Mountfort. They literally mounted this expedition, for the only access to this very remote part of Spain was on the back of mules. These three renowned birders rode to each of the major habitats of the reserve, concentrating on the avian aspects of Doñana, which were indeed spectacular, with flamingos, azure-winged magpies (*Cyanopica cyana*), and imperial eagles (*Aquila heliaca*) leading the list. In subsequent expeditions, they took along Eric Hosking, one of the world's greatest bird photographers, who was able to obtain pictures of several bird species that had never previously been photographed. With his help, Mountfort put together a classic account of this virtually unknown corner of Europe in their book, *Portrait of a Wilderness* (1958).

This book had a high impact on the increasingly conservation-aware public in Europe, and pressure began to rise for the protection of Coto Doñana. Eventually, in 1965 this resulted in the establishment of La Estación Biológica de Doñana, a protected area of 26 square miles (6,800 ha) with its field laboratory situated in the old Palacio building, and this formed the focus of a concerted research effort to understand further the varied ecosystems of the region. In 1969, the National Park of Doñana was created, protecting 135 square miles (35,000 ha) of this unique wetland. This has been an important step in the conservation of the area but has by no means ensured the survival of this wilderness. There is pressure from the tourist industry to develop the coastal strip and build concrete hotel complexes like those on the Mediterranean coast of Spain. Tourism also carries a demand for road development, which opens up the area for forestry, especially the expansion of eucalyptus forestry in this hot, dry region. The site also illustrates the problem of catchment management in wetland conservation, for the water supply demanded by the city of Seville and the agricultural developments upstream of Doñana are resulting in the extraction of water that deprives the wetlands of their required winter flooding. There is also the danger of industrial pollution, particularly from mining waste, some of which has recently spilled over into the rivers draining into the wetlands. The exploration and scientific appreciation of Coto Doñana may be relatively complete, but its continued survival is still in the balance.

■ WETLANDS IN LITERATURE— IDYLLIC IMAGES AND MOVING WATERS

The changing public perceptions of wetlands during the course of history are often reflected in the art and literature of the time. Perhaps, to some extent, attitudes may even have been determined by these expressive and influential media. Water is an essential requirement for life, so it is natural that wetlands should come to represent the sources of sustenance and life in the public imagination, but this is not universally

Still waters can instill a sense of peace and calm in the observer, but they can also be a source of fear.   *(Victor Kapas)*

true of the way in which they are portrayed, for wetlands can also be associated with disease and death, leading to a very negative representation in literature and in art.

The varying attitudes toward wetlands that are found in literature bear some relationship to the nature of the wetlands themselves, although there are numerous exceptions to any attempted classification. In general, early successional wetlands with open water that flows relatively freely (see photograph) are more attractive to the human imagination than later successional ecosystems with tall, dense, emergent vegetation and slow-moving or stagnant waters, or indeed to the raised and blanket bogs with open vistas, motionless murky pools, and an unknown depth of peat lying below the surface.

This preference for rheotrophy rather than ombrotrophy and for flowing rather than stagnant waters has been evident since the earliest literature and is not difficult to understand. The Bible, for example, uses water imagery both literally

and metaphorically in its teachings, as where Moses strikes the rock in the Sinai Desert at Meribah to bring forth water (Exodus 17:6) and an overheated deer seeks out flowing water for refreshment (Psalm 42:1). These are very real images of thirsty creatures in need of water, but they also have metaphoric value as pictures of spiritual need and fulfillment, a representation that is used again when Jesus promises "living water" to his listener (John 4:10). The use of the term *living* here has many spiritual connotations, but in its most obvious sense it clearly portrays moving water, a spring or a fountain, rather than a stationary or stagnant pool.

Moving waters, of course, whether in the dry lands of biblical and classical settings or in the more tropical or temperate areas of the world, are more likely to provide uncontaminated supplies that are safe to drink. This is especially true if they are encountered at the point where they first emerge from the ground. Flowing streams tend to contain lower concentrations of toxic chemicals, fewer parasites, and a lower content of algae, protozoa, and bacteria, all of which can cause disease. It is not surprising, then, that fountains and springs have been accorded a magical purity and power in so much literature. Edmund Spenser (1552–99) expresses the idea in *The Faerie Queene* (1589): "There was a springing well from which fast trickled forth a siluer flood full of great vertues, and for med'cine good." The association between such sites of moving water and the presence of oracles, sybils, and soothsayers, is a clear development of the purity image, and is frequent in classical literature. The notion is taken up by romantic writers of the 19th century, such as Sir Walter Scott (1771–1832), who often places admonitory apparitions and advisory spirits in the vicinity of springs and fountains, as in *The Bride of Lammermoor* (1819).

Fountains and springs, however, may not always be as pure as they seem. For example, Hylas, of classical mythology, was one visitor to such a fount who suffered as a consequence. The aquatic nymphs who resided in the pool (a favorite haunt for nymphs, incidentally) seduced him into joining them in the pool, and he was never seen again. The Roman poet Ovid tells the story in *Metamorphoses*, Book III, of vain Narcissus who had played with the affections of many (both female and male) and who subsequently fell under a curse from those whom he had hurt. One day he came upon a clear pool, with shining silvery waters, where he stooped to drink and immediately fell in love with his own reflection. Needless to say, such misplaced affection brought him little satisfaction. The mysterious properties of springs and fountains, therefore, are not always benign. In Spenser, once more, we have an example of an evil fountain in which wicked powers are associated with a reduced rate of water movement: "And bad the waters, which from her [the Nymph of the fountain] did flow. . . . Thenceforth her waters waxed dull and slow, and all that drunke thereof, did faint and feeble grow."

The Victorian writer and social reformer William Morris (1834–96) wrote some gripping allegorical novels in a fantasy mode, perhaps the most notable being *The Well at the World's End,* published in 1896. This is a medieval-style tale of the quest of a knight and his lady for the mythical well with strength-giving waters, but, in the final stages of the quest, they encounter a false well that had proved the downfall of many, beguiling them into the consumption of its putrid waters. As Ralph, the knight, stoops to drink from the well, the Lady Ursula cries out in terror: "O Ralph, do not do it! Seest thou not this water, that although it be bright and clear, so that we may see all the pebbles at the bottom, yet nevertheless when the wind eddies about, it makes no ripple on the face of the pool, and doubtless it is heavy with venom." Thus Ralph is saved by the lady to go on and find the True Well.

Whether resulting from the malign intentions of evil supernatural powers or from the mundane toxicity of certain environmental chemicals, the poisoned water source is clearly one of the great fears of humans when confronted with a wetland. This attitude reaches its extreme when faced with the saline wetlands of the dry lands, where thirst could so easily drive one to the ill-advised consumption of evil waters. The notion is expressed vividly in the painting *The Scapegoat* by William Holman Hunt (1827–1910), a friend of William Morris. The famous painting displays a forlorn, dying goat entrapped in mud and encrustations of salt. It is a dismal and harrowing picture of suffering. The scapegoat was an unfortunate animal upon which the sins of all the people of Israel were laid and which was driven out to the desert to die or, in this case, to the saline muds of the Dead Sea. The setting of this dismal picture of a dying goat was ably described in 1856 by the art critic John Ruskin (1819–1900): "The air in its [the Dead Sea's] neighborhood is stagnant and pestiferous, polluted by the decaying vegetation brought down by the Jordan in its floods; the bones of the beasts of burden . . . are bleached by the salt ooze, which, though tideless, rises and falls irregularly. Swarms of flies, fed on the carcasses, darken an atmosphere heavy at once with the poison of the marsh and the fever of the desert." The water no longer flows; it simply pulsates; and, with this loss of movement, the attractions of the mobile Jordan also fade. The wetland has lost its charm.

■ THE HORROR OF THE SWAMP THING

Death by drowning is a fear that afflicts a large proportion of humanity, and this phobia has undoubtedly colored our attitudes to wetlands. Stillness, depth, and rich weedy vegetation may appeal to some poets, as to the "Lake Poet," William Wordsworth (1770–1850), who saw in the stillness of the lake a greater assurance of life's stability and hope than in the tormented skies above. In "There was a Boy" (1798) he describes the reflection of the sky as an "uncertain heaven received into the bosom of the steady lake." A similar sentiment is expressed by the biblical psalmist when he writes of being led "beside the still waters" (Psalm 23:2), but most poets and writers see still waters as sinister and threatening behind an attractive and inviting facade.

In his poem "The Lady of Shalott" (1842), Alfred, Lord Tennyson (1809–92), describes the apparent idyll of a lowland swamp and stream where, "Willows whiten, aspens quiver, / Little breezes dusk and shiver, / Thro' the wave that runs for ever/ By the island in the river / Flowing down to Camelot." A sinister undercurrent also flows there, for we later read, "And at the closing of the day / She loosed the chain, and down she lay; / The broad stream bore her far away, / The Lady of Shalott." This scene is reminiscent of the wetland described in Act 4, Scene 7, of William Shakespeare's

(1564–1616) *Hamlet* (c. 1599) where, "There is a willow grows aslant a brook that shows his hoar leaves in the glassy stream." It is here, however, that the distraught Ophelia, while making garlands of crow flowers, nettles, daisies, and "dead men's fingers" climbed that willow and "fell into the weeping brook. Her clothes spread wide, and mermaid-like a while they bore her up;". . . . "But long it could not be till that her garments, heavy with their drink, pulled the poor wretch from her melodious lay to muddy death." Like the Lady of Shalott, she sang as she drowned.

Submerged and emergent vegetation adds to the terror of drowning in swamps and marshes. In imagination the plants can easily be assigned malign intent, entangling the flailing limbs of the victim. J. R. R. Tolkien (1892–1972), in his novel *The Lord of the Rings* (1954), plays on this fear as he describes the crossing of the Dead Marshes by intrepid hobbits on their journey to the Land of Mordor. They are driven close to panic in these wetlands where "the only green was the scum of livid weed on the dark greasy surfaces of the sullen waters. Dead grasses and rotting reeds loomed up in the mists like ragged shadows of long-forgotten summers."

A visual portrayal of the horror of swamps and the claustrophobia of subaqueous entanglements by weed and roots is unforgettably presented in the 1951 film version of *The African Queen* that starred Humphrey Bogart and Katharine Hepburn, in which the density of vegetation and the consequent lack of visibility and orientation add to the terrors of the experience of dragging a boat through a swamp. In the original novel of that title by C. S. Forester (1899–1966) published in 1935, we read that "it was a region in which water put up a good fight against the land which was slowly invading it. Through the mangrove roots which closed round them they could see black pools of water reaching far inwards; the mud in which the trees grew was half water, as black and nearly as liquid. The very air was dripping with moisture." Novelist Joseph Conrad (1857–1924) in *Heart of Darkness* (1902), his oppressive picture of west Africa, describes the view of these tropical swamps from a boat upon a river which was itself a "stream of death in life, whose banks were rotting into mud, whose waters, thickened into slime, invaded the contorted mangroves that seemed to writhe at us in the extremity of an impotent despair."

Although tropical swamps often figure in such portrayals, their temperate counterparts are by no means free from fears, as the hero of English novelist Charles Dickens's (1812–70) *Martin Chuzzlewit* (1844) discovered when he tried to settle in the coastal swamplands of eastern North America. "The trees had grown so thick and close that they shouldered one another out of their places, and the weakest, forced into shapes of strange distortion, languished like cripples." It was "a jungle deep and dark, with neither earth nor water at its roots, but a putrid matter, formed of

the pulpy offal of the two, and of their own corruption." Clearly, Dickens held the temperate swamps in low regard. The misty obscurity, so often associated with swamplands, adds to their mystical and fearsome aspect, as shown in the photograph of the Okefenokee Swamp in Florida.

The swamp's disorienting lack of visibility and tangled impenetrability contribute to the human fear of this habitat. There is a third source of fear, however, namely the possibility of a "swamp thing"—a harmful animal that can take advantage of human immobility. It may be large, or it may be small. In the film *The African Queen*, for example, the animals that cause the greatest distaste are small, bloodsucking leeches, and in some ways all the more terrifying because they lurk beneath the water among those same submerged plants and roots that impair our movements. The terror invoked here is essentially psychological, akin to arachnophobia (fear of spiders), but larger animals can also present a perceived or, indeed, a very real danger. Reptiles, particularly crocodiles, alligators, and snakes, can be encountered with little warning due to the restricted visibility, and this adds to the danger. In the reedy marshes of southern Iraq,

Wetlands, especially those with slow-moving or stagnant water, such as the Okefenokee Swamp shown here, can be eerie places. It is not surprising that many fearsome myths and legends are associated with wetland habitats. *(George Gentry, U.S. Fish and Wildlife Service)*

it is the wild pig that presents the greatest danger, while in the swamps of East Africa, the hippopotamus can present problems if encountered on one of its treks to and from the water's edge.

A more fantastic version of the "swamp thing" theme is that of the half-human, half-fish monster, as famously depicted in the 1954 film *Creature from the Black Lagoon*, which featured a large, bipedal, gilled creature discovered in the Amazon. In the 1982 film *Swamp Thing* (based on the 1950s comic book of the same name), the swamp thing takes a turn to the vegetable when a scientist abandoned by his colleagues in a swamp gradually turns into a sort of human plant.

One further reason why the swamp may generate unfriendly images in our minds is the fear of disease. Some parasites live in water or in water creatures, and several diseases are transmitted by mosquitoes (see "Wetlands and Disease," pages 197–199).

Swamps and other wetland types have also traditionally been associated with less material sources of concern—ghosts, spirits, demons, and magic—all of which played an important part in the medieval mind. It is no surprise that the wizard Merlin himself is said to have occupied an isolated island in the middle of the swamps of southwest England. When King Arthur was dying, his sword, Excalibur, according to the account in Sir Thomas Malory's (1414–71) *Le Morte d'Arthur* (1470), was cast back into the wetlands whence it came and was caught by a mysterious hand, shaken, and taken down into the mystical depths. Wetlands, it seems, are the perfect habitats for numinous beings whose presence may generate respect but also fear.

In view of all the negative aspects of swamps emphasized in literature, it is surprising to find that when the English author C. S. Lewis (1898–1963) set out to depict a perfect world, unsullied by evil, in his fantasy novel *Voyage to Venus* (1943), he used a floating swamp as his setting. In a constantly undulating, mobile world, Lewis's hero, Ransom, uses the rocking trees to steady himself and stand upright. He looks out over a carpet of "vegetable broth of gurgling tubes and exploding bladders," and on this fibrous platform of instability, Lewis perceives paradise.

■ BOGS AND THAT SINKING FEELING

If swamps have been harshly regarded by the authors in history, bogs have fared even less well. Perhaps the prospect of sinking into peat has even less appeal than that of drowning in water. The occasional discovery of human remains within peat, unearthed during the process of peat-digging, may have fueled the notion that bogs have the capacity to engulf those who are unwise enough to visit them. Or perhaps the sight and sound of a drowning sheep has been sufficient to generate such ideas. Whatever the basis for our fear of bogs, it is undeniable that literature, especially of the romantic and gothic variety, has exploited this fear extensively.

Bogs, particularly upland blanket bogs, are often represented as hellish and therefore provide an appropriate setting in which the villain of a plot can meet his timely end. The evil Carver Doone in English novelist R. D. Blackmore's (1825–1900) *Lorna Doone* (1869) sinks in and is finally lost beneath the peats of Exmoor in western Britain, despite the awkward fact that it is difficult to find peat deposits in that area where one can sink above one's knees. Mists and fogs, of course, are frequent over these upland mires, and these contribute much to their air of sinister mystery, often leading to the implication that the wetlands themselves generate them. *The Hound of the Baskervilles* (1901) written by the British physician, novelist, and detective-story writer Sir Arthur Conan Doyle (1859–1930) must be the best known of the gothic "bog novels" and certainly has the most splendidly exaggerated depiction of the habitat. Once again, the villain is lost in the bog, this time the "great Grimpen Mire" where Sherlock Holmes and his companions found that "rank reeds and lush, slimy water-plants sent an odour of decay and a heavy miasmatic vapour into our faces, while a false step plunged us more than once thigh-deep into the dark, quivering mire, which shook for yards in soft undulations around our feet."

The real problem with peat, it seems, is that it sucks one down. Although one undoubtedly sinks faster in water, it is perhaps a little easier in water to extract oneself or to swim, hence the tendency for authors to personify the peat bog and attribute evil motives to it. In the words of Conan Doyle, for example, "Its tenacious grip plucked at our heels as we walked, and when we sank into it it was as if some malignant hand was tugging us down into those obscene depths, so grim and purposeful was the clutch in which it held us." Conan Doyle leaves us in no doubt that the bog itself has predatory attitudes toward us.

English preacher and author John Bunyan (1628–88) appreciated the power of that sinking feeling and used this nightmare experience to great effect in his allegorical story *Pilgrim's Progress* (1684). One of the first tests for the pilgrim is the Slough of Despond, a quagmire from which he has great difficulty in extracting himself. The metaphorical and psychological significance of this mire becomes apparent when, later in the story, we meet Mr. Fearing, who had "a Slough of Despond in his mind—a slough that he carried everywhere with him."

Quite apart from the depressing nature of the peat bog, there is often a distinctive smell associated with these habitats, as Conan Doyle pointed out. Disturbing the peat or the murky pools or simply compressing the substrate underfoot can result in the release of bubbles of gas, containing both hydrogen sulfide and methane. The sulfurous fumes undoubtedly add to the association of bogs with hell. They also remind many authors of death and decomposition. Gabriel García Márquez refers to mangrove thickets as "swamps of putrefaction" in his novel *Love in the Time of Cholera* (1985), and Charles T. Powers writes of a location in Poland where there persists "an odor of mold and dampness that rises from the decay in lowland marshes and waterlogged fields" (from the novel *In the Memory of the Forest* [1997]).

These authors are right in their conclusion that the gases are the product of decomposition but wrong in giving the impression that the mires from which they emanate are the sources of excessive decay. In fact, nothing could be further from the truth. Decomposition in peatlands is suppressed by waterlogging, so the gases belched out by compressed peat are the outcome of partial decay and incomplete oxidation or organic detritus. The putrefaction idea, however, is deeply entrenched in the imaginative and poetic minds of many writers.

Swamps, with their tall vegetation and high levels of biomass, are often portrayed as enclosed, stifling habitats that engender claustrophobia. The open bogs, on the other hand, allow extensive vistas and wide skies that can have quite the opposite effect—they make the visitor feel small, insignificant, isolated, and lost. Loneliness, the total absence of human habitation, can be a torture to a social being. The lack of evidence of a human hand on the landscape—no field boundaries, no domestic animals, no sound of internal combustion engines—may be regarded as peaceful to some but strangely threatening to others. The very concept of wildness, or wilderness, affects different people in very different ways.

A taste for formal landscapes and human-induced order prevailed in Europe during the 17th century, and many regimented and geometric gardens were constructed by the rich landowners of that time. Nature was at its most acceptable when tamed by human hands. In the 18th century, attitudes began to change as a realization that irregular, natural features could add unexpected appeal to a landscape. English landscape gardeners, such as William Kemp (1685–1748) and Lancelot "Capability" Brown (1715–83) introduced irregularly shaped lakes and groves of trees to provide natural features into the gardens they planned. The paintings of the 17th-century French landscape artist Claude Lorraine (1600–82) had done much to popularize these natural, romantic, and "picturesque" (a word coined at the time of these changing attitudes) features. The quest for natural characteristics within formal gardens even extended to the planting of dead trees and artificially con-

structed ruins of "gothic" buildings ("follies"), showing how far people were prepared to go in order to insert an unpredictable but realistic element into the carefully planned landscape. Despite the appreciation of nature, culminating in the romantic movement, there remained in the mind of Victorian people a suspicion that untamed nature was not to be trusted and could demolish the human-imposed order that had required so much effort in its construction. When, in 1898, the English bishop F. W. Farrar (1831–1903) saw the silting and decay of the ancient harbour of Ephesus in Asia Minor he wrote, "the harbour is a reedy pool; the bittern booms amid its pestilent and stagnant marshes." The change from civilization to wilderness was undoubtedly a retrograde step.

In North America, however, a lonely but latterly influential voice was raised against this tide of environmental neatness. Henry David Thoreau (1817–62) wrote in the 1850s about his beloved American countryside in *Walden*. "Walden is a perfect forest lake. Sky water. It needs no fence. Nations come and go without defiling it. It is a mirror which no stone can crack, whose quicksilver will never wear off, whose gilding Nature continually repairs;. . . in which all impurity presented to it sinks." Idealistic, perhaps, especially with respect to the capacity of nature to cope with the human impact that was still to come, but an attitude that nevertheless valued nature for its own sake outside and beyond the influence of humanity.

What is it in our minds that allows us to appreciate a landscape in which our activities do not figure? Thoreau again has an answer: "It is vain to dream of a wildness distant from ourselves. There is none such. It is the bog in our brains and bowels, the primitive vigor of nature in us, that inspires that dream." An English poet, Gerard Manley Hopkins (1844–89), writing only a little later than Thoreau, expressed a similar view, generated no doubt by the "bog in his brain," and with an equally humid image, "What would the world be, once bereft of wet and wilderness? Let them be left, Oh let them be left, wilderness and wet; Long live the weeds and wilderness yet."

■ WETLANDS AND DISEASE

One of the reasons why wetlands have so often generated fears in human societies has been their association with certain diseases. Often the concerns have been unfounded, for such features as marsh mists and gases are certainly not a direct agency for the spread of diseases. In the century before the birth of Christ, the Roman Marcus Terentius Varro (116–27 B.C.E.) proposed that there are tiny, disease-carrying organisms in swampy places that are too small to be seen but are carried through the air and pass through human nostrils to cause disease. This suggestion is remarkable in its anticipation of medical microbiology but is not quite correct in its simplicity. Some disease-causing microscopic organisms are associated with wetlands, but their modes of transmission from swamp to people are much more complicated than Varro supposed. Among the most important and best known of these diseases are malaria and schistosomiasis, or bilharzia.

Malaria is one of the most widespread serious diseases in the world, affecting perhaps as many as 500 million people. It has been known throughout the history of humankind, although not always distinct from a number of fever-inducing diseases. The association between "swamp fever" and wetlands, however, certainly goes back 2,500 years. In the fifth century B.C.E., a student of Pythagoras, Empedocles, was evidently convinced of an association between this fever and the marshes surrounding the city of Selinus on the island of Sicily. As a result of the draining of these marshes, on the recommendation of Empedocles (490–430 B.C.E.), the city was freed from its malaria menace, and a special coin was struck to celebrate the event. The Greeks at this time were also aware that the most serious form of swamp fever occurred in late summer with the appearance of the

dog star, Sirius, and of Arcturus in the night sky. We now know that this was due to the predominance of an especially virulent form of malaria, caused by *Plasmodium falciparum*, at that time of year.

The symptoms and the course of the disease, with its intermittent fever and swollen spleen, became well known in the course of history. By the 18th century, English physicians were recommending the use of a tree bark from Peru, chinchona (quinine), as a means of treating the disease—a cure that had reached Europe from the native herbalists of South America via Jesuit priests. The mode of action or even the presence of a disease-causing organism in the blood of victims was not appreciated until 1880 when a French army surgeon observed moving organisms in the blood of a malaria sufferer. Even then, other medical scientists still maintained that the cause was a bacterium found in swamp water, but over the next few years the identification of the protozoan *Plasmodium* as the responsible agent was achieved.

The idea that the agents of malaria could be conveyed to humans by mosquitoes had occasionally been suggested

Mosquitoes and Malaria

Mosquitoes are found in many of the world's wetlands (see "Bog Invertebrates," pages 156–157), but not all are involved in the life cycle of the malarial parasite. Scientists have recognized about 3,300 species of mosquito, of which 410 belong to the genus *Anopheles*. Within this genus, 70 species of *Anopheles* are known to transmit the malarial parasite, *Plasmodium falciparum*. The larva of the mosquito is aquatic, hatching, spending its early life, and pupating in wetland pools. The adult insect, however, can disperse widely through the air. The males and females form courtship swarms at sunset, but the female is not able to lay eggs until she has consumed protein, and she obtains this from the blood of mammals (although some species parasitize birds or reptiles). The mouthparts of the female mosquito are modified into a long piercing tube with which she can cut into skin. She then forces the tube into a blood-carrying capillary beneath the skin and injects saliva containing anticlotting agents to ensure that the blood maintains its fluidity while being withdrawn. The mosquito takes only a small amount of blood, usually about 0.0002 ounces (5 mg), but this provides enough protein to manufacture up to 500 eggs.

If the mosquito has not previously taken a meal from a host contaminated with the malarial parasite, then the

effect on its victim is more of an irritation than a problem. But if the mosquito has fed on a person who has malaria parasites present in the blood, then it will have taken malarial parasites into its stomach, where the parasite completes its sexual reproduction and the zygotes take up residence in the salivary glands of the mosquito. When the female takes its next blood meal these will be injected into the host's bloodstream with the saliva, and the malarial parasite has successfully moved to a new host.

Once in the host's bloodstream, the parasite is carried to the liver, where it infects the liver cells. From here it enters red blood cells and replicates, eventually causing rupture of the blood cells and the release of parasites into the bloodstream. This release takes place every two to three days and causes the feverish symptoms of malaria to become apparent in the host. The disease cannot be passed from human to human except via mosquito transfer, so control of the disease has usually been focused on control of the vector, often involving the destruction of its wetland habitat. Blood diseases such as HIV cannot be transferred by a mosquito because the disease agent is not able to survive in the insect, but mosquitoes can act as vectors for some other diseases, such as yellow fever, filariasis, and dengue fever.

back as far as 500 C.E., but it was Dr. Josiah Nott (1804–73) of Alabama who in 1848 first formally proposed that both malaria and yellow fever (a viral disease) were transmitted by mosquitoes (see sidebar). The full story of the transmission and complex life cycle of the *Plasmodium* in mosquitoes and in humans was eventually worked out by the end of that century, largely through the work of a British parasitologist, Sir Patrick Manson (1844–1922). At last the connection between malaria and wetlands, suspected since the time of Empedocles, had been established. Unfortunately, of course, this led to further negative attitudes toward wetlands and their invertebrate inhabitants. The drainage of marshy areas and, when such pesticides became available, the widespread use of DDT as an antimosquito agent has led to widespread loss and damage to these habitats in many areas, particularly in the vicinity of areas of high populations, such as the Mediterranean basin in southern Europe. DDT is toxic to humans and to many predatory birds. Its use has been banned in the United States and in Europe, but it is still used in some parts of the world. Current research emphasizes the development of more effective drugs for humans that will prevent infection, but climate change may lead to an extension in the range of the infective mosquito, possibly resulting in increasing pressures for control by habitat destruction, especially in developed countries.

Schistosomiasis, or bilharzia, affects about 200 million people mainly in Africa and South America, and it is now known to be caused by tiny, wormlike flukes in the blood and organs of humans. The parasites take up residence in the intestines and bladder where they mate and lay eggs that are subsequently shed into the environment with feces and urine. When they encounter aquatic systems, carried along by untreated sewage, the eggs enter the guts of freshwater snails, inside which they hatch and spend the first part of their larval existence. They are shed from the snail back into the water, where they penetrate the exposed skin of any people who may be present in the water. The condition in humans damages the bladder wall and the liver, and it can prove fatal.

The papyrus records of ancient Egypt have references to this disease, which causes sweating and the passing of blood. It also reached epidemic proportions among Napoléon's soldiers during his Egyptian campaign of 1798. The worm was first described in 1851 by a young German doctor, Theodor Bilharz (1825–62), working in Cairo and was called *Schistosoma haematobium*. The life cycle of the liver fluke, which, like the schistosomes, has a snail as an intermediate host, was already known, so a search began for the particular species of snail that could transmit bilharzia. It was not until another military force, this time the British army, occupied Cairo in 1915, during World War I, that the life cycle was finally worked out in all its complexity.

The control of schistosomiasis has proved difficult, but not as difficult as that of malaria. Drugs are available to destroy the organism within the human body, so cure is possible, but prevention is obviously better. Draining wetlands has not been an option in this case because many of the wetlands involved are themselves important to local people for their utility as irrigation canals, transport waterways, washing areas, and so on. The alternative is the use of antimollusk pesticides (molluskicides) to reduce the populations of the snail vector (the intermediary host), but these pesticides must not be poisonous to humans, domestic animals, or even to fish if they are to be environmentally and economically acceptable. Education in hygiene, particularly the careful disposal of sewage, and about the dangers of entering polluted waterways with bare skin exposed is the most hopeful way forward, but it is clearly a difficult task.

Hippocrates (460–377 B.C.E.) wrote a book entitled *Airs, Waters and Places,* in which he set out some basic ideas concerning the relationship between environmental conditions and disease. He was certainly aware that illness often resulted from drinking water from the slow-moving waters of swampy wetlands. It was not until the 19th century, with the discovery of the organisms responsible for diseases, that the full significance of the aquatic component was apparent in the spread of such epidemics as cholera and typhoid. Medical scientists no longer believe that the inhalation of the humid "miasmas" of the wetland atmosphere is sufficient to lead to disease, but the fact that some diseases are closely connected with wetland environments cannot be denied.

■ CONCLUSIONS

Wetlands have a history that goes back almost as far as that of life itself. Some of the first incontrovertible evidence for life comes from shallow-water coastal wetlands, where blue-green bacteria formed stromatolites, in which their remains were preserved for more than 3.5 billion years. When plants first colonized the land it was in the brackish environments of coastal wetlands, more than 360 million years ago, so wetlands have been a key habitat in the development of life on Earth. The forested swamps of the Carboniferous period resembled the modern raised peat mires of Southeast Asia in their ombrotrophic hydrology, and the deep organic peat they formed eventually became the coal that fueled the industrial revolution. The subsequent wetlands in global history have continued to accumulate the fossil fuels that still form the major energy resource for human populations. Wetlands have thus had a major influence on human history and industrial development.

During the last 2 million years, the Earth has experienced a strongly fluctuating climate, and for much of that time the world has undergone glacial conditions interrupted by relatively brief interglacials. The world's pattern of wetlands has changed rapidly and recurrently with each cycle of glacial and interglacial climate, with tundra mires developing in increasingly low latitudes during periods of cold. Ancient lake sediments and peat deposits contain the fossils that have helped scientists to reconstruct these changing conditions. The oceanic raised bogs and blanket bogs, for example, have largely begun to form and expand during the last 5,000 years, as have the tundra mires of the northern parts of Canada. Even the Earth's subtropical arid lands have seen great changes. In the southwest United States, the glacial episodes were times of wetness, the pluvial periods, but in Africa and India they were times of drought. Wet conditions returned to these latter areas in the early Holocene, between 10,000 and 5,000 years ago, following which the desert regions have returned to their arid state.

The excellent preservative qualities of wetland sediments have resulted in the survival of many archaeological artifacts that would otherwise have been lost, especially those made of wood and other easily decomposed materials. Human settlements around the fringes of wetlands have been particularly well preserved and studied and have provided archaeologists with much information on the way of life among the prehistoric people of the world. Wooden trackways have indicated how people overcame the barriers to movement that wetlands sometimes presented, and wooden boats display the fact that in some respects the wetlands provided a means of movement and travel for prehistoric communities.

The use of wetlands as a way of accessing remote areas in historical times is illustrated by many stories of exploration. The European and American explorers of the 19th century expended much effort and many lives to discover the source of the Nile River in East Africa and that of the Congo River in Central and West Africa. In doing so they opened up the wetland riches of the African continent. Spanish explorers performed similar feats of daring in mapping the Amazon River in South America and revealing the extent and nature of the vast flooded forest. In Europe, the wetland regions, such as the Coto Doñana in southern Spain, inspired latter-day explorers and have contributed to the development of modern conservation strategies so vital for the survival of all the world's wetlands.

Wetlands have inspired and influenced people in many different ways, not least in the field of literary endeavor. Poetry and prose writings abound with wetland references, and a general pattern seems to underlie the various ways in which wetlands are presented. Flowing waters are often symbolic of goodness and health, while stagnant waters carry a sinister threat. Marshes and swamps are almost always evil places in literature, perhaps because they impair movement, give limited visibility, and always carry a threat of predatory or parasitic creatures lurking unseen either above or below the water. Peat bogs are seen as open but lonely places, reeking of decay and dangerous because of the soft peat that seems to suck down the unwary traveler. Many of these images are quite erroneous, but they have nevertheless influenced the public perception of wetlands, usually in a negative direction.

This negative attitude that wetlands often engender is partly justified by their association with disease. Stagnant waters can indeed carry disease, especially if they receive an input of human sewage. Wetlands may also support vectors of disease, such as the mosquito that carries the malarial parasite from one person to another.

When these negative aspects of wetlands are added to their limited value for grazing domestic stock or for growing crops, or even for human residence and village development, it is hardly surprising that these habitats have come to be regarded as wet deserts, wastelands that can only become useful if they are drained and destroyed, when they can be converted into a more valuable commodity, such as agricultural land. Wetlands do have their uses, however, for they are the home of much of the world's biodiversity, and they control a large proportion of the hydrology of the planet. The recent history of wetlands has begun to focus on these more positive aspects.

7

The Value of Wetlands

We live in a world where everything is considered to have a price, even people. Such a belief is ethically flawed, but is one that conservationists must face when arguing for the conservation of a particular site or region. A potential developer of an ecosystem may be able to put a precise economic value on the animals that can be harvested, or the timber that can be felled, or the peat that can be extracted from a wetland, but those concerned with its conservation are unable to match either the precision or the economic simplicity of such an argument. It is not surprising, therefore, that history is rich in accounts of the exploitation and spoilage of so much of our natural world.

■ VALUATION OF NATURAL ECOSYSTEMS

There have been attempts by conservationists to meet the economic argument head-on, as in the case of tropical forest exploitation. In one analysis of the Amazon basin, economists calculated the timber value per acre (0.4 ha) to be $400. The renewable resources of the same area (fruits, latex, herbs, etc.) were reckoned to be worth $280 per year. These figures, if correct, would certainly support renewable use rather than immediate timber exploitation, but they may not be entirely reliable, for the cost of transport of materials to market and the possible nonexistence of such a market detract from the achievable income from the forest. Similar work has been conducted on temperate wetlands, considering their value in terms of water supply, flood prevention, pollution control, recreation, and amenity. On these criteria, the coastal marshes of Georgia were valued at between $20,000 and $45,000 per acre (0.4 ha), and the wetlands of the lower Charles River floodplain near Boston, when considered for drainage, were set as high as $160,000 per acre, compared with about $4,000 per acre as the value of farmland at that time. In the case of wetlands, one could also rightly claim that their loss would

reduce the capacity of the Earth to absorb the additional carbon dioxide people inject into the atmosphere. Indeed, the exploitation of vegetation and peat would add to that atmospheric load of carbon. The economic consequences of such activities could be estimated.

An argument with fewer variables, however, is the need to maintain the Earth's biodiversity. Some people would argue that every species has an ethical right to existence and that our duty is to give them a chance. Extinction is a permanent and irreversible step, so the eradication of any organism cannot be viewed lightly. More pragmatic is the argument that other species are of use to people as food, as sources of drugs, or as a source of intellectual stimulation for study and are therefore worthy of preservation. Also, the need for humans to know that wilderness exists is another strong argument for habitat conservation. Visiting the wilderness can be a stimulating experience at all ages and, even if one cannot actually visit a wilderness, the knowledge that it exists is reassuring to many people. Wild habitats can also act as sensitive indicators of what humans are doing to their own global environment, a kind of litmus test that can give early warning of environmental dangers.

A large number of the world's people are now reasonably well convinced that some conservation measures are necessary, either for economic, cultural, or health reasons. What must now be debated (both globally and locally) is how much area can be assigned for this purpose and which areas are conservation priorities. There is clearly a limit to how much land (and water) can be used for wildlife conservation. The current world human population of more than 6.5 billion is reckoned to use or divert approximately 40 percent of the Earth's net primary production. Current predictions suggest that the global population will exceed 9 billion by 2050 and will continue to rise after that date. A doubling of the current population would leave very little room for the sustenance of wildlife habitats, for almost all productive land would be needed for agriculture or human residence. Meanwhile scientists need to set out criteria upon which conservation decisions can be made and comparisons between sites can be objectively established. Several have been suggested as noted below:

1. High diversity of species at a site is one obvious criterion for evaluation. "Hot spots" of biodiversity, where unusually rich assemblages of species are found, have been located in different parts of the world. Wetland examples are the Okavango swamps of Africa and Bharatpur in India. The reasons for their diversity include their high productivity, their value for migratory birds, their high microhabitat diversity, and their large areas. The presence of a range of successional stages of wetland adds to diversity, and sometimes a particular species acts as an "ecological engineer" in maintaining the successional cycle. The beaver of the boreal wetlands, the peccary of the Amazon, and the alligator of the Everglades all open up habitats and maintain early successional stages, which adds to potential biodiversity. So the maintenance of populations of these particular keystone species is essential for the survival of biodiversity.

2. How easy would it be to replace a particular ecosystem? This is an important question in evaluation. In art, the great paintings of the Old Masters are all the more valuable because they cannot be replaced if destroyed; the same logic applies to certain ecosystems. A raised bog, for example, may have taken 7,000 years to develop. It can be destroyed by peat extraction in a matter of months, but it can never be replaced in its original state. Other ecosystems, such as reed beds and swamps, can be more easily reconstructed.

3. Fragile ecosystems may be in greater need of protection than more robust ones. Most wetland vegetation is sensitive to trampling, but productive species, such as reeds, cattails, and papyrus, can recover quickly. Slow-growing species, such as the bog mosses, may take many years to recover from a single footprint, so such wetlands are in need of greater protection. Some sites are more vulnerable than others in the sense that they are in danger of exploitation, so vulnerability may also constitute a good reason for high priority in conservation.

4. An important concept is that of representativeness. Is the site under evaluation a particularly good example of this kind of ecosystem? This is where a thorough scheme for the classification of wetland types becomes important, for it ensures that conservationists can seek out good examples of each type for conservation priority. The geographical scatter of wetlands is also important here; planners must be sure that each wetland type is well represented in the conservation priorities of the different geographical regions within its range.

5. Naturalness—that is, how free from human operations a particular site may be—is also of interest in wetland evaluation. Such a consideration is of great importance in raised bogs, for example, where damage from human operations in the form of burning or peripheral peat extraction reduces the value of the site. It is a criterion that must be used with care, however. Some wetlands, such as reed beds, may have been subjected to centuries of human management, including harvesting and burning, and yet may have retained, perhaps even enhanced, their ecological value.

6. Rarity is an emotive subject that is often called on as an argument for conservation value. It is a concept that can be applied either to an organism or to an entire ecosystem, but it must be viewed within a broad geographical context. The presence of a rare animal at a site (particularly a large, attractive, furry one) has great public appeal in evaluation arguments. If that species is geographically widespread and abundant but locally scarce only because it happens to be at the edge of its distributional range, its presence should not be allowed to weigh too heavily. It is easy to lose sight of the importance of conserving the typical habitats of a region where such ecosystems are at their optimum and to concentrate instead on rare ones. Rare habitats or species are often of great local interest but are of lesser global value in the maintenance of biodiversity. There are some situations, however, where a species or habitat is confined to one geographical region, and in such a case rarity becomes an extremely important evaluation criterion.

Although economic evaluation may not be possible or meaningful for conservation agencies, there are criteria on which the value of a wetland can be assessed, and such criteria are particularly important in the comparison of different sites in order to select priorities for conservation.

■ FOOD FROM WETLANDS

Harvesting the products of wetlands provides a substantial source of income to people in many parts of the world, and the economic importance of wetlands, particularly at the local level, needs to be taken into account when considering their conservation and management. It is particularly important to determine what products of wetlands can be harvested in a sustainable fashion and to develop management techniques that will optimize yields with a minimum of damage to the habitat. Fish, for example, can be harvested in a sustainable fashion, while peat cannot.

Wetlands are a source of food for many people. Wild mammals that graze on the wetlands, particularly in tropical regions, can provide an important source of protein to local inhabitants. The lechwe antelope of the Okavango delta in Africa and the capybara of the flooded swamps and forests of the Amazon basin are both hunted and harvested by the local peoples. In the Sudd swamps of southern Sudan, it is thought that about a quarter of the meat eaten by local inhabitants is derived from hunting wild animals. The caribou of the boreal and arctic wetlands are harvested by humans and have effectively become domesticated in the nothern parts of Europe and Asia. From prehistoric times, the hunting and following of migrating herds has eventually developed into a symbiotic relationship between the reindeer and the Sami people.

Domesticated animals in many parts of the world exploit the high primary productivity of certain wetland environments. Even in developed regions such as North America and Europe, wet grasslands and coastal salt marshes are used for grazing cattle and sheep. The Sudd swamps of the Sudan are reckoned to hold about 800,000 cattle, and the wetlands of the Niger delta in West Africa may hold three million cattle, sheep, and goats. The future of harvesting animal protein from tropical wetland may benefit from a switch in emphasis from cattle to native animals, such as the lechwe. Cattle were originally forest animals and are poorly adapted to wetland life, although water buffalo are appropriate for the habitat, being better suited to the wetlands. Animals such as the lechwe can graze in the aquatic environment of marshes and swamps and are more resistant to the diseases and parasites associated with them. Perhaps it is time to domesticate new animal species to permit a sustainable harvest from wetlands rather than modifying these habitats to suit the requirements of the old domesticates, such as cattle.

Wildfowl have been harvested from wetlands since prehistoric times. Papyrus illustrations from ancient Egypt show fowlers at work with nets to trap wild ducks and geese. The use of firearms has, of course, permitted much more efficient and effective means of harvesting wetland birds, but these techniques have brought with them a clear need to limit the harvest to ensure its sustainability. Many wildfowl are migratory, so conservation must include protection of their breeding sites and permitting the birds to gain sufficient weight in their winter grounds to allow them to journey back to these sites in spring. Not only the harvest must be controlled but also the degree of disturbance to the feeding birds in winter (see "Wetlands and Bird Migration," pages 159–162).

Fisheries provide a major food and economic resource for many wetland peoples. In the Mekong delta of Southeast Asia, the Grand Lac is said to be one of the most productive freshwater fisheries in the world, and some of the rivers in the area produce 50–80 pounds of fish per acre (60–90 kg ha^{-1}), while the seasonally flooded regions nearby can yield three times that quantity. In Jamaica, the harvest of freshwater shrimps supports about 1,000 people, and in the Niger delta some 10,000 families are thought to depend mainly on the fish harvest. Almost all of the great wetlands of the world have fishing industries associated with them, from the carp fisheries of the Danube delta in Europe to the tilapia fisheries of Lake Victoria in East Africa, and from the bass fisheries of Lake Michigan and the salmon industry of the Pacific Northwest in the United States, to the carachin fisheries of the River Niger in West Africa.

Coastal wetlands are also vital for many types of fish and other forms of seafood. Crustaceans, including prawns, shrimps, and crabs, pass much of their early life cycle in coastal wetlands. Particularly rich and important in this respect are the mangroves of the tropics, where many shellfish, including various oysters, cockles, and clams, are harvested. Even the mudskipper provides the basis for an industry in the mangroves of Taiwan. Bamboo poles are used in many parts of Southeast Asia to provide settling locations for mussels, which can subsequently be harvested.

The provision of the facilities needed by an edible aquatic organism, such as the mussel, is the first stage in "fish farming"—an activity that is becoming increasingly important in many wetlands. Artificial ponds in Louisiana, for example, are used for crawfish farming, and hatcheries are used to boost the populations of trout and salmon in many parts of the world. Growing fish in cages or nets within a wetland can cause problems of eutrophication and poor hygiene if excess food falls to the bottom and decays. The rapid spread of disease among crowded, single-species populations of fish in such circumstances is a frequent problem. Wetland animal farming is not restricted to fish and crustaceans: Frogs, turtles, and alligators are also farmed in this way, providing an assured harvest of animal products at the end of the process.

Fishing is also one of the great recreational uses of wetlands (see "Wetlands and Recreation," pages 209–211) and is reckoned to be the most popular sport in the world. It is a form of wetland use that is not only sustainable but is also relatively easy to combine with wildlife conservation. Angling with a line, under controlled conditions, has limited impact on fish populations. It is a quiet activity and therefore creates only minimal disturbance for birds, mammals, and other wildlife.

Harvesting plant products from wetlands can take many forms. Wood for fuel or for construction is extracted from some wetlands. Reeds and papyrus are used for thatching roofs and even building the walls of huts. A variety of wetland plants, especially mangroves, provide dyes, drugs, tannins, sugar, tea, alcohol, and fibers. Even the flowers of some mangrove plants can be used to support beehives for honey and beeswax production. In some wetlands, the economically profitable harvesting of plants has led to their being managed almost as agricultural systems, as in the case of reed bed management in eastern England for thatch production. This form of wetland management can become harmful to other wetland wildlife, however, if management becomes so focused on one product that other wetland species are positively discouraged. This can be seen in the negative impact of cranberry cultivation industries on bogs, where the destruction of native vegetation is commonplace.

The sediments of wetlands, particularly the peats, are a very important resource, especially in the northerly mires (see "Peat as an Energy Resource," pages 211–213). Apart from the energy resource and the horticultural value of peats, their antibacterial properties have often led to their use as a primitive wound dressing (as in World War I) and as an absorbent and antiseptic diaper. Another use of peat, particularly in eastern Europe, is balneotherapy, wherein people immerse themselves in liquid peat as a tonic for the skin. No doubt the bacteriostatic and fungistatic properties of peat can effectively disinfect the skin surface and help in the treatment of some skin infections. The microbes that survive and live in peat, on the other hand, offer some unexpected opportunities for exploitation in industry. Methanogens are bacteria that operate under anaerobic conditions and partially decompose organic matter to produce the gas methane. Such bacteria could be very useful in the treatment of organic waste and in the process generate a useful biofuel. In 2006 a methanogen was first isolated and cultured from acid peat by microbiologists working on McLean bog, New York. Peatlands may thus contain as yet undiscovered attributes that could be of value to humans.

WATER CONSERVATION AND FLOOD CONTROL

Humans have an irrepressible desire to control nature. The mythical story of King Canute of England, who sat in his chair at the edge of the sea and defied the tide to engulf him, is a moral tale about human inefficacy in the face of nature's power. Controlling the flood, however, often appears to be a noble technological aim, since flooding, particularly in heavily populated areas such as the delta of the Ganges in Bangladesh and the delta of the Mississippi in the United States, can cause great damage and loss of human life, as in the case of the flooding associated with Hurricane Rita in New Orleans in 2005. In terms of ecology, one of the main problems is that wetlands have been seen as an impediment to the process of flood control rather than as a safety mechanism to absorb the consequences of flood.

Floodplains and deltas develop as a consequence of periodic overflowing of riverbanks. The floodwaters carry silts and detritus over surrounding land and create, or maintain, rheotrophic wetlands in their vicinity. These wetlands, in a natural situation, absorb the additional flow and are replenished in their nutrient contents as a result of the overflow.

Early agricultural societies often welcomed these events as part of Nature's cycle and as a source of irrigation water and fertilizer for the following season's crops, as in the case of the Nile floods in Egypt. Pastoral societies in temperate regions made use of seasonal floods by constructing "water meadows." Low-lying pastureland in the floodplain of a river was deliberately allowed to flood in early spring, bringing rich sediments but also raising the temperature of the soil and enabling the turf to begin an early growth of grasses and herbs for cattle and sheep. In southwest England, the "Somerset Levels" are winter-flooded wetlands that have been grazed in summer since time immemorial, hence the derivation of their name from the Anglo-Saxon term for summer lands, *sumorsaetan*.

A number of land-use changes have had a major impact upon these flooded areas and created a new set of problems. Agricultural reclamation and intensification, as a consequence of the richness of floodplain soils, have increased the tendency for year-round harvesting. Urban development of the floodplains has often taken place, especially in the lower reaches of rivers. The changes in land use in the upper catchments of rivers have often involved deforestation or wetland drainage, both of which can lead to an increased likelihood of flash floods downstream. The construction of dams and water-regulating systems to control flooding and to conserve water resources has affected the flood patterns on many rivers. The value of the rivers as transport systems and gateways to global trade through their estuaries has led to canalization of the rivers, creating straight, deep, embanked structures, often with higher water levels than the surrounding land. As a consequence of all these changes, the floods, which were once regarded as a blessing to the people of the floodplains, have now become their greatest fear.

In the Mississippi delta, for example, canals were first dug through the coastal wetlands to improve boat transport for early fur trappers and traders. Canals have subsequently been enlarged both to improve the transport systems and to permit access for oil and gas exploration in the coastal regions. In Barataria Bay in the Mississippi delta, about 10 percent of wetlands have already been lost to canal construction and further losses are taking place. These canals change the hydrology and erosion rates and also permit the penetration of saline waters farther inland, leading to ecological changes in vegetation and fisheries. One estimate claims that continued expansion of the canal system will cost the fishing industry about $1 billion in the next 20 years.

Conservationists have expressed concern about the loss of sediment to the floodplain wetlands resulting from canal construction, but it seems that storms such as Hurricane Rita compensate for any general loss of sediment. In the course of that one storm in 2005, a mud deposit on average two inches (5 cm) deep was deposited over 15,000 square miles (38,600 km²) of the coastal wetlands. This represents 130 million tons of sediment.

Higher up the river, canalization and the construction of levees (raised banks to enclose the water flow), together with urban development in the old floodplain region, has led to inevitable serious damage and loss of life during freak floods. Taming floods along such great rivers is attempted only at great risk, should the engineering fail. Similar problems have been facing Florida as its population has expanded in recent years and demand for water has increased (see "The Everglades," pages 222–224).

The kinds of problems experienced on the Mississippi are becoming increasingly frequent as the same course of development is repeated worldwide. On the Nile River, the building of the Aswan High Dam has permitted the control of the annual floods in lower (northern) Egypt, but this has been at the expense of the cycle of land renewal in the rich agricultural plains and has increased demand for artificial fertilizers. Sediment collection behind the Aswan High Dam has been more rapid than was anticipated, so there is also the danger that the system will become clogged with silt. Also, storing a large volume of open water in one of the hottest and most arid places on Earth has not proved wise. Evaporational losses of water have led to a much lower water level in Lake Nasser, behind the dam, than was expected.

In the Sudd swamps of Sudan, higher up the Nile, a flood-control program has been discussed for more than 60 years. The Sudd swamps, consisting of 4,250 square miles (11,000 km²) of wet floodplain, have proved a hindrance to navigation and a potential source of evaporative water losses. A canal (the Jonglei Canal) that bypasses the Sudd, linking the upper and lower reaches of the river, is more than 200 miles (350 km) long and has been an expensive project under construction since 1978. Various technological and financial problems, coupled with local civil unrest, have delayed its completion. Whether the economic gains from this venture in terms of transport facilities, water supply, and irrigation will exceed the costs to fisheries, pastoral migrations, floodplain productivity, and wildlife losses has yet to be seen.

Many parts of the world are looking closely at the opportunities available for modifying wetland patterns with the aim of increasing economic prosperity and human welfare. In China, the waters of the Yangtze River could be diverted into the north of the country to irrigate more arid lands. Several dams have been constructed in the Amazon region, often resulting in damage to fisheries and problems with the buildup of aquatic vegetation. Hydroelectricity obtained from dams along the Amazon River and its tributaries supplies 80 percent of Brazil's energy requirements. The Pantanal of South America, arguably the most important wetland on that continent, is under constant threat of water extraction and diversion. In Asia, the extraction of water from the rivers draining into the Aral Sea has led to the contraction and salinization of this enclosed water body, with severe economic consequences to the people who once made their living from its fisheries.

The diversion of waters from the Russian river Ob, taking its flow from the Arctic into the south, has long been debated. Hydrological changes on this scale could have an impact on the climate of wide areas, especially where the delicate balance of the Arctic Ocean is affected. Plans to dam The Mary River in Queensland, Australia, would destroy the last remaining pristine habitat of the rare Australian lungfish (*Neoceradotus forsteri*), an animal with a history stretching back 100 million years.

Wetlands retain water and store it in times of flood, protecting surrounding lands from catastrophe, and they also alter the quality of the water that passes through them. The fast-growing emergent plants of marshes and swamps demand large quantities of nutrients to build their biomass, and they take this from the water around their roots. The floating papyrus marshes of East Africa are capable of removing more than 50 percent of the phosphates and sulfates and the iron and manganese compounds, from the water that flows through them (see "Vegetation and Water Cleansing," pages 63–65), leaving the water less contaminated by these elements than when it entered the marsh. This significantly reduces aquatic pollution and eutrophication in many African lakes and also helps to alleviate the problems created by human agricultural fertilizers and sewage. The capacity of wetlands to purify water is one of their most useful properties as far as humankind is concerned and is a strong argument in favor of their conservation.

■ PEATLAND DRAINAGE AND AGRICULTURE

In 1919 the U.S. Department of Agriculture issued a map documenting wetlands in need of drainage. It showed the eastern coastal strip from New Jersey south to and including most of Florida; the coastal regions of the entire Gulf of Mexico; the Mississippi valley; and much of the Great Lakes region. By 1980, many of these areas had indeed been subjected to extensive drainage for agriculture. The peaty soils of these regions had been recognized as having great agricultural potential, once the initial expense of drainage had taken place, and this wetland history of the United States simply reflects what has been going on globally for several thousand years, ever since the agricultural idea took root in human minds.

Controlling floods and irrigating agricultural fields were primary concerns of ancient Mesopotamia and Egypt, but in temperate Europe the need to be rid of excess water was the greater concern. The fringes of the North Sea in northwest Europe were particularly rich in wetlands, and it was here that early modifications for agricultural advantage began in earnest. Many of these areas were flooded only in winter, so summer grazing by cattle was a regular form of land use, but extending the grazing period, or even expanding arable opportunities by means of drainage, was an attractive option. In eastern England, in the area known as the Fenlands, the process of reclamation had begun by the seventh century C.E., and by the year 1300, the cooperative effort of medieval farmers and the great monastic houses along the coastal regions had won for agriculture an area of 100 square miles (275 km^2) of maritime and reed bed wetlands. This was achieved by digging drains and then erecting embankments in a systematic series of land parcels, gradually pushing the borders of the wilderness eastward.

In the low-lying coastal areas of Holland and the Friesland region of north Germany, a similar program of reclamation had begun as long ago as 500 B.C.E., when settlers built their villages on artificially raised hummocks of transported soils. It was a simple progression then to begin linking these raised areas with embankments to hold back the flow of the surrounding waters. The simple use of gravity, coupled with animal- and human-powered scoop wheels to take water out of the enclosed areas, was all the technological equipment available for these first drainage exercises. The German lowlands were rich in raised bogs, so the peats from these sites could be harvested once the areas were protected from the sea. This cutting, coupled with the natural shrinkage of peats once drained, led to the lowering of the land surface behind the protecting embankments and left them open to great risk in the event of flood. One such flood is recorded for the year 1287, when a storm surge in the North Sea broke many of the banks, took back the reclaimed wetlands, and cost about 50,000 human lives.

Undeterred by tragedies, including recurrent bouts of plague, coastal peoples continued their relentless fight, cultivating the great raised bogs of the region, then burning and recultivating them as their capacity for supporting crops diminished. As their surface level sank and became wetter, it often became necessary to turn the land over to pasture. In Holland, the early 15th century saw the development of the wind pump, and this new technology led to great opportunities in land drainage. By 1650, the rate of land reclamation in that country had increased from about 1,250 acres (500 ha) per year to about 4,000 acres (1,600 ha) per year. A lull followed until the mid-19th century, when drainage began again in earnest.

Meanwhile, the drainage of the Fenland in England was dwindling as a result of various conflicts of interest. Much of the land was classified as waste and was therefore the property of the king. It was in his interest, therefore, to see the area reclaimed and made more productive. Local residents,

on the other hand, obtained a living from fish, eels, waterfowl, and reed harvesting and so could see little personal gain in the scheme. Merchants in the ports found that any reclamation simply left the former coastal ports stranded inland and trade diminished. What was needed for progress was a coordination of effort and a great investment of money. This came in 1630 when the earl of Bedford brought in Dutch experts to devise a drainage scheme. The English Civil War interrupted these plans, but eventually new drains were cut and new sluices established to hold back the sea. Land prices rose, but land surfaces fell as peats contracted and were oxidized. Rivers soon had to be embanked to contain them above the surrounding lands, which were falling at a rate of about a foot (30 cm) a century. Today, many of these rivers are 12 feet (4 m) above their surroundings and stand out as landmarks in the flat landscapes.

During one drainage operation in 1848 at a place called Holme Fen, an iron post was hammered down through the peat into the clays beneath, so that the rate of contraction of the peat surface could be assessed. In the first two years, the ground level fell by 6 feet (1.8 m) as water drained under gravity. The rate of shrinkage then tailed off until a pump was added to the system when the peat fell by a further two feet (0.7 m) in the following 20 years. The post is still present at the site, now protruding high above the surrounding land.

In 1787, the first steam engine for pumping water was installed in the Netherlands. It was as effective as about 20 wind pumps. The new technological advance gave further impetus to the flagging drainage project, and many new areas were drained during the 19th century in Holland and in England. Soil drainage was then further improved with the introduction of the tile drain, and the embankments became the weakest point in the system. Earth banks with wooden foundations have since given way to clay and then to concrete structures to withstand storm surges.

Wetland drainage in most other parts of the world, including the United States, has taken place largely since the beginning of the 20th century. Technologies developed over several hundred years in Europe were immediately available for application elsewhere. Tile drains, for example, were available after 1870 to drain the wet prairies, and by 1880 it is reckoned that there were more than 1,000 tile factories in the prairie wetland states. As a consequence of wetland drainage, land values in these regions rose by more than five times over the next 40 years, and deaths from malaria were cut to one-third of their former level (from 30 per 100,000 population in 1870) over the following 20 years.

The hardwood forests of the bottomlands of the Mississippi River valley were similarly subjected to intensive draining, mainly in the early part of the 20th century. As in the case of the English Fenlands, water had to be pumped out of the low-lying floodplain, over the embanked levees, and into the elevated river. This reclamation process left these lowlands vulnerable to severe flooding if the riverbanks overflow or burst.

TROPICAL WETLAND AGRICULTURE

Most tropical wetlands are low-lying ecosystems, whose drainage for conventional agriculture requires both technology and financial input beyond the capacity of most governments in tropical countries. Although complete drainage may be difficult, one crop that can grow effectively on reclaimed tropical peatlands is rice. Rice (*Oryza sativa*), which has been in cultivation in China for at least 5,000 years, is unusual among domesticated plants in its ability to grow in waterlogged conditions. It is a wetland plant with a hollow stem that serves to conduct oxygen to the roots for their respiration. The optimal growth of some varieties is within standing water, so the preparation of tropical peatlands for growing rice does not require total drainage of the site.

Some of the most extensive tropical peatlands are found in Southeast Asia in the coastal regions of Indonesia and Malaysia, and many of these locations have been cleared of their forest and planted with rice. The process of reclamation in this region began after World War I, when some of the coastal swamps of Kalimantan were converted to rice growing by the construction of canals for controlling water depth. The region was found to be favorable to this crop because, with the existing water supply, no expensive irrigation system was necessary. Water tables are usually lowered for harvesting, and reclaimed tropical peatlands exhibit some serious problems if allowed to dry. They contract (subside) rapidly as a result of decomposition—as much as five inches (12 cm) a year has been recorded. Water drains through the surface layers (acrotelm), leading to rapid losses of fertilizers into drainage streams. Wood in the peat (the main constituent of many tropical peats) can restrict root growth in the crop plant. The dark color of the peat can also lead to rapid changes in temperature in exposed peat by heat absorption. The peat surface is unable to bear heavy machinery for harvesting the crop, but despite all these problems, rice crops are still the major use of reclaimed peatland in Southeast Asia, especially for shallow peats less than three feet (1 m) thick. Perhaps it is fortunate for conservationists (and for wildlife) that the deep, ombrotrophic domes of coastal peatlands are considered unsuitable for agricultural development.

The reclaimed peat is often acid, pH 4.0 or less, but rice varieties are available that can cope with this degree of acidity. The addition of lime can remedy this situation, although in the volcanic regions of Southeast Asia a cheaper alternative is volcanic ash, which has proved quite effective in raising pH. However, it has the disadvantage of containing some undesirable elements, such as aluminum. The coastal location of these peats, together with their tendency to contract, can lead to increasing salinity with continued cultivation, but rice is able to grow even under such conditions. Where the water table can be lowered a little further, various other crops can be grown, including corn, cassava (manioc), soybean, and peanuts. Even such products as chili, ginger, cabbage, spinach, shallots, and peppercorns have been grown on drained tropical peatlands.

In the Sundarban wetlands of the Ganges delta in what is now Bangladesh, land reclamation has been proceeding since the mid-19th century. The task has proved difficult because of the rich growth of trees, the poor navigation facilities by water, and the presence of dangerous animals, including poisonous snakes, tigers, and crocodiles. Reclamation has continued over the last 100 years, and the land under cultivation in the Sundarbans grew from about 5,500 square miles (14,500 km^2) in 1880 to almost 8,500 square miles (22,000 km^2) in 1980. In the same period, the population of the area has grown from five million to 25 million, so the demand upon the agricultural resources of the area has become much greater. Most of the reclaimed land is used for wet rice cultivation.

In Burma, the delta of the Irrawaddy River once supported a similarly rich wetland forest and mangrove coastal zone. Tall evergreen trees, growing to heights of 130 feet (45 m), and nipa palms dominated the region, together with *Rhizophora* mangroves. Some clearance and reclamation for agriculture took place in the mid-19th century, but the rice cultivations provided ideal conditions for mosquitoes to breed, causing malaria to become even more prolific than previously, so reclamation slowed down. Changes in the law then encouraged squatters to settle in the unoccupied regions of the wetlands, and reclamation received a boost. Between 1880 and 1980, the area of reclaimed land in the Irrawaddy delta rose from 1,800 square miles (4,660 km^2) to more than 7,000 square miles (18,400 km^2). At the same time the population of the region grew from one million to five million.

In Africa, the pressure for agricultural reclamation of wetlands has been less intense. The lower density of population, when compared to the Indian subcontinent and Southeast Asia, together with the availability of drier cultivable land has served to protect many swamps. In addition, the tropical swamps have provided a source of several useful raw materials, such as papyrus thatch, rope fiber, and basket-weaving material, in addition to their fish resources. The edges of swamps are occasionally grazed and are sometimes burned to encourage new growth of reeds, but the concept of drainage is generally not attractive to most African people. During the last half-century, however, governments in Rwanda, Uganda, and the Congo have encouraged the reclamation of many valley-bottom swamps, through burning them and then establishing a series of drainage channels. The crops grown after drainage include sweet potatoes, sorghum, corn, beans, peas, potatoes, and sugarcane. As in so many other wetlands, the organic content of the resulting soil oxidizes rapidly after drainage and the ground surface sinks. Use of the drained resources for pastoralism, however, reduces this shrinkage effect. In Rwanda, this problem became extreme, with surface peat becoming oxidized to fine particles that could not be rewetted—a situation locally referred to as "black death." This can be avoided as long as water tables are retained within 18 inches (50 cm) of the peat surface.

The severe acidity of reclaimed peats in central Africa (often below pH 3.0) is another problem encountered in the development of agriculture on drained swamps. It probably arises as a result of the oxidation of sulfides in the peat to form sulfuric acid. The addition of lime to counteract this acidity, however, can result in the release of large quantities of ammonia gas, thus losing nitrogen from the system.

The use of highly water-demanding trees, such as eucalyptus, as an aid to peat drainage has been tried successfully in parts of Africa. It is also claimed that the presence of this tree deters the breeding of mosquitoes and hence helps control malaria.

■ WETLAND FORESTRY

Many wetlands bear forest vegetation that has economic value to human beings. From boreal forests to hardwood bottomlands and coastal mangroves, wetland trees have provided an immediate harvest of timber and pulp. True forestry, however, looks to the sustainability of timber production, as well as the maximization of timber growth rates, and this has often meant the management of wetland ecosystems following their initial harvest of trees. Efficiency of production, in the case of forestry, has often required similar basic treatments to those required by agriculture in the wetlands, namely, drainage and fertilization.

The first recorded drainage operations in peatlands with the observed outcome of increasing tree growth come from 17th-century Russia. In Britain, deliberate afforestation of a drained peatland was first attempted in 1730. By the 19th century, drained peatland forestry was widespread through Scandinavia, the Baltic states, and Russia. Foresters now estimate that 60,000 square miles (15 million ha) of peatland worldwide have been drained and planted with trees. The world's leading country in this practice is Finland, followed by Russia and Sweden. The United States has scattered drainage programs; the most extensive are around the Great Lakes, especially in northern Michigan, and the southeastern region, with emphasis on the bottomlands and swamps. In Canada, the forest resource and productivity is such that current demand is met without the need for the expense of extensive peatland drainage.

The objective of peatland drainage for forestry is to increase both the productivity and the biomass of the tree component of the vegetation. The fact that drainage is a means to this end is illustrated by a pine-covered bog in southern Finland. Before drainage, its biomass (in tons per acre—divide by 4 to obtain approximately tonnes per hectare) was 76; 50 years after drainage it was 308, and 74 years after drainage, 472. Annual productivity during that period rose from 9.2 tons per acre, to 17.6 and then to 26.4. These are typical figures, so there is no doubting the efficacy of drainage as a means of improving forest growth.

Drainage can be conducted either using surface ditches or buried tile drains, in which cylindrical tubes that carry water away from the site are buried several feet under the surface. These must open into larger drains to collect and remove their water content. The relative advantages of the two systems vary from one situation to another. In the wet prairies of the Midwest, for example, tile drains have long been favored, but in Scandinavia, where the peat is frozen for much of the year, surface ditches are normally used. The depth of a ditch need not be great in order to be effective, for most lateral drainage in a peatland occurs in the upper layers (the acrotelm—see "Hydrology of Peat," pages 24–27). The standard ditch as used in Finland is 2.5 feet (0.8 m) deep and 4.5 feet (1.36 m) wide at the surface. The optimal spacing between ditches will vary according to the type of peatland, but it ranges from 100 feet apart (30 m) for deep peats to 160 feet (50 m) apart for shallow peats. These ditches collect water moving laterally through the acrotelm and convey it rapidly to major drains and out of peatland. One consequence of this arrangement of drains is that peak flow following storms is greater, so account has to be taken of the possibility of flash flooding downstream.

Originally, ditches were dug by hand, but tractor diggers are now more frequently employed. The waste peat extracted from the ditch can be dumped alongside the furrow, since this elevates the immediate vicinity and provides a basis for the establishment of tree seedlings. Where peats are shallow, ditches may penetrate to the mineral soil beneath the peat, and this can lead to erosion of mineral materials by the runoff water. To avoid this sediment being carried out of the ecosystem and contaminating streams and rivers, sedimentation ponds can be constructed at the exit point of the drainage system.

One further problem is the extraction of mature timber. If the peat is deep, the load-bearing capacity of the ground is very limited, so it may be difficult to bring heavy machinery on to the site. Extraction trackways may need to be constructed, and this will cause blockage of ditches, which have to be repaired when the area is replanted.

The nutrient content of acid, ombrotrophic peats is generally low, so a high timber yield is only possible if the nutrients in short supply are artificially added. Foresters usually conduct chemical tests on the soil prior to planting, but sometimes the original vegetation can provide a clue to the nutrient composition. Sound ecological knowledge on the part of the initial surveyor can save much time, effort, and money later in the process. Rheotrophic, rich fen sites may need no additional fertilizer, while cotton grass bogs may need nitrogen, phosphorus, and potassium additions.

In Europe and Russia, Scotch pine (*Pinus sylvestris*) and Norway spruce (*Picea abies*) are frequently used in the afforestation of drained peatlands. In the exposed uplands of the British Isles, sitka spruce (*Picea sitchensis*) and lodgepole pine (*Pinus contorta*) are more often used. In Canada, the native black spruce (*Picea mariana*) is the main forestry species. The swamps of the southeastern United States have the bald cypress (*Taxodium distichum*), which has long been valued for its timber. Since the Civil War, most of the original forest has been harvested, but natural second-growth stands can still yield good crops of timber. One problem is that secondary growth is often dominated by water tupelo (*Nyssa aquatica*) and red maple (*Acer rubrum*), which are of less economic value than the cypress. The increasing value of bald cypress timber will encourage a more systematic program for the exploitation of this tree.

In tropical wetlands, forestry operations are often limited to the selective extraction of valuable timber, relying on natural regeneration to replace what is removed. In the great coastal swamps and raised bogs of Sarawak and Borneo, for example, the favored trees are *Gonostylus bancanus* and *Shorea* species. These are selectively felled and extracted from the forests on sleds and then by rail tracks, where the surface can support the load. Mangrove timber is also valuable for its durability and toughness and is widely harvested in the Sundarbans to provide poles and pilings for construction operations. Mangrove bark is also valuable as a rich source of tannins, used in the leather tanning industry. Natural regeneration is normally relied on for the sustainability of this timber.

■ WETLANDS AND RECREATION

Recreation in its broadest sense means to occupy oneself pleasurably. It involves entertainment but also an element of refreshment or renewal—literally, "re-creation." Humans use many natural and seminatural habitats for their recreational activities, wetlands among them. It is possible to divide recreational activities in the natural environment into two main types, "appreciative," which seeks only to look, and "consumptive," which demands something else from the experience that is taken from the environment.

The appreciative side of recreational activities can be quite diverse. It may be purely aesthetic, whereby pleasure is gained simply by looking at a landscape, painting it, or photographing it. It may be directed toward the wildlife content of the habitat, involving a desire to name and understand the different components and the ecological principles that underlie their relationships with one another and with their habitat. This type of interest can lead to deeper levels of scientific study and research that may demand materials from the habitat, specimens for identification and study, or samples for analysis. So, the "appreciative" attitude to nature can encompass approaches that range from the purely artistic to the purely scientific.

Consumptive recreation involves the extraction of something from the environment besides the pleasure of looking at it. There may be a symbolic element of domination in this approach, for example in trophy hunting or fishing. The dominance aspect may also take the form of pitting one's skills against nature, as in rock climbing, white-water rafting, or speedboat driving. Or the extraction element may be a simple desire to live with and from nature by hunting, fishing, or otherwise harvesting the products of nature. This approach can lead to wildlife protection and conservation concerns that are compatible with the scientific, "appreciative" approach.

In the case of wetlands, all of these different attitudes and approaches to their recreational use can be found. All of them, even the most apparently innocuous, can do harm to the wetlands that are the source of the pleasure, and for this reason, recreation ecology is becoming an increasingly important area of study. In deciding the management policy for a wetland site, and determining whether recreation (and what kind of recreation) will be acceptable, the evaluation criteria set out earlier need to be considered (see "Valuation of Natural Ecosystems," pages 201–202). A site of exceptionally high wildlife potential may be regarded as unsuitable for any form of consumptive recreation, and perhaps only the lightest forms of appreciative recreation. A site of poorer quality, on the other hand, or one that is less fragile may be able to cope with the consumptive element.

The history of wetland reserves often reflects the changing recreational demands of the public, as in the case of the Keoladeo Ghana National Park, near Bharatpur, India, shown in the photograph. In the 1890s, the maharaja of Bharatpur visited Britain and was impressed by the wildfowl shoots that he attended there, so he decided to set up his own wildfowl hunting preserve by extending the marshes on his land, not far from the city of Agra (famous as the site of the Taj Mahal). By constructing dams to hold back the water, and a series of tracks and causeways to make the interior of the wetland accessible to vehicles, he developed

The tropical wetland of Bharatpur, India. This is one of the richest of the world's wetlands for birds. *(Peter D. Moore)*

a wetland that remains one of the richest for bird life in the world. Its original purpose, of course, was the attraction of wildfowl so that they could be recreationally slaughtered by heavily armed tourists (limited to guests of the maharaja, and including such people as Viceroy of India, Lord Curzon, and the military leader Lord Kitchener). There is a record of 4,273 birds being shot at the site in a single day. This may seem excessive as a consumptive recreational activity, and it became clear through the 1950s and 1960s that such activities could not be reconciled with the increasing demand for appreciative tourism of a less destructive type. All hunting ceased in 1964, but the management of the wetland by artificial irrigation during the dry season has continued, and the causeways now permit birders from around the world to enter the very heart of this remarkable wetland and observe the wildlife at close quarters.

In Australia, the Kakadu National Park, not far from the town of Darwin in the north of the country, consists of river floodplains and swamps, coastal mangroves, and tidal flats. It is owned by the Aborigine peoples and is a site of great spiritual significance to them. For many thousands of years they have subsisted by hunting and gathering from the Kakadu, fishing, foraging, and hunting such birds as the magpie goose (*Anseranas semipalmata*). Nonindigenous visitors both from Australia and overseas have become increasingly frequent at this wetland, numbering currently more than 200,000 per year, and this has had an impact on the nature of the park. Commercial fishing (as opposed to angling), for example, a form of consumptive recreation that has now become little short of exploitative, has reduced the stocks on which the indigenous peoples depend. Even the "appreciative" tourists have inadvertently disturbed wildlife and transgressed the traditional aboriginal rules regarding access to certain religious sites. The further expansion of the Darwin airport will bring in yet more people and add to the problems of ecotourism in this wetland region.

The problem created by the very presence of human beings in high numbers in a reserve area is particularly acute in wetland sites. If water is used as a means of access, then bird and animal life at the aquatic edges of marshes and swamps are disturbed. If access is on foot, then the pressure of trampling can quickly destroy fragile wetland vegetation. Sometimes it is possible to construct walkways through wetland areas, which serve the purpose of controlling foot pressure on vegetation and also confine the visitors to certain locations, thereby limiting damage. The issue of visitor permits can provide income for reserve management but also leads to the temptation to exceed the carrying capacity of a site for such visitors.

The literal pressure of visitors on a reserve is particularly acute in the temperate raised bogs, where the *Sphagnum* bog mosses are especially sensitive to being crushed underfoot. Measurements of vegetation recovery in a hollow boot print on a bog have demonstrated that two or more years may be needed even for a cosmetic filling in of the hollow. Small changes in hydrology can also lead to permanent changes in the composition of the vegetation. Bogs, however, do not generally attract large numbers of visitors, so the damage is often limited to scientific and educational visits by parties of researchers or students. This is not always the case in Finland, where consumptive recreational activity takes the form of harvesting berries from bog plants. Various blueberry (*Vaccinium*) and cranberry species, cloudberry (*Rubus chamaemorus*), and others are gathered and used in the manufacture of a wide range of strongly alcoholic liqueurs. Finland's abundance of bogs and the fairly low population density, however, have resulted in this recreational activity being only a small threat to the health of the mires.

PEAT AS AN ENERGY RESOURCE

Peat consists of the incompletely decomposed remains of plants and animals (see "Peat Deposits: Origins and Contents," pages 31–32). It contains organic matter that is rich in energy but has proved relatively unpalatable to the microbes involved in decomposition in a waterlogged environment; hence lignin, cellulose, waxes, resins, and chitin abound. (These are all natural plant products that are slowest to decompose, so are abundant in peats.) When dried, or when permitted full access to the air, these materials can be decomposed or chemically oxidized quite rapidly, hence the shrinkage of drained peatlands. In a dry condition their energy content can also be released by the rapid oxidation involved in incineration; dry peats can be burned to release energy. The energy content of a *Sphagnum* peat is about 17–20 kilojoules per gram dry weight (about 4.5–5.0 Kcals g^{-1}).

Humans have used peat as an energy resource throughout their history wherever the material was easily available. In the coastal regions of western Europe, for example, peat blocks have long been cut by hand and then stacked to dry under the influence of gravity and the sun. Subsequently, they burn slowly to provide an efficient fire for space heating or for cooking. In areas where fuel wood was relatively scarce, such as Ireland, Scotland, and northern Norway, this form of fuel was and, locally, still is particularly important. Peat blocks could also be used as building materials, forming "turf walls" and even roofing materials. Peat roofs often become invaded by plants, forming a heath vegetation over the tops of these turf-roofed cottages.

In areas of the world where oil, coal, and gas are scarce and where peat is relatively abundant, the use of peat as an industrial energy resource becomes attractive. Russia, Ireland, and Finland are the most notable users of peat as an energy resource, although some is also consumed in this way in Germany, China, and France. In Finland, the energy value of the peat resource is its prime interest, with more than 90 percent of the harvested peat being used as fuel. In most countries, the harvesting of peat is associated more strongly with the horticultural industry (see "Peat for Horticulture," pages 213–214).

Peat can be extracted using drainage techniques, as in the case of agriculture, followed by stripping off the surface layer of vegetation, and then harvesting the peat using machinery. The cutting and stacking of blocks of peat is a technique still used, as in northern Germany (see the photograph on page 212), and especially in smaller operations on a local scale, but more frequent is the milling method, in which the exposed peat surface is allowed to dry, harrowed to break up the dried layer, and harvested using vacuum suction machines. The collected peat can then be compressed into briquettes and burned to produce electricity.

Sometimes the peat deposits cannot be effectively dried in order to collect it this way; occasionally it may even have to be harvested under water. Mechanical extractors are now available that can operate from floating barges. This method is likely to become important in tropical peatlands, where difficulties in developing adequate drainage systems has held back the process of peat exploitation. It will also have a negative impact on conservation, since peatlands that were formerly uneconomic for harvesting will now be threatened.

Other developments in the technology of peat use for energy involve more efficient methods of peat combustion. Traditional burning is inefficient, for it fails to use all the energy that is released. As in the case of coal, the conversion of peat to gas can provide more efficient energy use, and the production of methanol from peat might provide a flexible alternative and a competitive energy source.

Ireland is a small nation that depends heavily on peat as an energy source—14 percent of its energy is obtained from peat. Peat is very distinctly a nonrenewable resource, however, and future projections into the energy sources of Ireland suggest that its dependence on peat will fall, probably to about 4 percent by the year 2011. Ireland currently has five power stations that burn peat exclusively and in total extract 5 million tons of peat per year and employ more than 2,000 people. The nationally run Bord na Mona is responsible for this industry, and from its early days (founded in 1946) it began a vigorous policy of acquisition of the then-plentiful raised bogs of central Ireland. It now owns 205,000 acres (83,500 ha) of peatland. As the conservation importance of peatland and the increasing scarcity of peatland sites have become a matter of concern, Bord na Mona has developed a

Peat harvesting in northern Germany. Many of the extensive raised bogs of this area have been drained and mined for peat. *(Peter D. Moore)*

policy of cooperation with conservationists in selecting sites of particular ecological value that should receive protection. It has handed over certain raised bogs, such as Clara Bog near the Shannon River, to the care of conservation organizations. Following the extraction of peat, the cutaway bogs may be suitable for agriculture or forestry or may be rehabilitated as wetland habitats.

Russia and Finland harvest greater quantities of peat for energy than Ireland, but only 1.5 percent and 2 percent, respectively, of their total energy supply comes from this resource. In the United States, interest in the energy available from peat is obviously concentrated in those regions with a rich supply of peatlands, particularly Minnesota, where as much as 12 percent of the state's surface area could be exploited in this way. Other regions with energetic interests in peat include North Carolina, Maine, and Florida.

In several African countries, the use of peat for energy is being examined, since this would reduce the demand on fuel wood and slow the destruction of many forest habitats. In those African countries where 90 percent of the energy use comes from wood, the peat alternative is often an attractive one, if the peat can be harvested economically. Countries with very extensive peat reserves, such as those of the coastal regions of Sarawak and Kalimantan, are naturally eager to exploit these great assets. Unfortunately, the harvesting of peatlands in many developing countries is often considered without adequate attention to the possible environmental consequences. In Indonesia, the Directorate of Coal, the government body responsible for regulating peat harvesting, is taking ecological, especially hydrological, consequences into account in its selection of which areas to develop. In particular, the directorate is looking carefully

at peat masses with depths of 6–10 feet (2–3 m), which are generally too deep for agricultural drainage and development but may be suitable for peat harvesting.

It is quite understandable that countries with peat reserves (and often with limited alternative resources) should seek to take advantage of them, particularly with an aim to economic gain. The example set by the richer, developed nations would certainly tend to encourage this attitude. From the conservation point of view, however, the argument concerning the irreplaceability of these habitats must rate highly in considering peatland protection. The harvested peat has formed very slowly over many thousands of years; once removed it can effectively never be replaced. The decision to develop, therefore, is an irreversible one and must be made with care.

Peat is rich in energy because it is composed largely of organic materials derived ultimately from the process of photosynthesis. Peat can therefore be regarded as a kind of long-term storage reservoir for the energy of sunlight. The organic matter in which the energy is stored is also rich in carbon, derived ultimately from the atmospheric carbon dioxide that was fixed in photosynthesis by the plants of the past. When wetlands are drained, when peat is harvested and burned as an energy source, or when peat is collected and dug into garden soils, the organic carbon is converted back into carbon dioxide gas. Since this is a greenhouse gas that assists in retaining the radiant heat of the Earth, it is a major contributor to global warming, and any increase in its abundance in the atmosphere is environmentally detrimental. Peat exploitation, therefore, inevitably leads to the loss of a terrestrial reservoir and the enhancement of atmospheric CO_2 (see "Wetlands and the Atmosphere," pages 71–73).

If all the peat in the world were converted back to atmospheric CO_2, then it would raise atmospheric levels by about 20 percent, significantly affecting the energy balance of the Earth. The destruction of the world's peatlands would also remove an important sink for carbon because under present conditions the peatlands are taking more carbon out of the atmosphere than they are returning to it. It would be very unwise of people to destroy any natural mechanism for carbon removal, so the loss of peatlands and the drainage of wetlands is effectively one more self-destruct mechanism for humanity.

■ PEAT FOR HORTICULTURE

Although the use of peat as an energy resource is important in a few countries (mainly Russia and Ireland), its extraction as a material for horticultural use is far more widespread and a much greater global threat to the destruction of wetlands. The demand for peat, particularly in the United States and in Europe, as a soil "conditioner" has made the extraction industry a very profitable one. It is ironic, therefore, that gardeners, who generally have an interest in the natural world and the growth of plants, should bring great pressures to bear upon the survival of wetland habitats.

Most popular as a soil additive is moss peat, which is derived from the *Sphagnum* mosses. This type of peat is obtained by mining the upper parts of raised bogs after removing the surface vegetation. It is easily recognizable because the leaves and stems of the moss survive within the peat matrix. Of slightly lesser value are the sedge peats, which are more fibrous in structure and are derived from marsh and sedge swamp sediments. Also included under the sedge peat category are the peats derived from cotton grasses (*Eriophorum* species), which form distinctly stringy deposits of fine roots and stems.

There are many features of peat that make it effective as an improver of soil conditions for a very wide range of plants. Being an organic material, it provides soil detritivores (such as earthworms) and decomposers (the bacteria and fungi) with a source of energy. In this way, peat addition to a soil increases the activity of these animals, which has an impact on both the physics and the chemistry of the soil. Increased earthworm activity leads to a kind of plowing as the worms ingest soil from tunnels and expel it, often in casts (coiled extrusions of soil and fecal material) at the soil surface. Earthworms thus both aerate and mix the soil and create a structure of "crumbs"—small soil balls between which air can circulate and water can drain.

Microbial activity in peat can benefit plants in several ways. As the organic materials are broken down, elements needed by the plant, such as nitrogen, potassium, and magnesium, are liberated into the soil. Often this activity is concentrated in the region of the plant roots, so the released elements are immediately available to the growing plants. The encouragement of certain types of microbial activity can also result in the addition of new materials to the soil. Nitrogen-fixing bacteria, for example, use organic matter as a source of energy with which to fix nitrogen gas from the atmosphere. However, the value of peat as simply an organic resource in the soil is not significantly greater than that of any other type of organic compost. Indeed, it is usually less palatable to microbes than compost and so decays more slowly even in aerobic soils.

Peat is an effective means of holding water in soil. Being able to retain up to 20 times its own dry weight in water is a feature of the structure of peat, especially *Sphagnum* peats, which are rich in water-retaining pockets and microscopic holes (see "Plant Life in Bogs: Bog Mosses and Lichens," pages 147–151). Adding peat to a free-draining, sandy soil can greatly improve its ability to hold water that will subsequently be made available to plant roots during periods of drought. In heavy, clayey soils, on the other hand, the peat can lead to better structural formation, with more air spaces and freer drainage. Peats are usually packed for sale in a slightly moist state because the fine capillaries within it are very difficult to rewet once they become thoroughly dry. Water simply runs off the surface of fully dried peat and will not be absorbed by it unless a little detergent is used to reduce the surface tension of the water and allow it to penetrate the minute capillary tubes within the organic matter.

The physicochemical properties of peat are most distinctive and provide one of the most attractive features of peat as a horticultural tool. Complex organic materials (polyuronic acids) in the cell walls of plant tissues contained in peat have the capacity to attract and attach to themselves any free, positively charged ions (cations) circulating in the soil. So, calcium, potassium, ammonium, magnesium, and many other such ions are loosely bound to the surface of these compounds. The activities of microbes in the immediate vicinity of plant roots (the rhizosphere) displace these ions and liberate them into the environment once more, making them available to the plant roots themselves. Thus, the peat acts as a short-term storage location for these elements. By becoming absorbed on the peat surface, these ions are prevented from being leached out of the soil and lost to the ecosystem, but since the binding is relatively weak, they are still available to plants when needed. It is this cation-exchange capacity of peat that makes it such a very distinctive and valuable soil conditioner.

The most prized peats, such as the *Sphagnum* moss peats, are very acid and low in cations when they are first harvested from bogs. This may be a valuable feature if the horticulturist is interested in growing plants, such as rhododendrons, azaleas, or heathers, in acidic soils. Addition of raw peat will lower the pH (increase the acidity) of a soil and will favor these species, but if the peat is to be used as part of a general compost for broad use, then it must have chemicals, such as calcium, potassium, phosphorus, nitrogen, and so on, added to it.

Peats are sometimes used simply as a surface dressing to a soil, which is not entirely wise because they will dry out and be difficult to rewet and also because their exchange capacity cannot be fully utilized in this location. Cheaper alternatives, such as bark chippings and bracken detritus, are just as effective in this role.

Considerable effort is currently being expended in finding economically attractive alternatives to peat as a soil conditioner that will not involve the destruction of wetland habitats. Coir is the waste pith from coconut husks, which remains after the fibrous material has been extracted for rope manufacture. It has many of the properties of peat but has a higher pH, a richer content of minerals (especially potassium), and dries out very rapidly. It also seems to hold on to nitrogen more strongly than peat, and young plants grown in it often become deficient in nitrogen. Coir is a promising, renewable resource that may contribute in the future to the replacement of peat, but new handling techniques need to be developed, particularly in relation to watering and nutrient additions.

Perlite is a very light material derived from heating crushed volcanic rocks at high temperature. Particles expand to form light, airy blocks of material that are very effective as a potting substrate. It is better than peat in its aeration properties and just as effective in water retention and in dealing with added nutrients, especially slow-release fertilizers. At present, however, it is more expensive than peat and also has an artificial appearance that is likely to discourage its use in gardens. It is likely that the immediate replacements for peat will include a variety of organic waste materials of plant origin, from paper waste and wood waste to spent mushroom-growing compost and bracken mold. Any or all of these will reduce the pressures on our declining wetland resources.

■ CONCLUSIONS

It is very difficult to evaluate a natural ecosystem in economic terms, but it is a process that conservationists will inevitably have to face if they are to argue for the retention of natural reserves in competition with increasing economic pressures. The land surface of the Earth is limited and the human population is growing, so ecologists need to construct strong arguments if some areas of the world are to be retained in a relatively natural, biodiverse, but, from the human nutritional point of view, unproductive state. Conservationists also need to develop methods of management that can enhance the production of food,

timber, and other natural products while conserving as much as possible of the world's biodiversity.

In order to prioritize sites for conservation attention, the criteria for evaluating natural habitats need to be firmly established. Several such criteria have been put forward, including the degree of biodiversity, the difficulty involved in replacing a site if it were lost, the fragility and vulnerability of an ecosystem, the degree to which it is representative of its kind, the naturalness of the site, and its rarity or the rarity of the organisms it contains. Using these criteria it should be possible for ecologists to select for conservation those sites that are most valuable and most worthy of protection. Even in the absence of economic attributes, therefore, a natural ecosystem can be evaluated.

Wetlands also have a wide range of practical values for humanity, some of which are best appreciated in an undisturbed site, and others require the exploitation and even destruction of the wetland. The highly specialized microbes of wetland habitats may prove useful for the increasing problem of human waste management. Many of the world's wetlands have been a source of food since prehistoric times, containing fish, crustaceans, mammals, and plants that can be consumed. Some of these can be harvested in a sustainable manner, but generally the conversion of a wetland to an agricultural ecosystem usually involves its alteration beyond recognition. Relatively few domesticated animals are able to graze wetlands, the water buffalo being an exception. So the use of wetlands as grazing land also usually involves drainage. The same generally applies to crop husbandry; most domesticated plants need drier conditions than wetlands can supply, with the exception of rice, but even this crop requires severe modification of the structure and composition of a wetland if it is to be grown commercially. With increasing human populations and a rise in demand for food during the course of history, wetlands have inevitably been drained and converted into arable and pastoral systems throughout the world.

Wetlands, as their name implies, contain water, and freshwater is one of the most important requirements for human life. It is possible that the human population of the Earth will ultimately be limited by freshwater supply rather than by lack of food. Wetlands are important storage systems for this precious resource and can also protect low-lying land from floods in time of high precipitation or snowmelt. The drainage of wetlands results in soil shrinkage as water is lost and carbon is oxidized, which means that rivers need to be elevated and enclosed by levees to prevent overflow into surrounding lands, and this practice creates a very fragile and dangerous situation. Not only do wetlands protect lands from flooding, but they also cleanse the water that passes through them, extracting many of the excess nutrients that cause eutrophication.

Wetlands have often been regarded as wastelands, and their conversion to agriculture has usually been referred to as *reclamation,* but such reclamation has meant that the wetlands can no longer perform their natural function of water control and purification.

In some areas the climate or the soil conditions have meant that agricultural development is difficult even when a wetland has been drained. Under such conditions it may prove possible to use the site for growing trees, and the drainage of wetlands for forestry has been widespread. Trees that can cope with wet soil conditions and harsh climates have been transported around the world for use in forestry projects in drained peatlands. Some of the trees from the Pacific Northwest of America have proved extremely valuable in this respect, including Sitka spruce and lodgepole pine. Forestry has been more extensively used in the temperate zone than in the Tropics, but it will undoubtedly increase as natural forests in the tropical regions become depleted of their timber.

Many wetlands are valuable as locations for recreation. Recreation can be appreciative or consumptive. In the case of appreciative recreation, the participant is content to observe and is concerned to avoid any damage to the habitat, while consumptive recreation involves taking something from the site, perhaps hunting trophy animals or damaging the ground by excessive wear. Wetlands are relatively sensitive habitats compared with forests or alpine mountains, and even trampling through them can cause lasting harm, so they need to be handled with care. Boating and angling are often sustainable, but large-scale fishing and hunting need to be carefully controlled if they are to be sustainable in the long term.

Peatlands are particularly sensitive to trampling, but they have proved less popular as tourist resorts than open-water wetlands, so trampling intensity is not often a major threat. Peat extraction for energy or horticulture, however, has had a considerable effect on the peatlands of the temperate zone. Peat accumulates very slowly, so it is effectively irreplaceable, which means that any form of harvesting is not sustainable; peat is a nonrenewable resource. Peat contains energy, so it can be used as an energy source, being burned in power stations to generate electricity. Peat also has important properties as a soil conditioner, increasing the water-holding capacity of soils, enhancing their ability to hold nutrient elements, and providing soil microbes and detritivores with organic food. Peat has long been used for both energy production and horticulture, and the mechanization of peat extraction has led to accelerating rates of peatland loss in recent decades.

Wetlands, therefore, are valuable in a variety of ways, some of which have economic implications. Most of the economic values of wetlands, however, involve consumption and the destruction of an irreplaceable resource. The erosion of any asset that can never be regenerated should not be undertaken lightly, so humanity would be well advised to concentrate on those valuable attributes of wetlands that can be appreciated without involving their destruction.

8

The Future of Wetlands

Wetlands are extremely valuable for a very wide range of reasons, some financial and others aesthetic. They supply food and a means of transport, stabilize water supplies, and provide a means of recreation. But they are also highly threatened ecosystems. They face the threat of drainage and conversion to agriculture, pollution by communities who use them as a means of waste disposal, and disturbance by excessive recreational uses. But wetlands are also threatened by the global changes that are affecting all of the Earth's biomes, chief among which is climate change.

■ CLIMATE CHANGE AND WETLANDS

The increasing quantities of carbon dioxide, methane, and other greenhouse gases in the atmosphere will result in more of the Sun's heat being retained at or near the Earth's surface rather than being radiated back into space. The expected outcome is a gradual rise in the overall temperature of the Earth. Meteorological measurements are available from many parts of the Earth over the last 150 years, so it is possible to check whether this is indeed taking place, and evidence suggests that global temperature has risen by about 2°F (1.1°C) since 1860 (see figure on page 218). This may not sound very much, but when one considers that the difference between a glacial episode in the Earth's history and the intervening interglacial warm periods is only about 10°F (6°C), then the change can be seen as significant in climatic and ecological terms. If this trend continues, as is expected, there will be a number of impacts on the wetlands of the world.

Before examining the possible effects of global climate change on wetlands, however, one must consider in greater detail some complications in the general picture. The illustration shows the temperature curves for the Northern and Southern Hemispheres, and it can be seen that temperature changes have been greater and have exhibited stronger fluctu-

ations in the Northern Hemisphere. This is probably because more greenhouse gases are produced by human industrial activity in the northern part of the world, and there is only a slow exchange of air masses between the hemispheres across the equator. Since wetlands are also more abundant in the north (because of the greater preponderance of landmasses there), this is where one would expect to see the first impact of climate change on the wetlands.

A further complication resulting from atmospheric warming is that the seas will also become warmer and the rate of evaporation will therefore increase. A higher atmospheric humidity will add to the heat blanket of the Earth because water vapor is itself a greenhouse gas, but water vapor may also have the reverse effect by creating a greater cloud cover, which will reflect more of the incoming light energy and so shade the overheating Earth. This also raises the question of where the additional cloud will be found, for clouds produce more precipitation, which could in turn affect the distribution of wetlands. At present, predictions of patterns of future cloud cover are not very reliable, but the boreal wetlands may well receive more winter rainfall. Another factor to consider in this respect is the presence of sulfur in the atmosphere. Sulfur dioxide gas, which is produced by fossil-fuel burning, dissolves in water to produce droplets of sulfuric acid. These act as nuclei for further water condensation, so the presence of the sulfur actually generates cloud. There is an additional source of atmospheric sulfur from the sea, where algae produce the compound dimethyl sulfur, which enters the atmosphere and also acts as a focus for water droplet formation. Areas affected by pollution and oceanic regions, therefore, may be more prone to cloudy conditions in a greenhouse world.

Although most experts are now inclined to believe that the current rise in global temperature is a consequence of human activity by the release of greenhouse gases into the atmosphere, it must be remembered that this change is being imposed upon a much longer-scale fluctuation in global temperature as the cycle of glacial and interglacial periods, which has held sway over the past 2 million years, continues. Judging by past records of interglacial periods, we are currently past the temperature peak of the present warm phase and should be heading for the next ice age. Human industrial activities seem to have set this trend into reverse, but the effect may not persist or may be overtaken by other, shorter-term cycles imposed on us from outside the planet by changes in the Earth's orbit around the Sun or from closer to home as a consequence of high volcanic activity, which has a cooling effect on global climate.

Meteorologists are now developing computer models to predict just how all these variables will affect the global pattern of climate under the changed conditions. By the end of this century, the overall temperature of the world is expected to be 2 to 5°F (1–3°C) higher than today. The predictions are still crude because of the complicating factors that have been mentioned, but it seems that many of the regions of the world that are currently hot and dry will become even more extreme. This means that the temporary wetlands, typical of many desert margins in North Africa, the Middle East, and Australia, will be threatened. They may be replenished with water less often and will therefore dry out more rapidly. This will place pressures on the animals and plants that remain dormant between wet phases. Also, migrant species, particularly birds, will suffer if these wetlands become less reliable in their survival and duration. If the warmth and drought spreads yet farther, perhaps into such regions as the American grain belt of the Midwest, then such wetlands as the prairie potholes will also be threatened. Even the very extensive Pantanal wetlands of South America could experience reduced water supplies and longer droughts.

A close analysis of temperature change over the Earth's surface during the past 10 years shows that the increase in warmth has not been even. As might be expected, the hot regions of the world, such as the western Sahara, have become disproportionately hotter, but so, somewhat unexpectedly, have the polar regions. Global warming, it seems, may be observed first in the melting of ice sheets and of the arctic permafrost. Apart from the physical impact on soils, the temperature rise will affect the performance and the competitive balance between plant species. It is probable that the tree line will move north, reducing the area of open, barren tundra. It might be feared that some of the mire plants in these northern wetlands may be under threat as their available area is reduced, but one of the most impressive features of these plants is their great genetic diversity. The open tundra wetlands have expanded and contracted with glacial cycles over several million years, so the necessary genetic reserves in the vegetation to cope with the current changes are likely to be present.

The reduction in the stratospheric ozone shield at high latitudes may add further stress on the inhabitants of tundra mires, as increased levels of ultraviolet (UV) radiation are periodically able to pass through the atmosphere and reach the ground. Some experiments on the arctic bog moss *Sphagnum fuscum* have shown that artificially enhanced UV radiation over two years causes a stunting of the growth form, which results in a changed architecture in the canopy structure of the mires. This could affect the invertebrate fauna but could also affect litter production and decomposition. An important question is whether such change in temperature and UV radiation could cause changes in the overall balance of carbon in the northern mires. If these wetlands are indeed an important sink for atmospheric carbon and a buffer on the rising levels of carbon dioxide (see "Wetlands and the Carbon Cycle," pages 73–77), then a reduction in their capacity as a sink would have a negative feedback effect on the global carbon cycle. Warmer soils, poorer rates of litter formation, and stimulated microbial activity could convert these peatlands from carbon sinks to additional carbon sources, and the greenhouse spiral could be given a further upward thrust.

■ SEA LEVELS AND WETLANDS

One possible consequence of a warmer world is that more of the Earth's glaciers and ice sheets will be converted to water and enter the oceans. During the course of glacial/interglacial cycles, the level of the oceans has risen and fallen as the volume of water has increased and decreased. This has been due partly to the formation and the melting of ice sheets and glaciers and partly to the expansion and contraction of the oceans as their temperature has gone up and down. At the commencement of each interglacial episode, sea levels rise rapidly, and this is most effectively recorded in the heights of coral reef terraces in the tropical oceans.

At the beginning of the last complete interglacial episode 116,000 years ago, for example, the sea level as recorded by fossil coral reefs was about 300 feet (90 m) below present sea level. As the climate became warmer, the sea level rose by about 250 feet (75 m) within the course of 15,000 years. The difference between glacial and interglacial sea levels, therefore, is very considerable. Such changes affect the area of land over the Earth's surface and influence the hydrology of wetlands, especially those in coastal regions.

Coastlines change in the course of this long-term cycle, but reconstructing the coastlines of the past cannot be achieved simply by observing the contours of continental margins. The question of past coastlines is complicated by the fact that glaciers on the land surface actually depress the Earth's crust, so when they melt, the crust recovers by warping upward. This recovery process can take tens of thousands of years, and many parts of the world's land surface are still rising as a result of the melting of glaciers that were at their peak some 20,000 years ago during the height of the last glaciation.

The British Isles provide a good example of the complexity of interaction between rising sea levels and rising

Changes in temperature over the past 150 years in the Northern Hemisphere, the Southern Hemisphere, and globally. Note the stronger changes in the Northern Hemisphere, possibly because of the higher levels of industrial activity in the north. The baseline is the 1961 to 1980 average.

land surface after the loss of ice loads. Most of the ice during the last glacial episode in Britain lay over Scotland, northern Ireland, and Wales, while the southeast remained ice-free. The sea level at the peak of glaciation was about 300 feet (90 m) lower than at present, so the British Isles were linked to one another and to the mainland of Europe by land bridges. As sea level rose during the present interglacial, these land bridges were cut by rising water, but the north of Ireland and Scotland also became elevated as the land surface warped upward. At present, the northwestern parts of the country

are rising faster than sea level, at a relative rate of about 0.08 inches (2 mm) per year, while in the southeast, where the crustal material is relatively static, sea level is rising relative

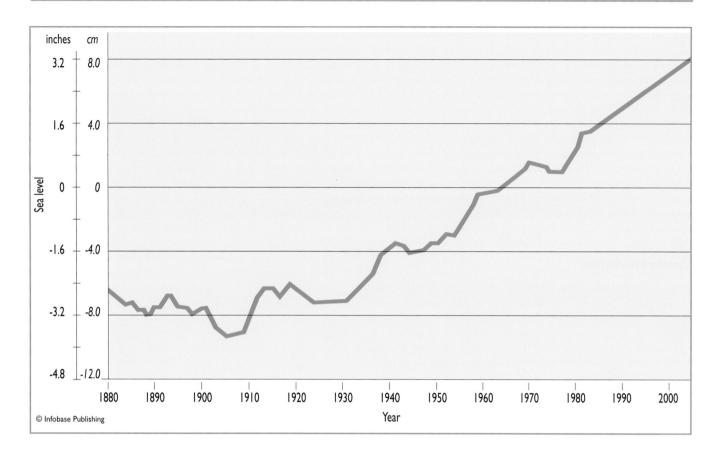

Changing global sea level over the past 120 years.

to the land surface at about 0.04 inches (1 mm) per year. So the different behaviors of coastlines must be taken into account if the consequences of the projected rise in sea level with global warming are to be fully appreciated.

As in the case of global temperature rise (see diagrams on page 218), measurements around the world have established that global sea levels are indeed rising, as shown in the accompanying diagram. The complications of crustal warping make such measurements very difficult and unreliable, but the picture that has emerged from numerous studies around the world suggests that global sea level has risen by about 4 inches (10 cm) over the past 100 years. This corresponds precisely with the annual figure for sea-level rose quoted above for southeast England. Projecting into the future depends on how fast global temperature is likely to rise, so any estimate will have a wide margin of error. The most dependable figures currently available suggest that by the year 2050, sea level will be 8–12 inches (20–30 cm) higher than at present. Some more extreme proposals place this rise at 24 inches (61 cm) higher. By the end of the century, the rise could be as high as 3 feet (91 cm).

Even the moderate figures for the year 2050 make it evident that many coastal wetlands around the world will be put at risk. Most particularly, the saline coastal wetlands, such as the salt marshes, will be strongly affected by such a change. A higher sea level will result in the erosion of sediments, together with vegetation, at the lower end of the marsh sequence. The upper zones of the marsh should theoretically be forced upward as mud is redistributed and the level of the upper marsh is raised. In practice, however, this may not occur because of the nature of the artificial coastlines that human activity has often constructed at the upper end of salt marshes, especially in developed regions of the world, such as eastern North America and western Europe. Protective barriers have often been erected to allow the enclosed land behind such structures to be drained and reclaimed for agriculture or for building. These walls prevent the salt marsh from moving inland laterally, and the result is that the marshes will become confined to a narrow strip between the sea and the human barriers.

In some cases, the barriers themselves will become threatened and may prove too expensive to maintain. When this happens, the simplest solution is a managed retreat, allowing the sea access to low-lying regions once reclaimed from marine incursion. This will be unpopular with landowners, especially in low-lying areas such as the Netherlands, but may become an economic necessity.

Sea-level rises in the past have left their mark in wetland stratigraphy, where marine sediments are found as a distinct layer within the peat profile of a bog. Such episodes

Coastal raised bog on the North Sea coast of Germany. Rising sea level has caused the salt marsh in the foreground to erode and flood the freshwater peat of the bog. *(Peter D. Moore)*

may become more common in coastal peatlands, such as the raised bogs of northern Germany. Already some of the peat bogs that border the North Sea are being eroded and overwhelmed by marine incursions, as shown in the photograph. Saline saltmarsh vegetation is extending into the acidic wetlands as the relative level of the sea to the land changes. Such changes have taken place in the past, but the rate of change may well prove much greater during this century, and wetlands may have problems in coping with such rapid changes.

Rising sea levels in the Arctic will reduce the tundra area from the north, while temperature rises will put this habitat under pressure from the south. In warmer climates, such as Florida and in the eastern regions of China and South Korea, forested coastal wetlands will change in their character, as an influx of saline water penetrates into the inner parts of these ecosystems. The extensive marshes of southern Iraq and the complex of coastal wetlands on the Tana delta in Kenya will also be changed in character by sea-level alterations. The tropical mangrove swamps will be particularly threatened, not only by the rise in water level but also by the likely increased incidence of storms. The possible destruction of the protective mangrove strip along such

coasts as Bangladesh will have serious consequences in the form of storm damage and flooding in the low-lying regions behind them. Other tropical areas to be influenced strongly will be the mouth of the Amazon in South America and the forested raised bogs of the coastal regions of Sumatra and Kalimantan in Indonesia.

Another coastal region likely to be affected by sea level change is the delta of the Nile River in Egypt. The Nile Delta consists of two main branches, spreading in a fan. The seaward edge of the Delta region between these branches consists of sandy beaches and salt marsh, becoming increasingly fresh as one moves inland, with abundant cattail marshes and shallow water lagoons. In the past, the pressure of floodwater down the Nile and the constant extension of the deposited river-borne sediments into the Mediterranean Sea have resulted in gradual replacement of saline wetlands by fresh marshes. This successional replacement can be traced by examination of the sediment stratigraphy beneath the current cattail marshes. Many of these fresh marshes are reclaimed for agriculture as they become shallower with the constant input of silt from the flooding river. But conditions in the Nile Delta have been changing rapidly in recent years. The building of the High Dam at Aswan in southern Egypt (see map on page 186) was intended to control the annual floods of the Nile and to provide a source of water for irrigation at all times of year, but the dam has also acted as a trap for sediments, and Lake Nasser is rapidly becoming silted. This also means that there is less eroded sediment carried down to the delta, so that its northward extension has ceased. New land is no longer becoming available in this northern region of Egypt, and the expectation of rising sea levels in the Mediterranean will make matters even worse because erosion and marine flooding of the Delta wetlands and the reclaimed agricultural land behind them will become increasingly likely.

Sea-level rise is inevitable as the world's temperature rises, but it may not occur as fast as has been anticipated. There are suggestions that the higher sea temperature in the Antarctic, for example, could result in higher precipitation over the Antarctic ice cap, which would compensate for the ice melting around its edges. There are still many unknowns in the equation, and it is consoling to recall that wetlands have survived very considerable sea-level changes in the past. This time the additional constraints placed upon them by human land use may render the wetlands more prone to long-term damage.

■ WETLAND DRAINAGE AND DESTRUCTION

Most of the natural habitats of the Earth are being eroded by human exploitation, and wetlands are no exception. There are many features of wetlands that make them particularly liable to destruction. They are often perceived by the general public as being at best a useless wilderness area and at worst as being a threat to health and safety (see "Wetlands and Disease," pages 197–199). Once drained, however, they can produce valuable agricultural land, and during the process of preparation they may yield peats that have an economic value (see "Peat for Horticulture," pages 213–214). There is also a general feeling that wetlands should be controlled because of the perceived threat of floods and danger to human property and life (see "Water Conservation and Flood Control," pages 204–205). When all these attitudes are put together, it is not surprising that wetlands are one of the most threatened of the Earth's remaining natural habitats.

Wetland destruction, damage, and loss are caused by different activities with different aims in various parts of the world. The main causes of destruction include: drainage for agriculture and forestry; drainage for mosquito control; stream channelization and flood control; waste disposal; development for urban settlements or roads; irrigation schemes diverting water; groundwater abstraction; pesticide leakage, fertilizer and sewage runoff; mining for peat, gravel, or minerals; and sea-level rise. Overall, it is calculated that about 50 percent of the world's wetlands have been destroyed by humans for one or several of the above reasons.

Figures for the United States are more detailed than those available from many other parts of the world, and they seem to fit well with this overall picture. When Christopher Columbus arrived on the western side of the Atlantic, there were probably about 670,000 square miles (1.7 million km^2) of wetland in what is now the United States. Since that time, about 360,000 square miles (870,000 km^2) have been lost because of human action, an area equivalent to about 54 percent of U.S. wetlands. Of this, about 80 percent of the loss was due to land drainage for agriculture. George Washington himself, in 1763, set up a company with the aim of draining the Great Dismal Swamp in Virginia and North Carolina. This policy of wetland destruction continued throughout history, encouraged further by the Swamplands Act of 1850, wherein Congress mandated the drainage of 20 million acres of wetland.

The idea that wetlands might be worth conserving has been slow to catch the public imagination, so the process of destruction has continued to the present day. Between the mid-1950s and mid-1970s, about 5 percent of the remaining wetlands in the United States were destroyed, including 5 percent of marine intertidal wetlands, almost 7 percent of estuarine intertidal wetlands, more than 14 percent of freshwater marshes, and more than 10 percent of forested swampland. So, the threat to wetlands evidently continues. The purposes of these so-called reclamations are varied. Of the freshwater wetlands, 80 percent are converted to agriculture and 6 percent to urban use. Of the saline wetlands, only 2 percent are used for agriculture, 22 percent for urban development, and 56 percent are dredged to make deep, navigable channels.

In some parts of the United States, the proportional loss of wetlands has been even greater. In the central valley of California in 1850, for example, winter floods covered the entire floodplain and led to the development of 62,000 square miles (160,000 km^2) of wetland. By 1939, drainage and conversion to agriculture had eliminated 85 percent of these wetlands. The process has continued and now there remain only about 580 square miles (1,500 km^2) of wetland in the region—rather less than one percent of the original. In North Carolina, the Pocosin wetlands once covered an area of 2.5 million acres (1 million ha), but by 1980 this had been reduced to just 670,000 acres (280,000 ha). The area has mainly been cleared for agriculture, but the consequent changes in hydrology have greatly increased the danger of flooding by runoff water.

Once the land has been drained and the rivers channelized, it often becomes necessary to irrigate floodplain areas to maintain their agricultural potential, and in many of the hotter regions of the world this has led to the evaporation, rather than the through-flow, of water. What is left behind is an accumulation of salts dissolved in the water, and the soils become salinized. Changing the pattern of river flows, together with the agricultural conversion of wetlands, has been a particular feature of the Indian subcontinent, especially in the region of the Indus River in Pakistan. Over the past century, the Indus has been diverted, controlled, and used to irrigate one of the largest reclamation schemes in the world. This newly created agricultural area has proved vital for the maintenance of Pakistan's 138 million people, but it is also now suffering from the inevitable salinization. Each year a further 150 square miles (400 km^2) of the agricultural land of the region is lost because of saline soils developing; already some 22,000 square miles (57,000 km^2) have been affected. A further effect of removing freshwater from the Indus for irrigation is that there is lower flow of water through the estuary, and this leads to the penetration of salt water from the sea farther inland. The remaining wetlands of the Indus delta (mainly mangroves) are now threatened by the increasingly saline nature of the waters within them.

In New Zealand, the remaining wetlands occupy only 10 percent of the area they covered when Europeans first arrived on the islands. Not only do these remaining wetlands suffer from drainage and reclamation but also, as is the case in Australia, from invasive aquatic plants.

The peatlands of northern Europe are being lost at an alarming rate, particularly in view of the fact that they cannot be reconstructed once lost. In England, the wetland area fell by 52 percent between 1947 and 1982, and in Belgium only about 10 percent of the original wetland area remains.

In Africa, many wetlands have survived because of the lack of equipment and funding for major reclamation schemes, but this may well change as investment and agricultural development take place. There are many plans for the construction of polders (drained areas protected from inundation by banks), similar to the schemes long used in the Netherlands. Riverine floodplains and deltas will be particularly attractive for this form of development, which is projected for the Tana delta in Kenya, the Niger delta in Nigeria, and the Kafue Flats of Zambia, among other localities. It is to be hoped that the costs to fisheries, as well as to wildlife conservation, will be taken into account before these plans go forward. Rice is the projected main crop, but the costs of rice production under these circumstances would be very high.

■ THE EVERGLADES

Although North America has experienced severe losses in its wetland heritage as a result of agricultural development, it is most fortunate to possess one of the largest areas of freshwater wetland in the world, the Florida Everglades, sometimes called the River of Grass. These occupy a massive 4,000 square miles (10,000 km^2), from Lake Okeechobee in the north, through the full length of Florida in the southwest. To the west lies Big Cypress Swamp, and around the southern coastal regions are the mangroves that border the Gulf Coast. These features are shown on the map on page 223.

Ever since Florida became a state in 1845, there have been plans and attempts to drain this vast wetland. Early efforts in the decade following statehood concentrated on the Kissimmee River region to the north of Lake Okeechobee, and 50,000 acres (more than 20,000 ha) of this system were drained and planted with sugarcane and rice by about 1865. It was not until the 20th century, however, that serious attempts were made to drain the region of marshland south of the lake. In 1909, the first major canal was completed, linking Lake Okeechobee with Miami, and by 1917 four additional canals had been dug through this area. The region was devastated by hurricanes in 1926 and again in 1928, which caused extensive flooding and fatalities, with more than 2,500 lives lost. For several weeks after these catastrophes, the decomposed bodies of unidentified drowned people were found throughout the region and often had to be piled together in heaps for burning. This calamity led to a concentration of effort aimed at flood control by improving the discharge channels from Lake Okeechobee to the East Coast. The building of levees and hurricane gates, coupled with the lowering of the water table, also permitted the reclamation of the land south of the lake for agriculture. This proved successful, with sugarcane production doubling from 410,000 tons in 1931 to 873,000 tons in 1941.

The region was again hit by hurricanes in 1947 and 1948, and these, coupled with the neglect of the drainage project resulting from the financial depression of the 1930s and the problems caused by the invasion of water hyacinth that blocked the drainage canals, meant that there was no further advance in the draining of the Everglades. Again, most effort was concentrated on protecting the rapidly urbanizing eastern coastal strip from the effect of flooding. This was tackled in 1952–54 by construction of a levee running north and south between nine and 18 feet in height (3–6 m), for 100 miles down the eastern side of the Everglades region. Water conservation areas and pump stations were installed to the west of the levee to cope with periods of high water runoff. These areas held surface water and were also able to supply the urban water demands of the east coast settlements, together with those of the Everglades National Park that had been established to the south.

in the Florida interior. These advances, particularly the enhanced diversion of water to the east in 1963, were not without cost to the wetlands. Many of the wetland habitats, particularly the transition regions of the marshes and glades, were lost and with them their populations of breeding birds. The flow pattern within the region was modified by the systems of canals and levees, leading to alterations in the extent and timing of floods. The general lowering of water tables also led to peat oxidation and the subsidence of land surfaces, as in the fenlands of eastern England (see "Peatland Drainage and Agriculture," pages 205–207). Records from the 1980s indicate that the discharge of water from the area is currently much greater (perhaps three times as great) into the Atlantic Ocean as that through the southern exit rivers. This excess water could be diverted into the Everglades National Park.

Through the 1970s and 1980s, the demand for water from agriculture and the expanding cities of the Atlantic coast has led to an increasingly unpredictable supply of water to the Everglade wetlands, with consequent droughts and wildlife losses. Less than half of the original wetland is still in existence, and only about one-fifth (mainly in the southern end) lies within the protection of the national park. There is now considerable support for Everglades conservation, and there has been extensive legal wrangling over water management in the region. The example of the Everglades illustrates very clearly that conservation of a wetland reserve must take into account and assume some control over the entire hydrological catchment that serves it. Often, the most valuable wetlands lie in the lower reaches of drainage systems, and the abstraction of water, or the diversion of streams, in the upper reaches of the catchment starves them of the water they require for survival.

The Everglades region has very little slope overall, so the movement of water from north to south is extremely slow, and this may add to the problems of management. It has been calculated that it takes eight months for a molecule of water to travel from Lake Okeechobee to the Gulf of Mexico, if it ever gets that far, given the diversion of water into urban use, agriculture, or the ever-present likelihood of evaporation. Ideally, the current period during which water levels are high (called the *hydroperiod*) in the eastern marshes of the Everglades of up to two months in the year should be extended to three–seven months, but this can only be achieved by making the agricultural and urban demands less dependent on the water that belongs to the Everglades and leave more behind to raise the water table of those wetlands.

The population of south Florida currently stands at about six million, mainly along the eastern coast, but it is predicted to rise to between 12 and 15 million by the year 2050. This will place even greater strains on the water resources of the state, so careful management is necessary if

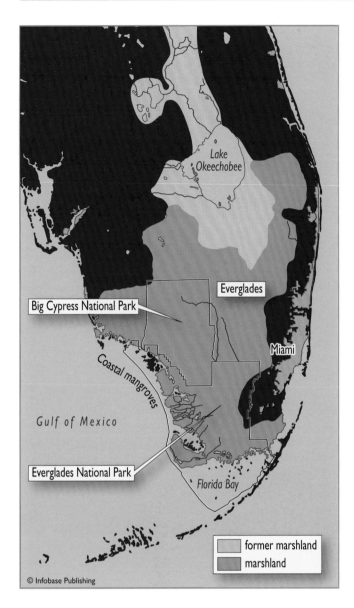

Map of the Florida wetlands showing their former extent and the current location of the Everglades National Park and Big Cypress National Park

During the 1950s, cattle farming was the main form of land use in the drained area, but cattle, grazing on the saw grass peats in the early days of ranching, were found to develop copper deficiencies. More nourishing grasses were introduced as forage for the cattle, including St. Augustine grass (*Stenotaphrum secundatum*). By the 1960s, cattle ranching was giving way to sugarcane and vegetable farming, and in recent times, rice is sometimes grown in rotation with these other crops.

The establishment of flood control and the lowering of water tables on the northern side of the Everglades permitted the development of higher population densities on the eastern seaboard and also the intensification of agriculture the people are to be supplied with water and the Everglades are to survive as well. A plan known as the "Restudy" is now being implemented. It is designed to rearrange and partially restore the water flow patterns of the region, resulting in less water being diverted directly to the sea and lost and more being allowed to move through the marsh and swamplands. The Florida Everglades Forever Act, signed in May 1994, also provides for water treatment areas, where water polluted with fertilizers such as nitrates and phosphates passes through filtration marshes. The vegetation of these marshes extracts the offending elements before the water enters into more sensitive areas.

Steps are thus being taken to prevent the loss of this remarkable ecosystem that was so recently in danger of destruction as a result of human misuse. There still remains, however, the long-term threat of damage by rising ocean levels, which are anticipated over the next hundred years. An increase in global sea level of only a few inches could have profound consequences for these low-lying wetlands by increasing their salinity and therefore changing their flora and fauna. The battle may at long last be swinging in the Everglades' favor, but the war is not over yet.

■ WETLAND POLLUTION BY PESTICIDES

Pollution can take a number of different forms. It may result from adding a chemical (often of human origin) to an ecosystem that is foreign to that system and toxic to some or all of the organisms found there. On the other hand, pollution may result from adding an element that is widely present in the environment, such as aluminum, but in such high concentrations that it proves toxic, or pollution may be caused by the addition of a perfectly innocuous element, perhaps even one that is required by organisms, such as nitrogen, but in concentrations that disturb the balance of the ecosystem. All of these forms of pollution affect freshwater wetlands in different ways.

The rheotrophic wetland ecosystem is a natural collector of all materials that are soluble in water or easily carried by it. The flow of water from a catchment brings these materials together, and the activities of plants and animals in the wetland often lead to the concentration and retention of materials either in living bodies or in the wetland sediments. Such materials may be derived from anywhere within the hydrological catchment of the system, and human activities many miles away can have their effects on wetland chemistry. The use of pesticides and herbicides in the upper reaches of a catchment, for example, can easily affect the quality of water and the community composition of the wetlands downstream. For this reason, wetlands have

often served as biological indicators of the health of entire landscapes, providing early warning of dangers to wildlife and humans from toxic chemicals.

DDT is a good example of a man-made chemical whose history is closely linked to wetlands. This compound, an organic molecule containing atoms of chlorine, was first synthesized in 1874, but its insecticidal significance was not appreciated until the outbreak of World War II in 1939. At that time insecticidal compounds were being tested as a means of controlling lice, fleas, ticks, and mosquitoes among soldiers living under conditions of overcrowding and crude hygiene. DDT emerged as a "wonder compound," proving highly toxic to insects but with no apparent danger to humans. Its chemical toxicity was regarded as similar to that of aspirin—effectively negligible. The tests conducted were based on acute rather than chronic exposure. There is no doubt that the chemical saved large numbers of lives during the war by controlling the vectors of serious diseases, but its success led to progressively more extensive use in agriculture and in malaria-control programs following the end of hostilities.

One property of DDT that soon emerged was its toxicity to fish. This became apparent as a result of spraying wetlands during the course of mosquito control. Such control programs were used not only in those parts of the world where malaria occurred but also where mosquitoes were a mere inconvenience to those visiting wetland areas for recreation. A good example is Clear Lake in California, where anglers found the density of biting insects intolerable. In 1949, a mosquito-control plan was put into effect, using the related chemical DDD, which was considered less toxic to fish than DDT (an important concern, of course, to anglers). DDD was applied in sufficient quantities to cause a final water concentration of 0.015 parts per million—so low that fish toxicity was considered very unlikely. The plan worked well, and midges and mosquitoes were virtually eliminated within a year, although they recovered over the following few seasons in the absence of further applications of the pesticide. It became necessary, therefore, to apply further doses of DDD, but by 1954, other problems became apparent that were feared to be linked to the chemical. The breeding success of western grebes fell to low levels, and then dead grebes began to be found in large numbers during the winter. Extensive research and chemical analysis of various components of the lake's biota (the flora and fauna of the area) led to the conclusion that DDD was becoming concentrated along the food chain from plankton to fish-eating grebe, leading to concentrations of 1,600 parts per million in the fat tissues of the bird, which was sufficient to prove fatal.

In this instance, an artificial chemical of low toxicity built up within the wetland food chains to result in fatal consequences in top carnivores. The same process was then observed in many other food chains, and the top carnivores were always the first affected, whether bald eagles, peregrines, or grebes. High levels of DDT in birds often led to the thinning of eggshells, together with behavioral abnormalities, which together resulted in breeding failure. Such observations gave rise to Rachel Carson's book *Silent Spring* and to a great surge in environmental concern, since it was quickly recognized that humans are at the head of many food chains and are themselves distinctly at risk from accumulative toxins in the diet. It is of note, however, that wetlands, by accumulating pesticides, provided the first warning signals of this kind of problem, which is a convincing argument for the value of wetlands to humans (see sidebar).

For many years, DDT has been banned as a general pesticide in the developed world, though its value for pest control in developing countries, especially tropical countries, has resulted in its continued use in many areas. Malaria remains an extremely serious disease of the Tropics, and in 2006 the use of DDT as a control agent against the mosquito vectors of the disease was being extended, despite the environmental risks.

DDT and DDD are by no means the only pesticides that have been (and in some countries continue to be) a threat to wetland wildlife. The organophosphorus compounds form another group whose members are marked by very high toxicity. These were also introduced into agriculture after World War II to control insect pests of crops and ticks on sheep.

Wetlands as Pesticide Accumulators

One of the reasons why wetlands are such efficient detectors of persistent chemicals in the environment is because they contain animals that process large quantities of water, extracting materials from it. Mollusks and crustaceans, together with fish, need to pass large volumes of water through their bodies to obtain oxygen (and sometimes food, in the case of filter feeders). For this reason, a fish can be much more sensitive to toxins present at low concentration than people would be if they simply drank the water. The concentration of pesticides in humans does rise, however, if they feed on contaminated mollusks, fish, or birds from wetland sites in which toxic chemicals are still used for mosquito control. This is a major hazard among West African peoples, in whose countries DDT is still used for insect control. The dangers of poisoning and the loss of wildlife must then be set against the risk of malaria, if mosquitoes are tolerated.

One of the most toxic pesticides ever used in agriculture is TEPP, one ounce of which is sufficient to kill 500 people. Parathion is less toxic but still lethal to wildlife in relatively low doses. It was used for mosquito control in California and resulted in the death of a range of birds, including game birds. Its association with these deaths, however, was more difficult to prove than was the case with DDT and DDD, because the compound is less stable and therefore more difficult to detect in a corpse.

Other chlorinated hydrocarbons, including dieldrin, aldrin, heptachlor, and endosulfan, have all been used in a similar manner, especially as seed dressings to prevent insect attack. All have been shown to have toxic effects on birds, however, and are persistent in the environment, finding their way into soils and wetland sediments where they can continue to cycle into food webs for many years (at least 10 years in the case of dieldrin).

WETLAND POLLUTION BY ORGANICS AND METALS

Not all pollution results from the addition of man-made toxins to wetland environments. Pollution can be caused by naturally occurring compounds or elements becoming concentrated in certain ecosystems. Even an excess of organic matter entering a wetland in suspended form can result in damage to the ecosystem by overloading part of the natural cycle of decomposition. This can occur, for example, as a result of the discharge of untreated sewage into a wetland—a common condition of many villages in tropical countries.

Sewage treatment involves the destruction by bacteria of much of the organic content of the original material, so the discharge from a treatment plant into a waterway will be relatively low in organic matter, although it may contain soluble chemicals. Untreated sewage, however, is rich in organics, and this stimulates microbial activity in the aquatic system where it ends up. In fast-flowing wetlands, this may not present a problem, for the organic matter is scoured through rapidly and the oxygen supply is usually maintained at a high level. In sluggish systems, the decomposition of organic matter may result in oxygen deficiency in the waters as it becomes used up by the bacteria. Water polluted with organic matter thus has a high demand for oxygen, and this requirement is conventionally measured by incubating the sample to determine how much oxygen is consumed by decomposition in a five-day test period. This is known as the *biochemical oxygen demand* (BOD).

To give an idea of the amount of oxygen that can be consumed by sewage, each human produces in a single day enough sewage to require four ounces (112 g) of oxygen for its decomposition. It would take more than 2,600 gallons (10,000 liters) of oxygen-saturated water to supply enough oxygen to accomplish this. It is evident, therefore, that untreated sewage presents a very serious oxygen demand on a wetland ecosystem, and that such pollution can quickly lead to anaerobic environments and the death of oxygen-requiring organisms. The problem is not entirely confined to sewage pollution. Excess food from fish farms, slurry from farms and agricultural applications, and excessive organic production caused by high nutrient influx to a wetland (see "Eutrophication," pages 61–63), can all lead to oxygen depletion in an aquatic ecosystem.

Organic matter is not the only natural material that can cause problems in wetlands. Many of the elements that are found in nature at low concentrations become toxic at higher concentrations, and these are sometimes found in those wetlands affected by certain types of pollution. Industrial activity and mining in a catchment can lead to the discharge of effluent rich in some of these potentially toxic elements, including lead, copper, zinc, cadmium, arsenic, and mercury. Some plants are sensitive to certain of these elements. Algae, for example, are generally sensitive to copper even at low concentrations (about 0.5 ppm). Fish, on the other hand, can normally survive concentrations of up to one ppm of copper. Many mosses are also copper sensitive, including *Sphagnum,* which means that it is unwise to use copper-based fungicides on boards laid across bog surfaces to reduce pressures from human trampling.

Aluminum is a relatively common element in nature, but it is insoluble in neutral and alkaline conditions. At low pH (acid conditions), however, it becomes soluble and can be leached from catchment soils into wetlands in concentrations sufficient to cause the death of fish. Both Canada and Scandinavia have experienced problems of this sort, especially following the spring snowmelt, when acid water containing sulfuric and nitric acids formed from oxides of sulfur and nitrogen (derived from industrial air pollution) is released from the snow and moves through the soil, dissolving aluminum on its way toward streams and wetlands. Some fish, such as trout, are particularly sensitive to dissolved aluminum in the water, and some plants are also affected.

Lead is also soluble in acidic conditions and, until relatively recently, has been widely discharged into the atmosphere by the combustion of gasoline, where it has long been used as an additive. Mining in the catchments of wetlands can also be a source of lead to wetlands, as can sewage treatment, which may result in concentrations of up to 8,000 ppm. The use and frequent discarding of lead weights by anglers can also add significant quantities of this metal to wetlands. Many deaths of swans in Britain have been blamed on this source of pollution, as the birds deliberately scavenge for the lead pellets in the mud to add to the grit in their crops. Plants are generally unaffected by lead until it reaches relatively high concentrations in the soil (about one percent), above which few species are able to survive. In animals, it is an accumulative poison. Human adults can cope with blood levels of 0.08 percent, but children are more sensitive, and the human fetus even more so. Minute quantities of lead in fetal blood may cause damage to developing nerve and brain tissues, possibly by interfering with blood oxygen transport.

Mercury is another heavy metal that can be an important pollutant of wetland ecosystems. It has been used as a seed dressing and also as a fungal inhibitor in the paper industry. It can also be a pollutant from mining activities. Although metallic mercury is toxic, the products of bacterial activity in the anaerobic sediments of wetlands can create methyl mercury, which is far more poisonous. Unfortunately, the methyl form of mercury is also more soluble, so it moves more easily throughout a wetland. As in the case of lead, it is an accumulative poison, and like DDT it concentrates in the food chain. The most famous situation in which mercury from a wetland caused human calamity took place in Minamata, Japan, between 1953 and 1961. Mercuric sulfate discharged from an industrial plant in a coastal region was concentrated by mollusks, which were then eaten by the local human population. A total of 65 people died from mercury poisoning before the source was discovered.

Metal toxins are carried by the air as dust particles, hence they can be washed out by the rain onto the surfaces of wetlands far from industrial activity. Even the ombrotrophic bogs can be reached by pollutants in this way because of the dependence of these mires on rainwater. Gaseous pollutants, such as the oxides of sulfur and nitrogen, are also transported to the ombrotrophic mires, where they may have a negative impact on the growth of sensitive species, such as the *Sphagnum* mosses, or may stimulate the growth of some species by supplying additional nitrogen to this nutrient-poor wetland. Some wetland plants, such as the grass *Molinia caerulea,* that are more nutrient-demanding have undoubtedly benefited in recent times from aerial eutrophication on the blanket bogs of western Europe. The death of *Sphagnum* as a consequence of sulfur pollution, on the other hand, has weakened the surface structure of many bogs in polluted areas and has rendered them liable to erosion.

Chemical analyses of blanket bog peats reveal the abundance of these elements in recent deposits. Many attempts have been made to trace the history of metal pollution in industrial areas by documenting the changing concentrations of these elements with depth down the peat profile. The mobility of the metals in the peat have often made this task difficult; it results in the movement of some metals both up and down the profile depending on the degree of aeration in the peat. Copper is the one metallic element that seems to remain fixed in the profile, and this offers the best opportunity for historical environmental reconstructions recording the history of pollution in a region.

PEAT EROSION

Damaged peatlands can become extremely dangerous. Raised and blanket bogs often develop in upland locations, in depressions, and over slopes where their stability can be very delicately balanced. In these situations, a bog, which is constantly growing through the accretion of new organic matter from the top and inhibited decomposition below, is liable to become increasingly unstable as its bulk increases and the force of gravity constantly encourages downslope movement. Added to this is the fact that the peat itself is capable of retaining large masses of water in its structure—up to nine times its own dry weight. When totally saturated, therefore, peat on slopes becomes even more unstable. If, at this stage, the impact of human activity is added, such as the cutting of peat faces on the lower parts of slopes, then the stability threshold may be crossed and the entire bog may become mobile, with catastrophic consequences.

In the northeast region of Killarney, Ireland, in 1896, just such a catastrophic bog burst took place three days after Christmas. Cornelius Donnelly, his wife and six children, together with their livestock, were swept away and buried by a tide of liquid peat flowing down the hillside on which the family lived. The tragedy followed a particularly heavy rainfall but was partly due to the excavation of peat from the scarp slope above their residence, resulting in a line of weakness developing. Such flows were not uncommon in 19th-century Ireland, where peat masses are recorded moving as far as 14 miles (23 km) from their source and creating havoc in their wakes. In recent times, it has been possible to study these peat movements in scientific detail. By driving wooden poles deep into the peat but not so deep that they enter the soil beneath, it is possible to survey the movement of bogs on slopes. At one site in northern England, for example, a lateral movement rate of two inches (5 cm) per year has been recorded. If this mobile peat mass were to encounter a break in the slope of the bedrock, or if the cutting of peat were to take place at its lower face, the entire bog could destabilize and burst downhill.

Mass peat movements of this kind are less frequent, however, than the gradual erosion of upland peats by gullying, especially in areas of blanket peat development. Gullies develop in a peatland as a result of drainage streams cutting back into the peat, causing the collapse of vegetation at the surface and exposing the dark-colored older peats to the alternating wetting, drying, heating, and cooling of the atmosphere. This process is gradual and progressive and can result, in time, in the removal of entire blanket bogs, leaving only degraded hummocks of peat resting on the underlying soil. Although water runoff is the main medium for erosion, the exposed peat surfaces can also become weakened and exfoliated by frost and torn apart by wind.

Historical records suggest that these erosion processes have become more frequent in recent times, and this raises the question of what factors actually contribute to the onset of erosion. Is it a natural consequence of continued accumulation of peat, leading ultimately to an inevitable instability, or are there natural climatic factors or human-induced changes that have led to peatland collapse? There are many theories about why upland peatland erosion takes place and why it appears to be increasingly frequent.

Climate change is an obvious contender as a possible cause of the onset of erosion. Peatland ecologists, however, cannot agree whether erosion is more likely under wet conditions or dry conditions. A drying of the bog surface can cause instability in the vegetation cover, leading to the death of some species and the creation of open gaps. If this is followed by very wet conditions, it is reasonable to suppose that the surface "skin" of turf could be disrupted and that erosion would set in. Drought can also render a bog liable to fire, and intense burning can eradicate the entire surface of a bog and greatly encourage erosion. There is a study from northern England that records the history of an upland blanket bog that was regularly burned to encourage the growth of fresh shoots of heather (*Calluna vulgaris*), on which grouse feed. The maintenance of grouse populations was the main management aim, in the hope of conserving the hunting potential of the area, but a period of poor management resulting from a lack of gamekeepers during World War I led to a relaxed burning regime and the accumulation of dry litter. Inevitably, the increasing flammability of the peatland led to a catastrophic fire in 1921, and the peatland subsequently became badly eroded and has never fully recovered.

Another possible cause of erosion in upland peatlands is intensive grazing by domesticated animals. Sheep, in particular, can damage the surface vegetation of a mire both by removing plant material through feeding and by their trampling activities. Sheep grazing on wetlands must step carefully, and they usually develop tracks along which they are assured of firm ground and safety, but these tracks damage the thin surface layers that give stability to sloping mires, and the small gullies produced by sheep tracks carry surface water runoff after storms that often develop into erosion gullies.

Damage to surface vegetation can also result from pollution. Since the blanket and raised bogs are ombrotrophic, pollution in this case can only enter the ecosystem from the air, often washed out by rainfall. It is noticeable that bogs in the vicinity of large industrial complexes are more liable to erosion than those in isolated rural areas. It is known that some of the important peat-forming plants of these mires, including the *Sphagnum* bog mosses, are sensitive to some heavy metals that can be transported through the air from industrial or mining sources. Copper and lead are particularly important in this respect, and their abundance in rainfall has certainly increased over the period when peat erosion has become more common and widespread. Examination of peat profiles often shows a correlation between the disappearance of *Sphagnum* mosses from the fossil record and an increasing abundance of heavy metals, but the correlation may be coincidental, and the real cause may be some other human-induced factor.

One such possibility, which has received increased attention, is the aerial eutrophication of upland wetlands by such elements as nitrogen in the form of nitrates and ammonium ions in the rainfall. The fertilization of bog habitats, which are by their very nature mineral-poor and infertile, can lead to extensive changes in their vegetation. This process may well lead to the kind of ecological instability that results in erosion.

The prevention of erosion, therefore, may be more difficult than has been anticipated. It is possible to control the intensity of sheep grazing, and, to some extent, fire frequency by means of controlled burning. The control of aerial eutrophication of bogs is much more difficult and involves international cooperation. If this is indeed a major cause of erosion, then remediation will prove difficult.

◼ WETLAND CONSERVATION

Wetlands have been particularly subject to damage and destruction in the last few centuries, when mechanical aids for drainage and excavation have become increasingly available. The conservation of wetlands, therefore, requires protection of the remaining intact areas, management to ensure that valued features are retained or enhanced, and reconstruction or repair of lost or injured wetlands in order to expand this shrinking and threatened habitat.

Wetland protection, like so many other conservation problems, is international rather than national in its extent. The development of protective policies by individual governments is obviously commendable, but international agreement is essential if this global resource is to be saved. In this respect, the first priority is the recognition of sites that are of clear international significance and ensuring their protection before destructive processes lead to their loss. Perhaps the most important initiative toward international protection began in the 1960s with the launch of the International Biological Program (IBP), which was intended to coordinate studies of the productivity and energy relations of the world's ecosystems, including the wetlands. The importance of wetlands as highly productive and biodiverse systems became apparent from these international studies, and this led ultimately to the establishment of a Convention on Wetlands of International Importance, the first meeting of which took place in 1971 in the town of Ramsar in Iran; hence, it became known as the Ramsar Convention. The convention has led to the naming of internationally agreed-on important wetlands sites (Ramsar sites) that are worthy of special protection and conservation, and such recognition has helped to assure these sites of a future that is free from destructive development.

Conservation, however, does not mean a simple designation of a valuable site and then exclusion of human activities; such a policy would be practically impossible, even if it were desirable. Humans are part of the global ecological scene, and their relations with wetlands are often very close. The diverse nature of many wetland areas has often resulted from human activity, and some, from the blanket bogs of Wales to the flooded swamps of Bharatpur, India, may even owe their existence to such activity. The long-term survival of wetlands is much more likely to be assured if local human populations can appreciate that economic advantage or an improved quality of life is likely to emerge from their retention. Wetlands as a means of flood control and as a source of fish, fuel, and recreation are therefore more likely to be attractive and to be perceived as useful by local human populations.

Wetland conservation requires active intervention and participation in ecological processes, usually referred to as ecosystem *management*. Precisely what form these activities take depends on the management objectives, such as water storage or raising fish populations, reed harvesting, or the fostering of wildlife biodiversity. Management, therefore, is a purposeful intervention aimed at achieving particular targets for the wetland, and it often involves manipulating the ecological processes within the ecosystem, particularly the process of succession.

It has been seen that succession, the development of successive stages over time in a wetland ecosystem, underlies all wetland habitats. Often the progress of succession is reasonably predictable and is related to such factors as water table depth and the dominant vegetation type (see "Succession: Wetland Changes in Time," pages 14–15), and management practices are very frequently designed to modify or even reverse this process in order to retain a particular stage in the wetland development, such as open water, or reed beds and marsh. In large wetland sites, the overall management policy, particularly if aimed at the maintenance of wildlife biodiversity, may be one of creating and retaining a mosaic of as many different habitats as possible. This may require some drastic intervention, for the natural progress of succession leads toward a degree of uniformity, often with the soil surface elevated above the water table. To slow this process down, the water table may have to be artificially raised, or the soil surface artificially lowered. The latter is often less problematic and less expensive than the former, so conservationists may dig pools in desiccating valley mires or use even quicker but more spectacular methods, such as explosives, to achieve the same result.

Many wetlands have already been severely damaged by human action, but they are not beyond restoration. The process of wetland rehabilitation, therefore, can be attempted even where surface vegetation has been lost and drainage carried out. The main considerations involved in wetland rehabilitation are the reconstruction of the hydrological regime, the reestablishment of nutrient and energetic balances, and, where necessary, the reintroduction of lost species into the ecosystem.

The reinstatement of a lost or damaged hydrological system in wetlands usually involves reversing any drainage that has taken place. Drainage ditches will need to be blocked, perhaps even dams built, to prevent water from escaping the system and to create the waterlogged conditions that are essential for wetlands and are the precursor to the development of peatlands. A return to former conditions of nutrient inputs and circulation may involve controlling land use in surrounding catchments, particularly in the case of rheotrophic mires. The use of local land for arable agriculture, for example, can lead to site eutrophication, as can the establishment of fish farms in the upper reaches of the feeder streams. These factors are easier to control than aerial inputs of nutrients, which may come from a range of sources over a wide catchment and will affect the development of ombrotrophic as well as rheotrophic mires.

Controlling the species composition of the community is best achieved simply by manipulating its physical conditions and relying on natural reinvasion of a site. This is likely to work effectively with mobile organisms like insects and birds. Even fish are remarkably adept at moving from one wetland to another, though they may require reintroduction in more isolated systems. Plant life in wetlands is also relatively mobile, often being carried on (or in) birds, or by aerial dispersal (see "Plant Dispersal in Wetlands," pages 162–164). Plant seeds too are quite resilient and may be found in a dormant state within residual peats and in sediments that have survived at a damaged site. If reintroduction is necessary, it is advisable to use a local source so that the genetic quality of the plants used approximates that of the original populations.

Although the assumption is that the wetland is to be restored to a former condition, the question of which prior stage is preferable can be a difficult one. In the case of raised bog restoration, for example, it is not normally a manager's intention to return to such an early stage in succession as reed beds, marsh, or forested swamp, all of which existed in the bog's former history. The aim is usually to reestablish a relatively late stage of succession in this case. Just how far the clock must be turned back is one of the most important initial questions facing the wetland conservationist.

■ MARSH AND FEN RESTORATION

On the whole, it is easier to restore habitats that belong to an early stage of ecological succession rather than those which belong to a late one. For example, it is easier to construct an open-water habitat than a raised bog, but, even so, the manipulation of these habitats to reinstate conditions and vegetation that have been lost can still present problems. Reed beds, marshes, fens, and swamps are all rheotrophic mires (see "Temperate Reed Beds and Marshes," pages 83–86); their water supply comes from both rainfall and catchment drainage. Damaged mires of this kind have often been denied the input of catchment drainage water or have had this water diverted around them or away from them. The first stage of restoration and maintenance, therefore, is establishing a reliable water supply.

In the case of the prairie pothole mires that have been drained and converted to agriculture, rehabilitation must begin with the reestablishment of a raised water table. The success of this often depends on the type of drainage that was used in the first place. Ditch drainage is easier to correct than tile drainage. Also, where tile drainage has been used, residual fragments of wetland vegetation have often been lost, so raising water tables may not result in a rapid recolonization by aquatic vegetation. In a survey of natural and reflooded prairie potholes, 43 percent of the wetland plant species of the area were restricted to the natural wetlands and had not been able to reinvade the restored sites. This may prove to be a situation where human reintroductions are required.

The chemical composition of the water supply will also modify the type of mire that develops following fenland restoration. Nutrient-rich waters, especially nitrate- and phosphate-rich waters from an agricultural landscape or even from a sewage outfall, can be used in wetland management but will give rise to productive, fast-growing wetland plants, such as the reed (*Phragmites australis*) in temperate regions and papyrus (*Cyperus papyrus*) in tropical Africa. The reed beds and marshes thus produced will be low in plant species diversity but may be precisely what is required for animal diversity. Even reed beds dominated by a single plant species can support a wealth of insect and bird life, and this may be precisely what the wetland manager requires. The problem then is how to maintain these conditions and prevent them from proceeding to the next stage in succession, namely the invasion of woody species.

Halting a wetland succession is extremely difficult because invasion by the next colonist species must be stopped, preferably by preventing the appropriate physical and biological conditions for its establishment from ever arriving. In the case of tree invasion into reed beds, a number of measures can be taken to make the habitat less conducive to successional advance. Cutting the reeds at the end of the growing season (after the bird nesting time is over) reduces the organic input to the sediments (which makes the water more shallow) and also severely restricts the growth of any invasive trees and shrubs. Trees grow from the tips of their stems, whereas grasses (including reeds) grow from their bases and recover rapidly following cutting. The ancient reed beds of Europe were maintained in that condition for centuries by a continued program of harvesting reed beds for thatching materials.

Reed beds may also be burned in winter to reduce biomass, again harming potentially invasive tree species more than the resilient reed. This can cause local eutrophication of water, but this is of little consequence if the reed bed habitat is to be retained. The practice of reed bed burning as a means of management dates back at least 10,000 years in Europe. In the forested swamps of western Java, charcoal has been found dating back as far as 130,000 years, and it is at least possible that *Homo erectus* was responsible for burning these swamps even before our own species evolved. In recent times, burning has been used as a management tool in cattail marshes, where the effect is to reduce cattail domi-

nance and increase the diversity of other plant species. Thus, controlled burning may provide a useful wetland management tool once again.

An alternative approach to management is the alteration of the water table, either by lowering the soil surface through excavation or by raising water levels through construction of levees and dams. Closely spaced dams are more effective than widely spaced ones (see diagram). Pumping water into the conserved area is an expensive process and may be used as a last resort, but alternative methods of marsh management are clearly preferable where possible.

Drainage ditches in a peatland cause the lowering of the water table and consequent damage to the surface vegetation. Rehabilitation is best achieved by raising the water table, which involves damming the ditches. Widely spaced dams are not very effective in raising the water table, but closely spaced dams, as shown in the lower diagram, are very effective in raising the water table.

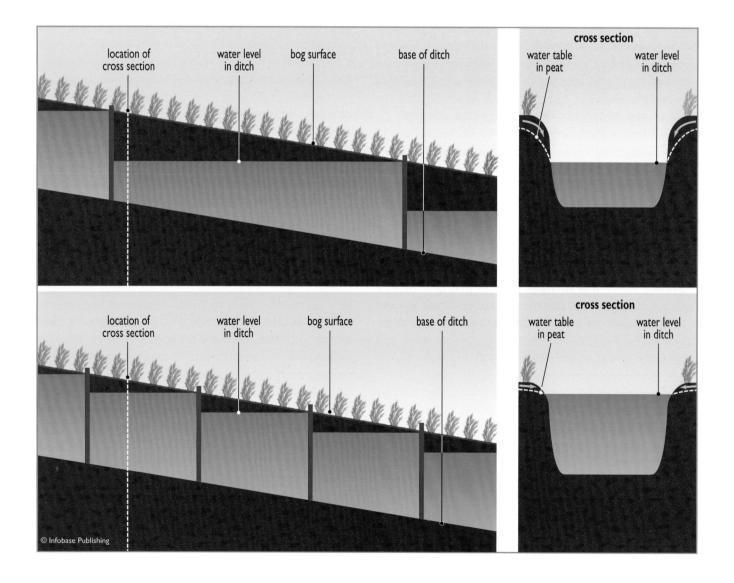

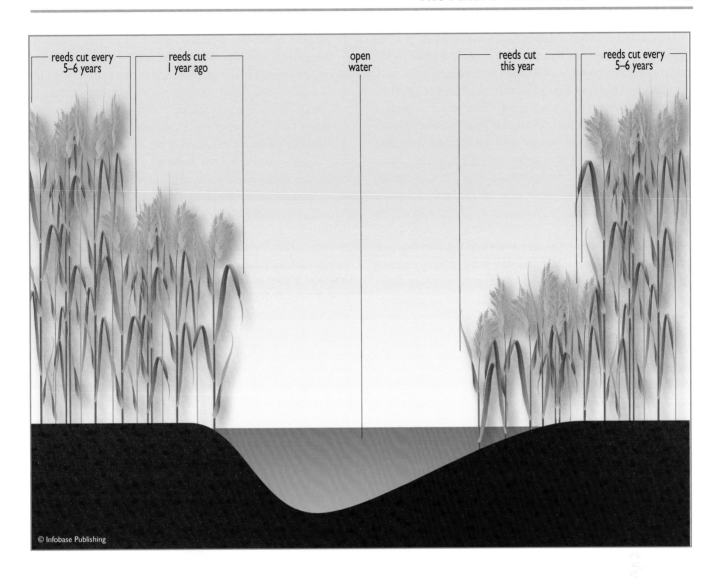

| reeds cut every 5–6 years | reeds cut 1 year ago | open water | reeds cut this year | reeds cut every 5–6 years |

A reed-cutting regime, such as the one shown here, enhances the range of habitats available for birds and other animals. The zone closest to the drainage channel is cut more frequently than more distant reeds.

Most forms of management demand active, and sometimes destructive, measures for a short period of time but result in long-term benefits. In order to ensure that the destructive stages do not result in undue damage to the entire ecosystem, it is advisable to apply management in a "patchy" manner. If the area of reeds cut each year at a site is restricted to about 10–15 percent of the area involved, for example, the animals and birds that require winter cover will still have plenty of suitable habitat available. Birds such as the bitterns require open areas with standing water, and it is quite feasible to manage patches precisely for these requirements, having open pools with reeds around the edge cut every other year, while the reed beds more distant from the pool may be cut (or burned) less frequently. This kind of rotational management can even be applied on the sides of the ditches that conduct water into and through a marsh, as shown in the diagram. Habitats, therefore, should be compartmentalized for management with a mosaic of recovery stages constructed. In this way both habitat and species diversity can be maximized.

Management of a rheotrophic fen wetland for plant species diversity requires a different method than that used for bird species. Natural fens (see "Fens," pages 87–89) can vary in their chemical components from calcareous (rich) to acid (poor), but they are rarely well supplied with sources of nitrogen. They are not as productive as reed beds, are lower in stature, and stand in shallow water. The water supply source needs to be monitored, as an influx of nitrate-rich water could result in alteration of the composition of the fen vegetation to favor competitive, fast-growing species of plants. Pumping polluted river water, rich in sulfur, into fens can create further chemical problems. The sulfur, on contacting the soils of the fenland system, exchanges for phosphates which are then released into the ecosystem, creating a type of internal eutrophication. Simply raising water levels in a soil can help to rid the system of nitrogen because denitrification is increased by anaerobic conditions in the soil.

Like reed beds, fens are liable to become invaded by trees and shrubs if they are not managed. They may be mown as a means of preventing this, and many fens have traditionally been mown to produce winter feed for animals. Frequent mowing (more than once a year) can lead to a high diversity of plant species and may also reduce the nitrogen capital of the ecosystem by constantly removing it in the biomass. Alternatively, of course, animals may be grazed directly on fens. This also serves to reduce biomass accumulation, but defecation ensures that nitrogen is recycled rather than lost to the system. Where calcium-rich, nitrogen-poor waters are available for fen rehabilitation, the greatest plant diversity, particularly among the mosses, is likely to be found.

■ BOG REHABILITATION

Repairing damaged bogs is a far more difficult task than the repair of fens, swamps, and marshes. This is because a true bog is elevated above the groundwater on a mound of peat, and once this peat is removed, it takes thousands of years to renew. So, once a bog has had its peat mound extracted, it is almost impossible to restore it to its former condition. There are, however, certain management options that can at least remediate the damage or make a cosmetic repair possible.

When peat is extracted for horticulture, the most valuable material is the *Sphagnum* peat that often occupies the upper part of the peat dome of raised bogs. This is the prime target of the extractor. Some of these upper peats are also rich in the fibrous remains of sedges and cotton grasses which are of some horticultural value but generally less than pure *Sphagnum* peat. Below these layers in many bogs, however, are peats that are rich in the woody remains of trees from the forested wetlands that preceded the bogs in the course of succession (see "Peat for Horticulture," pages 213–214). The trunks and branches preserved in these peats make harvesting using the milling method difficult and produce a low-grade horticultural peat. The peat extraction, therefore, often ceases at this stage, and the site is either used for agriculture or forestry or becomes available for wetland rehabilitation. The management policy for the repair of a wetland that has reached this stage of exploitation will depend on the level of groundwater left in the mire.

Most bogs are drained by ditches prior to peat extraction, so the first stage in wetland rehabilitation for a cutover bog is to block the remaining drainage ditches and reestablish a higher water table. It is probable, however, if cutting has proceeded as far as the wood layers, that the supply of water to the remaining peats will come from the flow of catchment drainage. Any mires that can be reestablished, therefore, will be rheotrophic marshes, swamps, or fens, depending on the water level achieved. Ditch blockage is most effectively and most easily carried out using peat, especially compacted peat from lower layers which has a very low hydraulic conductivity (see "Hydrology of Peat," pages 24–27). Seepage is obviously the main problem to be avoided, especially in the case of mires where the input of water is not high and where the hydrological balance of the entire system is delicate. There is, of course, a danger that seepage will occur through the sediments that underlie the peat, especially if the sediments are sandy in nature. Sediments of low hydraulic conductivity, such as impermeable clays, form ideal bases for sites where rehabilitation is attempted. In regions where rainfall is low and catchments are small, it may even be necessary to pump water into the mire to keep its water table high during dry periods.

It may be helpful to create pools and reservoirs in the regenerating peatland system by further local peat excavation. These can increase the water storage potential of a mire, becoming flooded in wet winter periods and supplying the peatland system with water through the dry periods. They also create microhabitats within the wetland that are valuable to invertebrate animals and to birds. It is also necessary to consider the hydrology of areas surrounding the mire and to ensure that water tables are kept as high as possible in these "buffer" regions. This can reduce the lateral seepage losses of water from the wetland ecosystem.

Various outcomes are possible for such a rehabilitation scheme. If the water table is maintained at a high level, then open water and marsh habitats may be re-created, providing a wetland of high wildlife potential. If, on the other hand, it proves impossible to maintain high water tables, then invasion by pioneer trees, such as birch, or by grasses, may take place. This is seldom regarded as a desirable outcome because such habitats are often relatively poor in species and do not supply the conditions required by more specialized wetland plants and animals. Trees are water-demanding, so they place a strong hydrological demand on the ecosystem, further lowering water tables and leading to increased decomposition in the remaining peat. Mechanical removal of trees may be necessary, but this can be difficult since trees often regenerate from their stumps. The use of selective herbicides, applied directly to the cut stumps of trees, is often the only feasible option. Even when this is carried out, however, the final structure of the vegetation obtained is not likely to mimic that of the original raised bog. Restored vegetation is often taller and does not have the complexity of hollows and hummocks that are typical of a *Sphagnum* bog surface. Some wetland birds are adaptable and able to cope with the new conditions, but others, such as the golden plover, are very fastidious in their habitat requirements and are unlikely to recolonize bogs of this type.

Not all bogs are cut back to basal peat layers over their entire surface. In older peat-cutting systems, a complex of ditches, channels, and pits, together with intervening ridges along which peat was transported, may have left a very varied topography on an abandoned bog. The flooding of old peat cuttings creates a series of deep, acid, peaty pools that quickly become colonized by floating aquatics and by a diverse assemblage of insects. Old ridges and paths are also a source of many plant species that reinvade the abandoned peat surfaces, and the superficial layers of cut peat may contain fragments of vegetative parts of plants and seeds that can regenerate and germinate to provide a new plant cover.

The critical group of species in bog rehabilitation is the bog moss, *Sphagnum*. The most important of all peat-builders (see "Plant Life in Bogs: Bog Mosses and Lichens," pages 147–151), these mosses modify their local environment to maintain the acidic conditions necessary for bog growth. They are also sensitive to disturbance and tend to be slower in the process of recolonization than their competitors among the grasses, sedges, and ericaceous shrubs. It may even become necessary to assist them in recovering their former dominance in a damaged bog. Experiments on the "seeding" of *Sphagnum* onto damaged bogs in Canada and England have shown that the best results are derived from using fragments of whole plants obtained by passing the moss through a blender, then scattering the pieces over the peat surface or the flooded pools. In the case of the latter habitat, the presence of dead twigs and branches has been found helpful, giving support to the floating carpet of *Sphagnum* that develops. Surprisingly, in view of the low nutrient requirements of most of the *Sphagnum* mosses, a fertilization of peat surfaces with phosphates provides improved results in the reestablishment of the moss.

The time taken for a bog to recover from peat extraction and other types of damage (such as severe drought and fire) varies with the degree of that damage. Totally cutover bogs are effectively irreparable, when viewed on a human time scale. Recovery times measured in thousands of years are unrealistic as far as humans are concerned, so such bogs are effectively lost forever. Flooded peat cuttings, however, and damaged peat surfaces, can be repaired within a matter of decades. Peatlands that suffer regular fire will change in their vegetation constitution. Some plants are more fire resistant

than others; some cotton grasses, such as *Eriophorum vaginatum,* are fire resistant because they form dense, fibrous tussocks, which are not easily burned (see diagram). Bog mosses are more easily damaged and take longer to recover.

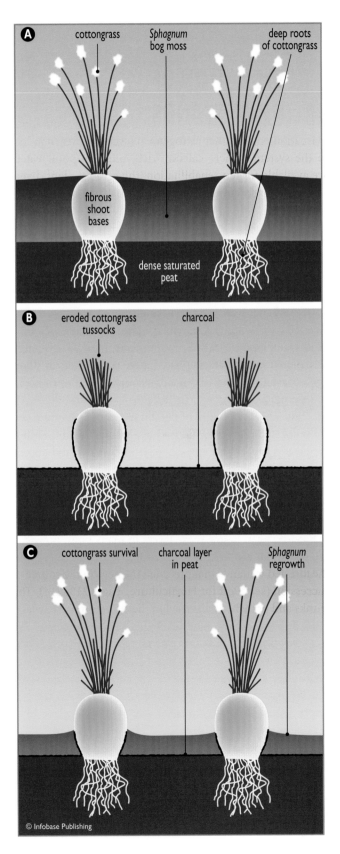

Recovery of wetlands from fire. (A) Before the fire, carpets of *Sphagnum* moss are interspersed with tussocks of cotton grass (*Eriophorum* sp.). (B) Following the fire, the *Sphagnum* is destroyed, but the fibrous tussocks of cotton grass survive the high temperatures and project above the damaged surface of the mire. (C) If the surface of the mire becomes wet again, then *Sphagnum* redevelops, but the cotton grass tussocks continue to project above the surface for many years.

It may take 20 to 50 years for a peat-producing vegetation type to reestablish after severe disturbance, and even longer for an active carpet of *Sphagnum* to invade, but the restoration of such bogs is at least possible and certainly worth the management effort. It is much easier and cheaper, however, to avoid such damage in the first place.

■ RESTORATION OF SALINE WETLANDS

The estuaries, salt marshes, and mangroves of the world have suffered extensive destruction as a consequence of coastal development and coastal pollution. For centuries, global trade was dependent on sea transport, so the demand for ports and docks in estuaries and harbors led to these sites becoming the focus for industrial and urban development. Polluted rivers draining into the sea have also brought to the coastal wetlands a heavy load of toxic materials. In recent years, sea transport of materials that are potentially damaging to natural habitats, such as oil and chemicals, in increasingly capacious vessels has also led to a rising risk of catastrophic pollution incidents around our coasts, with the wetlands being in the front line of damage (see photograph). Hence, restoration of damaged coastal wetlands has become a major area of research for ecologists.

Efficient functioning of coastal wetlands is of great significance to the fisheries industry as well as to wildlife enthusiasts, and capacity of coastal wetlands to filter and absorb many of the river-borne pollutants that would otherwise find their way into the seas also makes them worthy of careful conservation and management. The fact that between 1950 and 1970 some 360,000 acres (146,000 ha) of estuarine wetlands were lost in the United States alone is clearly a cause for concern, especially when the economic value of these wetlands has been estimated at between $16,000 and $70,000 per hectare. Destruction has largely taken the form of dredging and filling of shallow mudflats and marshes, especially since the 1950s. Such activities have been in decline since the 1970s, and in New England, especially, there is now a strong move toward salt marsh rehabilitation.

Where salt marshes have been filled and drained for agricultural reclamation, the natural salt marsh vegetation has often been replaced by more brackish or freshwater varieties. The reed *Phragmites australis* has often taken over from the salt marsh cord grass, *Spartina alterniflora*, along the east coast of North America. The stems of reeds decompose more slowly, and the dead remains have often posed a fire risk in coastal areas. In a coastal valley marsh on the Hammock River in Connecticut, for example, a tidal gate was established in 1913 to control the input of seawater, and the resulting fall in salinity led to the salt marsh being replaced by reeds. It has been easy to reverse this trend by simply opening the gate once more, but the cost has been an increased risk of marine flooding.

More serious problems arise where dumping of spoil or waste materials has taken place over a salt marsh. Again, it is often the reed that assumes dominance in such situations, but the cure has to be more drastic, such as dredging away all waste cover to expose the original marsh surface. The next stage is to replant *Spartina* to stabilize the mud and to commence the process of revegetation. Fortunately,

Polluted coastal wetland near a petroleum refinery in eastern England   *(Peter D. Moore)*

Spartina is very easily propagated by vegetative means, so the establishment of an initial plant cover is not difficult. Other salt marsh species, particularly those associated with low marsh, such as *Salicornia,* invade spontaneously. With an expansion rate of four times its initial cover each year, the process of marsh recovery under the influence of *Spartina* is relatively rapid and rehabilitation is quickly effected. Totally new marshes can also be created in this way, if the appropriate mud cover and slope can be achieved mechanically, so that the area to be created lies between the high and low water marks.

A further type of salt marsh rehabilitation is often required in the vicinity of major oil terminals, where damage can result from accidental oil spillage, or deliberate and illegal tank cleaning at sea. The most immediate and emotive consequence of such spills is the death of animals, particularly seabirds and wildfowl, that become contaminated. Less obvious are the filter-feeding animals, such as mussels, that become clogged with oil. The plants of a salt marsh are generally quite resilient with respect to minor oil spills, but very severe oil contamination can kill even *Spartina.* In such a situation, it may be necessary to replant mudflats to restabilize them. Experiments suggest that the presence of the plants actually enhances the microbial decomposition of oil residues in the mud, thus providing a self-cleansing system for the marsh.

Salt marshes in northern Europe have a very long history of grazing by cattle and sheep, and ecologists have become concerned with the relative advantages of using grazing animals as a management tool in the community. In experiments where marshes that had long been grazed were protected from animals, the general outcome was the excessive growth of a tall, aggressive grass, the sea couch grass, *Elytrigia atherica,* and an overall reduction in plant diversity. Light grazing, it seems, curtails the growth of the most robust species and permits the diversification of the salt marsh turf. Excessive grazing, however, reduces biodiversity by eliminating sensitive species, especially those that grow slowly. The density of grazing animals, therefore, needs to be carefully controlled. Some wild grazers can cause extensive destruction of salt marshes, such as snow geese, which have caused devastation in the salt marshes bordering Hudson's Bay in Canada.

In the Tropics, the coastal mangrove swamps have suffered similar fates to those of the temperate salt marshes—pollution, agricultural reclamation, and despoilation being rife. In the Sundarbans of India and Bangladesh, the recent history of mangrove destruction has led to an experimental program of restoration and reforestation. Reclaimed areas of grassland and scrub in the Indian region of Orissa have been planted with hundreds of thousands of mangrove trees comprising 18 different species. The initial establishment has proved very effective and relatively cheap, but the lon-ger-term care, such as the exclusion of domesticated grazing animals and the trampling effects of local people collecting shrimps and crabs, has proved more difficult to manage. There is also, of course, the problem of predatory animals, such as crocodiles and tigers, being brought close to domestic settlements by the expansion of the mangrove cover. The additional protection the villages received from storms and wave action must be balanced against these risks.

The extensive mangrove forests of the Mekong delta in Vietnam were an unfortunate casualty of the Vietnam War in the 1960s. Defoliants dropped in the area by U.S. military forces reduced about 50 percent of the mangroves to what was then classified as an unproductive wasteland. The area south of Saigon, for example, had virtually no regrowth of mangrove trees or salt water fern, *Archrosticum aureaum,* even several years after the war had finished. The soils have become acid because of the sulfur generated by oxidation as the water table has fallen, and large numbers of crabs are now found there, which may also be slowing down the process of revegetation. There is now a program for reforestation of the mangroves, through which about 75,000 acres (30,000 ha) have been planted with trees. In this way, the protective function as well as the economic value of these coastal wetlands will be restored. Already it is possible for the local inhabitants to begin to use the mangrove forest for fishing and for the sustainable production of fuel wood. Meanwhile, the abundant wildlife of the Mekong delta is gradually recovering.

■ DO WETLANDS HAVE A FUTURE?

It is evident that wetlands are valuable, fragile, and endangered. They are valuable because of their intrinsic economic contribution in numerous ways to the welfare of many peoples in the world, because they harbor a unique and distinctive section of the world's biodiversity, and because they provide an aesthetic and recreational resource for large numbers of people. They are fragile because of their living components, since both plants and animals are particularly sensitive to environmental changes. Wetlands are also fragile because disturbance that may be far removed from the wetland itself, often higher in the catchment, can have an impact on the lower reaches of drainage basins and hence on the wetland ecosystem. Wetlands are endangered because their value is not appreciated, their resources (such as water and peat) can quickly be turned into cash, and the land they occupy has the potential to provide productive pasture, forestry, or arable opportunities, with evident economic advantages. When these three factors are considered together, it is not surprising that so much of the world's wetlands have already been lost.

Efforts to protect wetlands are globally patchy, and good intentions are not always effective when measured in actual results. In the United States, there are extensive legal measures in place that assist in wetland conservation. The Wetlands Loan Act of 1961, for example, enables the Fish and Wildlife Service to provide interest-free federal loans for the private acquisition of wetlands, and the Coastal Zone Management Act of 1972 empowers the Office of Coastal Management to provide funding for state wetland initiatives. Efforts to encourage the rehabilitation of prairie pothole mires and the bottomland hardwood forests are now actively being pursued.

On the other hand, there are tax benefits to farmers who drain, clear, and develop wetlands, as well as a Payment-In-Kind (PIK) Program that encourages agriculturalists to bring unfarmed areas (including wetlands) into cultivation. The actual outcome of these conflicting signals is that wetlands are still being converted to farm use in the United States at a rate of more than 600,000 acres (250,00 ha) per year.

In Europe, the protection of wetlands is even less secure than in the United States. Hunting and wildfowling enthusiasts in Europe have never developed a fully conservationist approach, and their efforts have not been adequately coordinated. As a result they have had less political impact than in North America. In recent years, interest in birding has risen steeply, and birders may develop sufficient political muscle to permit a more powerful lobby for wetland conservation. In Europe, however, there is not as much appreciation of "wilderness" habitats. Thousands of years of relatively high density of human occupation and land use has resulted in a cultural landscape entirely dominated by past and present human activities. In this situation, continuation of intensifying agricultural productivity has prevailed as a general European aim. Political fragmentation makes it difficult to obtain precise figures for the rate of wetland loss in all the countries of Europe, but the rate of drainage in England and France alone exceeds that for the whole of the United States of America.

Many of the world's most important wetlands lie in the developing countries of the Tropics, and here the threat is particularly great. As in the case of the tropical rain forests, the developed nations are calling on developing nations to protect and care for wetlands, despite their own history of habitat destruction for the sake of economic gain. Among the most serious threats to wetlands are the South American governments, most of whom are not signatories to the Ramsar Convention. The floodplain (llanos) and delta of the Orinoco River in Venezuela are increasingly being cleared for cattle grazing, and the enormous flooded forests and grasslands of the Pantanal in southern Brazil are equally threatened. The lower reaches of the Paraná River, which has a basin of more than 1 million square miles (2.8 million km²), forms one of the most densely populated areas of South America.

Rapid deforestation of the upper basin of the Paraná has greatly increased flood frequency, causing, during the catastrophic floods of 1983, more than $1 billion worth of damage. Perhaps sheer economic necessity will provide at least temporary protection to the wetlands of this area.

In Africa, the Niger delta is also used for cattle and sheep grazing. Nomadic pastoralists take their flocks through the area following the receding floodwaters and obtaining nutrition from the first flush of plant growth in February and March. By July, they have moved on to upper grasslands, while agricultural peoples grow their crops on the watered, fertile soils before the coming of the floods once more in August. This ancient cycle, in which wetlands, floods, and human subsistence can operate in a sustainable manner, is threatened by development plans to control the floodwaters by a system of dams, which will mean an end to the old ways of life and the establishment of irrigated agriculture. Despite the problems already experienced in the control of the Nile floods in Egypt, such plans abound in Africa, with no lack of international monetary support.

In the Far East, the coastal swamps and bog forests of Indonesia are in great danger as high population density and the attendant demand for agricultural and other resources increases. Deforestation, peat extraction, and the development of rice cultivation on the coastal plains that currently support these enormous tropical wetlands seem inevitable.

The outlook for wetlands is far from rosy, but neither is it entirely a matter for despair. Among the bright spots in the history and the future of wetland conservation is the Ramsar Convention, which has resulted in a list of several hundred internationally important wetland sites worldwide, and their presence on this list, though not a guarantee of protection, does convey some international prestige that argues well in their defense. Bird life plays an important part in the identification of internationally important sites, mainly because of the high public interest in this class of animals and the fact that survey data are more reliable for birds than for any other group. Wetlands with more than 10,000 wildfowl, or 20,000 waders, or with greater than one percent of the world's breeding population of a species, become eligible for nomination on this list. Although these criteria are somewhat arbitrary, they form the basic requirements for the highest grade of evaluation, and to this core group of wetlands it is hoped that more extensive and representative sites can be added.

Wetlands are a global resource and a global responsibility. They cover only about six percent of the world's surface, but they harbor a substantial and distinctive proportion of the Earth's biodiversity. Their value in flood control, the provision of water, and even the absorption of pollutants, both dissolved and gaseous, is only now being fully appreciated. For too long in human history they have been regarded simply as wastelands, and it is to be hoped that their true value as part of our planet's heritage will be recognized before too much of this distinctive habitat is lost.

CONCLUSIONS

Wetlands, unlike the other major terrestrial biomes of the world, are scattered over many latitudes and climatic zones. They are often fragmented into small areas, and they often lie in regions, such as valley floors, where human settlement and agriculture is a priority. As a consequence, wetlands are under extreme threat from human population expansion and demand for land. They are also at risk from climate change, so their security and survival is very precarious.

Climate change, especially global warming, will have its greatest impact on the high latitudes. The tundra and boreal wetlands will undoubtedly suffer as the polar regions become warmer, partly because microbial activity will increase as soils become warmer and drier, and partly because more robust vegetation, including shrubs and trees, will invade from lower latitudes and colonize mires. The vegetation of the Arctic is already changing, and this is expected to continue. The chances of survival of tundra wetlands may be increased if precipitation becomes greater as a result of climate change, and current models predict that this is very possible.

Sea levels will rise as a consequence of global warming, and this presents a threat to coastal wetlands, especially salt marshes and mangrove swamps. Coastal defenses, such as seawalls, levees, and barriers, have been erected in many developed parts of the world to protect towns and agricultural land against incursion by the sea, but marshes on the seaward side of these defenses will become crushed into increasingly narrow strips as sea level rises. Mangrove swamps are tropical and subtropical coastal wetlands, and their survival will depend on their ability to move inland with the retreating shoreline. In Southeast Asia, some tropical bogs lie close to sea level and will also be placed under threat.

Human drainage and destruction of wetlands has been proceeding for many thousands of years, and the process has accelerated very rapidly in recent decades as agricultural demand has increased with growing populations, and the development of sophisticated technical equipment has allowed drainage to proceed faster. Only very recently have people begun to realize the value of wetlands as mechanisms for water control and to appreciate the advisability of con-serving them. The history of the Everglades in Florida provides a clear illustration of the errors that have been made in wetland management and also the opportunities still available to rectify these errors.

Wetlands have long provided a convenient way of refuse disposal. Most contain flowing water, which has been perceived as the ideal medium for carrying sewage and other waste away from human settlements. As societies became increasingly industrialized, the wetlands were considered the simplest solution to the problem of chemical waste disposal, so the lowland wetlands of the world became increasingly polluted as a result of human activity. The intensification of agriculture added to the problem as fertilizers and pesticides were added to the effluent. The dangers of these processes are now appreciated, and it is realized that pollution of wetlands results in human health being put at risk as well as the survival of the wetland. What is not always appreciated is that air pollution by pesticides and by exhaust fumes from vehicles can also affect wetlands, even the ombrotrophic mires that lack any incoming streams and have escaped the worst effects of aquatic pollution in the past.

Air pollution is also responsible for the erosion of some upland peatlands, especially those downwind of industrial activities and dense urban areas. Pollutants kill some sensitive plant species, especially the bog mosses, and the outcome is the loss of surface vegetation and the erosion of underlying peat by drainage water. Sheep grazing can add to the pressure and encourage the development of erosion gullies.

Wetlands that have been damaged by drainage or erosion can be restored to their former state if they are properly managed. The rehabilitation of mires demands careful control of their hydrology by manipulating water levels, usually raising them by establishing dams. Early stages in wetland succession, such as marshes and fens, are relatively easy to reestablish, but the ancient peatlands, including raised bogs, are very much more difficult to recover. For this reason, the remaining examples of these types of wetland need to be protected carefully.

Undoubtedly, wetlands do have a future. Enlightened people are now aware of just how dependent human communities are on the wetlands, so the processes of destruction and pollution that have accelerated in recent decades hopefully can be controlled, and some of the damaged wetlands can be restored to something approximating their former state.

9

General Conclusions

Wetlands are unlike all of the other terrestrial biomes of the Earth because they are not restricted to any particular region. They are scattered throughout the world, wherever the supply and retention of water in a landscape permits their development. As a consequence, wetlands are much more fragmented that any other biome, and fragmentation brings both advantages and disadvantages from an ecological point of view.

The wide geographical scatter of fragmented wetlands has meant that, over the course of geological time, animals and plants from a very diverse range of taxonomic and ecological backgrounds have been able to invade them and adapt to the distinctive conditions they find in the wetland habitat. Some wetlands are dominated by Arctic lichens, bog mosses, or tough, woody dwarf shrubs, while others bear tall tropical trees, even palms. The northern mires are grazed by reindeer, while the tropical marshes support antelopes and water buffaloes. Arctic foxes and wolves prey on the tundra mire wildlife, but in the tropical forested wetlands big cats, such as tigers and jaguars, are the top predators. The scattering of wetlands through such diverse latitudes and climates has thus resulted in an immense biological diversity becoming associated with them. Although individual wetland types, when considered in isolation, may not contain the great biodiversity of the tropical rain forests, when the wetlands of the world are considered as a group, their variety is impressive.

Geology, and consequently chemistry, also vary greatly in different parts of continents, resulting in a range of different landscapes developing, and each of these topographic landscape types develops its own characteristic set of wetland types. Some wetlands are rich in plant nutrients, while others are poor. Some wetlands have an abundant supply of calcium carbonate, especially if they develop in limestone regions, and this greatly affects their flora and fauna. The supply of nutrients to wetlands is also dependent on the nature of the human land use in the surrounding catchment. Farming and forestry often involve soil disturbance, which leads to soil being eroded into low-lying wetlands, and the application of fertilizers usually results in an excessive runoff of nutrient elements that find their way into the wetlands. So the nature of a wetland is affected by the activities of the people who surround it.

Wetlands are also dynamic ecosystems that are in a constant state of change. Many of these changes are fairly predictable, as in the case of infilling lakes as sediment accumulates, helped by the growth of aquatic plants. Reed beds and marshes extend from the margins into the shallow waters and establish new types of wetland with different assemblages of plants and animals. The process continues, and marshes often give way to forested swamps as the soil surface becomes elevated by deposition and conditions become suitable for tree invasion. Where precipitation is high enough, even these swamps may become supplanted by fast-growing humble bog mosses that prevent the regeneration of trees and gradually overcome their taller competitors by locking up nutrients and acidifying the waters. The peat domes of raised bogs are among the most impressive of all wetland habitats and, though relatively poor in terms of biodiversity, they contain some remarkable creatures, including plants that eat animals, and spiders that create their own tunnels through the canopy of bog mosses.

The wide and diverse scatter of wetlands through so many climatic and geological settings thus provides them with a great range of plants and animals that cope with the distinctive problems associated with life in the wet. Fragmentation has also given wetlands a certain degree of protection. A local catastrophe, whether natural, such as a hurricane or a fire, or human induced, such as a pollution event, may be confined because of the relative isolation of wetland units. The survival of neighboring wetland units that are untouched by the catastrophe allows species to reinvade the damaged wetland and assist in its recovery. A scattering of sites means that all of the wetland eggs are not in a single basket.

Fragmentation also has its disadvantages. Habitats that are split into smaller units by alien terrain can become vulnerable as pressures develop around their edges. Small wetland sites are more easily drained for agriculture, often commencing with the extension of agricultural land around their perimeters. Edges of all habitats are often the most dangerous places for local inhabitants because there they are less well protected by their normal surroundings and become easier targets for predators. A wetland that is broken into small units has a greater length of perimeter habitat per unit area than the equivalent area of a single unit. Local extinction of wetland species thus becomes more likely in a fragmented situation. The problems of wetland fragmentation are becoming more apparent as human populations expand and the demands for agricultural and urban development place greater pressures on natural landscapes.

For the conservationist, the biodiversity of wetlands is reason enough for their protection, but many people find value only in those things that bring direct health or economic benefit. The survey of wetlands presented in this book has illustrated that very wide range of advantages that wetlands bring for humanity. They control the water cycle of the Earth, and the world's supply of water may soon limit the development of human societies in many parts of the world. Without wetlands the residence time of freshwater on the surface of the land and in its buried aquifers would be reduced, leaving people with less water for their needs. Protecting wetlands is thus a wise move to ensure a continued supply of our most vital resource after food, namely, water.

Wetlands are easily contaminated by agricultural and industrial waste products and leakage. This damages the wetland, but it also has its impact on human society. The products of the wetland, including fish or ducks, become polluted and may even be unfit for human consumption. The water itself will require costly treatment if it is to be drinkable. There are many places even within the developed world where the leakage of seemingly harmless agricultural fertilizers, such as nitrates, into wetlands is so excessive that local tap water is unfit for use with young babies. As is so often the case, pollution of the environment rebounds directly onto the local human population. Some wetlands, especially marshes and reed beds, accumulate toxins and thus cleanse the water that passes through them, helping to purify the waters that are drawn on by people.

Wetlands also record the history of their local environments in their sediments, and this source of information is becoming increasingly important to archaeologists and students of climatic history. The loss of a peat bog is also the loss of a unique archive of data concerning environmental history. As scientists try to understand and predict current changes in global climate, the records from the past take on increasing importance, forming a baseline against which present change must be evaluated. The record of the fossils from lakes and peat bogs also provides a means of studying the history of our own species. Land-use change, especially those that date from before the time of recorded history, can best be studied through the indirect evidence of fossils. Sometimes the remains of dwellings, boats, trackways, and even human bodies emerge from the wetland sediments and provide a glimpse of the way we once lived. The picture that emerges from these ancient times is one of a long and wet relationship between people and wetlands. Sometimes they formed barriers that could be crossed only by boats or elaborate wooden trackways, and sometimes they formed the routes for exploration, discovery, migration of peoples, or simply the access to food.

The development of human society has been closely linked to wetlands in many parts of the world, from the Arctic to the Tropics. In recent times people have become separated from wilderness areas and have increasingly been concentrated into urban habitats, where every aspect of the environment has a human signature on it. People are only comfortable when they see the marks of other humans around them and become insecure and frightened when separated from signs of civilization. The prehistoric attitude of oneness with the wilderness has been completely lost, and this alienation has cost the environment dearly because it has engendered a desire to bring all things under human domination and control.

Wetlands are no exception, indeed they are generally perceived as particularly useless and even threatening to humanity. Perhaps in recent times the materialistic and utilitarian attitudes of humankind are gradually being replaced by an appreciation of the natural world and a return to the admiration of wilderness for its own sake. The rise of ecotourism is a sign of this change in attitude as worn urban dwellers seek the inner peace that wild places offer.

Year by year scientists are uncovering more and more remarkable facts about the ways in which wetlands function and also about the sophisticated adaptations of the strange creatures that inhabit these waterlogged habitats. No doubt many of these discoveries will prove of value in the advancement of academic studies as well as in industry and medical science, but among the more spiritually perceptive, such discoveries will also encourage a sense of wonder at the extraordinary complexity and order of the world.

Glossary

aapa mire   sloping FENS found in the boreal regions of North America (where they are sometimes referred to as STRING BOGS) and Eurasia. A series of ridges (STRINGS) and linear pools (FLARKS) run across the slope of the fen, giving these mires a distinctively striped appearance when seen from the air

acrotelm   the upper layers of peat at the surface of a mire that consist of uncompacted dead organic matter. Water moves with ease in the acrotelm—it has a high HYDRAULIC CONDUCTIVITY. These layers are periodically aerated when the water table falls. Decomposition is therefore faster in the acrotelm than in the CATOTELM below

active layer   the upper soil layers in Arctic PERMAFROST environments that melt in the summer and freeze in the winter

aerosol   particles suspended in the atmosphere as a result of their very small size

aestivation   a period of dormancy associated with the unfavorable conditions of summer drought (equivalent to hibernation in winter)

albedo   the reflectivity of a surface to light

allochthonous   of material that has originated away from the site in which it eventually settles (such as leaves carried into a lake). *See* AUTOCHTHONOUS

allogenic   forces outside a particular ecosystem which may cause internal changes; for instance, rising sea level can influence water tables in freshwater wetlands farther inland. It is therefore considered an allogenic factor

anaerobic   lacking oxygen. *See* ANOXIC

anion   elements or groups of elements carrying a negative charge, e.g. NO_3^-, HPO_3^-

anoxic   lacking oxygen. *See* ANAEROBIC

aquifer   a body of rock that is porous and permeable to water underlain by an impermeable layer, resulting in the storage of water beneath the ground

ATP   adeonosine triphosphate. The molecule in cells that acts as a temporary storage system for energy

autochthonous   material that has originated in the site where it is deposited, e.g., bog moss peat in a bog. *See* ALLOCHTHONOUS

autogenic   forces within an ecosystem that result in changes taking place, e.g., the growth of reeds in a marsh result in increased sediment deposition. *See* FACILITATION

autotrophic   organisms that are capable of constructing complex organic molecules from inorganic sources, such as green plants

biochemical (or biological) oxygen demand   the amount of dissolved oxygen used up during the breakdown of organic pollutants in a water body. It is used as an index of organic pollution in water

biodiversity   the full range of living things found in an area, together with the variety of genetic constitutions within the species present and the range of microhabitats available at the site

biomass   the quantity of living material within an ecosystem, including those parts of living organisms that are part of them but are, strictly speaking, nonliving (e.g., wood, hair, teeth, claws), but excluding separate dead materials on the ground or in the soil

biome   a large-scale community that is defined on the form of its vegetation. Biomes include tundra, boreal forest, desert, tropical rain forest, and others

biosphere   those parts of the Earth and its atmosphere in which living things are able to exist

biota   the sum of living organisms, plants, animals, and microbes

blanket bogs or **blanket mires**   rain-fed wetlands (true BOGS) found in regions of high PRECIPITATION, mostly in cool, temperate, oceanic regions, but also on some tropical mountains. Blanket peat deposits spread over valley floors, hilltop plateaus, and over all but the steepest slopes

blue-green bacteria (Cyanobacteria) once, wrongly, called blue-green algae. Microscopic, colonial, photosynthetic microbes that are able to fix nitrogen. They play important ecological roles in some wetlands as a consequence of their nitrogen-fixing ability

bog a wetland ecosystem in which the water supply is entirely from rainfall (OMBROTROPHIC). Such wetlands are acidic and poor in nutrient elements. They accumulate a pure peat with little mineral matter, so are prized for horticulture. They include raised bogs, blanket bogs, and bog forests

bog burst a term applied to catastrophic and sudden erosion of peat masses that results when they develop on unstable slopes and then become charged with excessive loads of rainwater

bog forests acidic, rain-fed, domed tropical mires (true BOGS) that accumulate deep peat deposits in some equatorial coastal regions, particularly in Southeast Asia. They are regarded as the closest modern equivalent to the Carboniferous coal-forming swamps

bog mosses a distinctive group of mosses, all belonging to the genus Sphagnum. They have the capacity to hold up to 20 times their own weight in water and are also able to retain CATIONS. Most species are associated with acidic mires

boreal northern, usually referring to the northern temperate regions of North America and Eurasia, which are typically vegetated by evergreen coniferous forests and wetlands. Named after Boreas, the Greek god of the north wind

calcareous rich in calcium carbonate (lime)

capillaries fine tubes, as in the structure of partially compacted peat

carr a Scandinavian word generally used in Europe for forested wetlands with flowing water (rheotrophic). Equivalent to the North American swamp

catchment a term meaning a region drained by a stream or river system (equivalent to WATERSHED)

cation elements or groups of elements with a positive charge, e.g. Na^+, NH_4^+, Ca^{++}

cation exchange the capacity of certain materials (such as peat and clay) to attract and retain CATIONS and to exchange them for hydrogen in the process of LEACHING

catotelm the deeper, compacted layers of peat, which are permanently waterlogged and ANOXIC and which have a very low HYDRAULIC CONDUCTIVITY and so are virtually impermeable to water. Decomposition within the catotelm is very slow, in contrast to decomposition within the ACROTELM

charcoal incompletely burned pieces of organic material (usually plant) which are virtually inert and hence become incorporated into lake sediments and peat deposits, where they provide useful indications of former fires. Fine charcoal may cause changes in the drainage properties of soils and lead to waterlogging and mire formation, as in the case of many VALLEY MIRES and BLANKET MIRES

climax the supposed final, equilibrium stage of an ecological succession. Many question whether real stability is ever achieved

coal ancient peats that have been physically and chemically altered as a consequence of long periods of compression, sometimes at high temperature

community an assemblage of different plant and animal species, all found living and interacting together. Although they may give the appearance of stability, communities are constantly changing as species respond in different ways to such environmental alterations as climate change

competition an interaction between two individuals of the same or different species arising from the need of both for a particular resource that is in short supply. Competition usually results in harm to one or both of the competitors

consumer an organism that relies on other organisms for its food (HETEROTROPHIC). Consumers may be primary, secondary, tertiary, and so on, depending on their position in a FOOD WEB

continental climate a climate characterized by hot summers and cold winters, often with low precipitation, resulting from a weak influence of the world's oceans

coppicing a management system applied to certain trees and shrubs in which the stems and branches are cut back to a "stool" only a few inches above ground level. Buds on the stool ensure that the plant regrows, producing new shoots for future harvests. A cutting cycle of between 10 and 20 years is usually applied. This system of wood harvesting has been used for at least 6,000 years in Europe

Coriolis effect the tendency of free-moving objects, such as air masses, to be deflected by the rotation of the Earth on its axis. Deflection is to the right in the Northern Hemisphere and to the left in the Southern Hemisphere

Crassulacean Acid Metabolism (CAM) a photosynthetic mechanism in which carbon dioxide is temporarily fixed in the form of organic acids (often during the night) and is later released within the plant cells to be fixed again by conventional metabolic processes

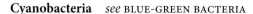

Cyanobacteria *see* BLUE-GREEN BACTERIA

deciduous a plant that loses all its leaves during an unfavorable season, which may be particularly cold or dry

decomposer a MICROBE involved in the process of DECOMPOSITION

decomposition the process by which organic matter is reduced in complexity as microbes avail themselves of its energy content, usually by a process of oxidation. As the organic materials are respired to carbon dioxide, other elements such as phosphorus and nitrogen are returned to the environment where they are available to living organisms once more. It is therefore an important aspect of a NUTRIENT CYCLE

deterministic a process in which the outcome is predictable and does not allow for chance (STOCHASTIC) events

detritivore an animal (usually invertebrate) that feeds on dead organic matter

diatoms a group of one-celled photosynthetic organisms that form an important part of the phytoplankton in wetland habitats. They construct cases (FRUSTULES) made of silica, which survive in lake sediments and can be useful in the reconstruction of past conditions, such as the acidity of water bodies. They are thus useful in the study of environmental history

dissociation the separation of two elements from one another in solution, forming charged IONS

diversity a term that includes both the variety of elements in an assemblage and the relative evenness of their representation

domed mire *see* RAISED BOG

dune slack the wet FEN ecosystems found in the hollows between sand dunes. As a result of the lime in broken mollusk shells within the sand, these slacks are often fed by lime-rich groundwater. The word *slack* is an old Viking word, as in "to slake one's thirst"

eccentric bog a type of raised mire that develops on gentle slopes, with the result that the pool systems, instead of being uniformly concentric in arrangement, have their focus in the upper section of the mire, and the pools assume a crescentic form downslope

ecological niche *see* NICHE

ecosystem an ecological unit of study encompassing the living organisms together with the nonliving environment within a particular habitat

emergent aquatics wetland plants growing in water but having their shoots projecting above the water surface

erosion the degradation and removal of materials from one location to another, often by means of water or wind

eukaryote an organism with cells containing a distinct nucleus enclosed in a double membrane. *See* PROKARYOTE

eutrophication an increase of fertility within a habitat, often resulting from pollution by nitrates or phosphate, or by the leaching of these materials into water bodies from surrounding land. This increase in fertility results in enhanced plant (often algal) growth, followed by death, decay, and oxygen depletion

evaporation the conversion of a liquid to its gaseous phase. Often applied to water being lost from terrestrial and aquatic surfaces

evaporite a sediment rich in salts resulting from the evaporation of warm shallow lakes

evapotranspiration a combination of evaporation from surfaces and the loss of water vapor from plant leaves

evergreen a leaf or a plant that remains green and potentially photosynthetically active through the year. Evergreen leaves do eventually fall but may last for several seasons before they do so

exfoliation the erosion of rocks by the flaking off of surface layers, often caused by frost

facilitation one of the forces that drives ecological succession. When a plant grows in a particular location, it may alter its local environment in such a way that enables other plants to invade. When a water lily grows in a lake, for example, its leafstalks slow the movement of water and encourage the settlement of suspended sediments. The lake becomes shallower as a consequence, and other species of plant are able to invade, eventually supplanting the water lily

fen a wetland dominated by herbaceous plants, fed by the flow of groundwater, and having the summer water table at or below the soil surface

flark the elongate pool of an AAPA MIRE that runs along the contour of the slope

floodplain the low-lying, alluvial lands running alongside rivers over which the river water expands during time of excessive discharge

food web the complex interaction of animal feeding patterns in an ecosystem

fossil ancient remains, usually applied to the buried remnants of a once-living organism, but the term can be applied to ancient buried soils or even to the organic remains we call fossil fuels

fragility the degree of ease with which an organism or a habitat may be damaged

frustule the silica shells of DIATOMS. The frustule has two "valves" which fit together like the overlapping base and lid of a box

fundamental niche   the potential of an organism to perform certain functions or to live in certain areas. Such potential is not always achieved because of competitive interactions with other organisms. *See* REALIZED NICHE

gene pool   the sum of genetic variations found within a population of an organism

gley   a soil that forms under waterlogged conditions, which is therefore ANOXIC

greenhouse effect   the warming of the Earth's surface as a result of short-wave radiation passing through the atmosphere, then converted to long-wave radiation as a result of interception and reflection by the Earth. Long-wave radiation is more likely to be absorbed by the atmosphere after reflection because of the presence of GREENHOUSE GASES

greenhouse gas   an atmospheric gas that absorbs long-wave radiation and therefore contributes to the warming of the Earth's surface by the greenhouse effect. Greenhouse gases include carbon dioxide, water vapor, methane, chlorofluorocarbons (CFCs), ozone, and oxides of nitrogen

groundwater   water that soaks through the soils and rocks, as opposed to water derived from precipitation

habitat structure   the architecture of vegetation in a habitat

halophyte   a plant that is adapted to life in saline conditions as a result of its physical form or its physiology, or both

hydraulic conductivity   a measure of the ease with which water moves through a material. A high hydraulic conductivity means that water moves easily through that material

hydric sediments   deposits laid down beneath water bodies, such as lakes

hydrogen bonding   a bond between two molecules in which one of the components is a hydrogen atom, often linked to oxygen or a halogen. The hydrogen bonding between water molecules contributes to the great cohesive strength of this material, which is vital for the uplift of water in tall plants

hydrology   the study of the movement of water in its cycles around ecosystems and around the planet

heterotrophic   organisms that are unable to fix their own energy and are therefore ultimately dependent upon green plants or other AUTOTROPHIC organisms as an energy source

hydroseral succession   a process of ecosystem development that originates with open water habitats

igneous rocks   rocks created as a result of volcanic activity

inbreeding   a population in which genetic exchange from the outside is severely restricted, resulting in lack of genetic variation. *See* OUTBREEDING

insectivorous   an organism that feeds on insects and other invertebrates. The term may be applied to certain plants that trap insects and digest them as a source of energy and nutrient elements

interception   the activity of plant canopies in preventing rainwater from reaching the ground directly. Intercepted water may continue on its way to the ground, or may be evaporated back into the atmosphere

ion   a charged element or group of elements. *See* ANION and CATION

Intertropical Convergence Zone (ITCZ)   the band of the low latitude zone of the Earth where air masses from the north and the south converge. Its position varies with season, migrating poleward during the summer season in each hemisphere. It is a region of low atmospheric pressure and consequently high precipitation

invertebrate   an animal lacking a backbone, including, for example, insects, mollusks, and crustaceans

jet stream   a rapid movement of air from west to east resulting from the rotation of the Earth

kettle hole   a hollow in glacial detritus deposits resulting from the melting of a block of ice in that position

kleptoparasitism   a form of parasitism in which theft is involved, as in the removal of insect prey from carnivorous plants by ants

lagg   the area of FEN and CARR that surrounds a RAISED BOG as GROUNDWATER flows around the edge of the peat dome

latent heat of evaporation   the energy needed to convert liquid into vapor at the same temperature

latitude   conceptual lines running around the world that are named according to the angle subtended to the equatorial plane. Thus the equator is regarded as 0° and the poles as 90° North or South. The equatorial regions thus lie in the low latitudes and the polar regions in the high latitudes

leaching   the process of removal of IONS from soils and sediments as water (particularly acidic water) passes through them

lignite   soft, brown coal intermediate between PEAT and COAL in its development

limestone   sedimentary rocks containing a high proportion of calcium carbonate (lime)

limnic sediments deposits formed by sedimentation in water bodies. They are commonly rich in mineral, inorganic materials derived from watershed EROSION

longitude conceptual lines running from pole to pole and intersecting the equator. They are numbered from 0° at the Greenwich Meridian in southeastern England running east and west to 180° running through the Pacific Ocean

lycopsids a group of plants related to modern horsetails (Equisetum) that once included large wetland species that dominated the coal-forming swamps of Carboniferous times

macrofossils FOSSILS that are large enough to be examined without the use of a microscope; sometimes referred to as megafossils

macrophyte large aquatic plant that can be observed without the use of a microscope

management in the context of wetlands ecology, refers to the process of manipulation by humans in order to achieve a particular end (e.g., flooding, mowing, burning, harvesting, and so on)

mangal forested coastal ecosystems in the Tropics. The mangrove trees characteristically have upwardly bending roots that extend above the water level and act as respiratory organs

mangrove a term that is applied both to the MANGAL habitat and to the trees that typify this habitat

marsh this term can cause some confusion because it is used in different senses in America and in Europe. In its American sense, it means herb-dominated wetlands with a high water table, generally above the peat or sediment surface. In Europe it is used for terrestrial wetlands on moist mineral soils, often maintained in a short-turf herbaceous condition by grazing and trampling

Mediterranean climate a climate characterized by hot, dry summers and mild, moist winters. Found in the Mediterranean basin, California, Chile, South Africa, and southwest Australia

megafossils *see* MACROFOSSILS

meromictic lake a lake in which there is strong stratification of water layers, often as a result of temperature differences. Such lakes usually produce strongly stratified, laminated sediments. *See* VARVES

metamorphic rocks rocks that have been modified in their structure and composition as a result of high temperature and pressure in the vicinity of volcanic or tectonic activity

methanogenic bacteria bacteria that produce methane gas as a result of their metabolism

microbes bacteria and fungi

microclimate the small-scale climate within habitats, such as beneath forest canopies or in the shade of desert rocks

microfossils FOSSILS that can be observed only with the aid of a microscope, such as POLLEN GRAINS, diatom FRUSTULES, plankton remains, and so on

migration the seasonal movements of animal populations, e.g., geese, caribou, or plankton in a lake

minerals inorganic compounds that in combination make up the composition of rocks. The term is also used of the elements needed for plant and animal nutrition

mire general term for any peat-forming wetland ecosystem

mire complex a wetland that consists of a series of different mire types

net primary production the observed accumulation of organic matter mainly by green plants after they have used some of the products of photosynthesis in their own respiration and metabolism

niche the role a species plays in an ECOSYSTEM. It consists both of spatial elements (where the species lives) and the way in which it makes its living (feeding requirements, growth patterns, reproductive behavior, etc.)

nutrient cycle the cyclic pattern of element movements between different parts of an ECOSYSTEM, together with the balance of input and output to and from the ecosystem

occult precipitation precipitation that is not registered by a standard rain gauge because it arrives as mist, condensing on surfaces, including vegetation canopies

oceanic climate a climate in which summer temperatures are cool and winter temperatures mild and often accompanied by high precipitation. Such conditions are most often encountered in regions close to the oceans

oceanic conveyor belt the movement of the oceanic waters of the world in a pattern that distributes energy from the equatorial regions to the polar regions. Warm, low-density water moves northward into polar regions, where it cools and becomes denser, returning to southern regions as deep-water currents moving in the opposite direction to those at the surface (*see* THERMOHALINE CIRCULATION)

ombrotrophic fed by rainfall. BOGS are ombrotrophic mires, receiving their water and nutrient input solely from the atmospheric precipitation

osmosis the movement of water molecules from low SOLUTE concentration to high solute concentration through a semipermeable membrane that prevents the diffusion of the solute

outbreeding   a population in reproductive contact with other populations of the same species and able to exchange genetic material. Such populations generally contain more variety in their GENE POOL than INBREEDING populations

ox-bow lake   a crescent-shaped body of water produced from an old river channel as a result of a new route being cut, concluding with the isolation of the old channel

paleoecology   the study of the ecology of past communities using a variety of chemical and biological techniques

palsa   a wetland type found only within the Arctic Circle. Elevated peat masses expand as a result of the development of frozen cores within them. They pass through a cycle of growth and then collapse, forming open pools

paludification   a process in which an ecosystem becomes inundated with water

palynology   the study of pollen grains and spores

peat   organic accumulations in wetlands resulting from the incomplete decomposition of vegetation litter

peat extraction   the harvesting of peat by humans either for energy production or for horticultural applications

peat profile   the cross section of a peat deposit that provides an opportunity for the study of peat STRATIGRAPHY and hence the reconstruction of past plant communities

permafrost   permanently frozen subsoil. The upper layer (ACTIVE LAYER) thaws during the summer and freezes in winter

pH   an index of acidity and alkalinity. Low pH means high concentrations of hydrogen ions (hence acidity). A pH of 7 indicates neutrality. The pH scale is logarithmic, which means that a pH of 4 is 10 times as acidic as pH 5

photosynthetic bacteria   bacteria possessing pigments that are able to trap light energy and conduct photosynthesis. Some types are green and others purple

physiological drought   a condition where water is present in a habitat but is unavailable to a plant because of low temperature

pioneer   an initial colonist in a developing habitat

pluvial   a time of wet climate in low latitudes corresponding to times of glaciation in higher latitudes. The southwest of North America experienced such conditions during the height of the last glacial episode in the north

pneumatophore (or **pneumorrhiza**)   root structures on MANGROVE trees that project above the mud and act as a means of gaseous exchange with the atmosphere. They are needed because of the anaerobic conditions in waterlogged mud which prevent roots from respiring

polar front   the boundary where tropical air masses encounter cooler polar air masses, resulting in unstable weather conditions and the formation of depressions

pole forest   the uniform tree cover found on the elevated peat domes of tropical coastal BOG FORESTS of Southeast Asia

pollen analysis   the identification and counting of fossil pollen grains stratified in peat deposits and lake sediments

pollen grains   cells containing the male genetic information of flowering plants and conifers. The outer coat is robust and survives well in wetland sediments. The distinctive structure and sculpturing of the coats permits their identification in a fossil form

polygon mire   patterned wetlands of the Arctic regions in which raised polygonal sections are separated by water-filled channels. These mires are particularly impressive when, seen from the air, their intricate patterns can be appreciated

population   a collection of individuals of a particular species

pothole mires   an extensive series of scattered wetlands found in the temperate continental regions. The North American pothole mires are important duck breeding areas

precipitation   aerial deposition of water as rain, dew, snow, or in an OCCULT form

primary productivity   the rate at which new organic matter is added to an ecosystem, usually as a result of green plant photosynthesis

prokaryote   a simple organism having cells that lack a true nucleus

protist   simple organism belonging to the Kingdom Protoctista, consisting of unicellular or colonial organisms including amoebas, flagellates, diatoms, foraminifera, and many others

pyramid of biomass   a principle that applies to terrestrial ecosystems stating that the succeeding trophic levels of an ECOSYSTEM have a lower total BIOMASS than preceding levels

quaking bog   a wetland in which floating vegetation extends over a lake basin from the edges, eventually forming a complete cover. The acid, floating surface may become covered by trees and quakes when walked on. Also called Schwingmoor

raised bog   a mire in which the accumulation of peat results in the formation of a central dome that raises the peat-forming vegetation above the influence of groundwater flow. The surface of the central dome thus receives all its water input from precipitation (OMBROTROPHIC)

rand the sloping periphery of a RAISED BOG

realized niche the actual spatial and functional role of a species in an ecosystem when subjected to competition from other species. *See* FUNDAMENTAL NICHE

reclamation the conversion of a habitat to a condition appropriate for such human activity as agriculture or forestry

redox potential a scale indicating the potential for oxidation and reduction in an environment

reed bed a wetland dominated by the single species, the reed (*Phragmites australis*)

regeneration complex the arrangement of hummocks and pools on the surface of a RAISED BOG, which was once believed to arise from a sequential replacement of hummocks by pools and vice versa

rehabilitation the conversion of a damaged ecosystem back to its original condition

relict a species or a population left behind following the fragmentation and loss of a previously extensive range

replaceability the ease, or difficulty, with which a particular habitat could be replaced if it were to be lost

representativeness the degree to which a site illustrates the major features characteristic of its habitat type

resource allocation the division of the products of photosynthesis among different parts of a plant, such as leaves, stem, and roots

resource partitioning the manner in which different species assume different roles (NICHES) in an ecosystem and thus divide the resources between them

respiration the oxidation of organic materials resulting in the release of energy. Waste products include carbon dioxide, methane, or ethyl alcohol, depending on the availability of oxygen (*see* REDOX POTENTIAL) and the type of organism involved

rheotrophic a wetland that receives its nutrient elements from groundwater flow as well as from precipitation. In rheotrophic mires the groundwater flow is usually responsible for the bulk of the nutrient input

rhizopods one-celled microscopic animals resembling *Amoeba,* but with a protective shell around their bodies. These shells are often preserved as fossils within peat deposits

ridge-raised bog occasionally RAISED BOGS occupying adjacent hollows may grow to such an extent that they become fused by a linking area of peat covering the ridge that separates them. Such joined raised mires are termed ridge-raised bogs; sometimes called saddle mires

salinization the increasing salt content in the wetlands of hot dry regions when they have no exit drainage. Salinization is a consequence of the evaporation of water leaving behind the salts contained in the incoming water

salt marsh coastal intertidal wetlands dominated by herbaceous plants

saltation the movement of a particle, such as a sand grain, by bouncing over a surface

saturated vapor pressure the total amount of water vapor that a volume of air can hold at a given temperature

Schwingmoor *see* QUAKING BOG

sediment focusing a process that occurs in lake basins where sediments slump from the steeper parts of the lake basin and accumulate in the deeper regions

sedimentary rocks rocks formed by the gradual accumulation of eroded materials, either under water or on land, eventually forming a compressed, stratified mass of material

soligenous of mires that receive water input from groundwater sources, often fed by spring lines. Such mires are usually rich in mineral nutrients and are RHEOTROPHIC

solute a material that dissolves in a SOLVENT

solvent a liquid in which materials (SOLUTES) can dissolve

species richness the number of species of organisms within a given area; an important component of BIODIVERSITY

Sphagnum *see* BOG MOSSES

spores the dispersal propagules of algae, mosses, liverworts, ferns, and fungi

spring mire a peat-forming wetland that develops over springs, often having layers of mineral sediments within its profile as a consequence of water injection under pressure

stochastic a chance event; one that cannot be predicted. *See* DETERMINISTIC

stomata the pores through which a plant exchanges gases with its environment and through which it loses water by TRANSPIRATION

stratification the layering of lake sediments and peats as a consequence of their accumulation in a time-related sequence

stratigraphy the study of layering in sediments and the description of sediment profiles. This may provide information on the developmental sequence of a mire over time

stratosphere the part of the Earth's atmosphere lying above the TROPOSPHERE, from about nine to 30 miles (15–50 km)

string bog *see* AAPA MIRE. The term *bog* is not strictly accurate in this context because these mires are flow-fed (RHEOTROPHIC), hence aapa mire is preferred

strings the raised ridges running along the contours of AAPA MIRES (STRING BOGS). The linear pools between them are termed FLARKS

stromatolite rock-like mounds formed by CYANOBACTERIA in shallow seas. They are found in fossils dating as far back as Precambrian times

sublimation the direct formation of a solid from a gas

submerged aquatic freshwater MACROPHYTE that lives in a submerged (rather than EMERGENT) state. Some carry flowers that extend above the water surface

succession the process of ecosystem development. The stages of succession are often predictable as they follow a directional sequence. The process usually involves an increase in the BIOMASS of the ecosystem, although the development of RAISED BOG from CARR is an exception to this. Succession is driven by immigration of new species, FACILITATION by environmental alteration, competitive struggles, and eventually some degree of equilibration at the CLIMAX stage

sulfide zone the waterlogged, ANAEROBIC CATOTELM in a peat profile. So named because a silver wire inserted into this zone in the pear profile becomes rapidly blackened by silver sulfide

swamp a vegetated wetland in which the summer water table remains above the sediment surface so that there is always a covering of water. In North America, the term is restricted to forested wetlands of this kind, while in Europe the term is normally used only for herbaceous reed beds and cattail marshes

telmatic sediments deposits formed at the WATER TABLE, including some peats

tephra the glasslike dust particles emitted from erupting volcanoes. Layers of tephra in peat STRATIGRAPHY can provide time markers, since the dates of eruptions are well known and the chemistry of tephra often indicates the precise volcano involved

terrestrial occurring on land, as opposed to aquatic

terrestrialization the process of SUCCESSION whereby aquatic ecosystems gradually become infilled

terrestric sediments materials that are deposited above the prevailing WATER TABLE, e.g., the PEATS of RAISED BOGS

texture refers to the proportions of different sized particles in soil. A soil containing a relatively even contribution from sand, silt, and clay is called a loam

thermohaline circulation the movement of water masses around the oceans of the world in a circulatory system that varies in the density of waters, caused by a combination of temperature and salt content. *See* OCEANIC CONVEYOR BELT

topogenous of a mire that receives water by runoff from surrounding slopes. Such mires are flow-fed (RHEOTROPHIC)

topography the form of a landscape

transpiration the loss of water vapor from the leaves of TERRESTRIAL plants through the stomata, or pores, in the leaf surface

troposphere the lower layer of the Earth's atmosphere, up to about nine miles (15 km)

tundra the open vegetation of cold, arctic conditions. Trees are absent, apart from dwarf species of willow and birch

valley mire strictly a mire complex, consisting of a central stream and surrounding fen or carr vegetation, and lateral poor fens in which the flow of water is slow and the nutrient supply is restricted. Often called valley bog because of the acidity and nutrient poverty of the lateral regions, but the wetland normally remains RHEOTROPHIC, so it is not strictly a BOG

varves distinct layers in a lake sediment resulting from annual variations in the conditions of sedimentation and productivity in the lake

vertebrate an animal with a backbone

vessels specialized cells in wood that are responsible for the transport of water up a stem or trunk

vulnerability the degree to which a site is threatened, as when a wetland is in danger of drainage for alternative uses, such as agriculture or forestry. Contrast FRAGILITY

water level the height of water above the surface of a soil

water table the level at which water is maintained within the soil of an ecosystem

watershed the region from which water drains into a particular stream or wetland (equivalent to CATCHMENT). The term is also used of the ridge separating two catchments, literally the region where water may be shed in either of two directions

weathering the breakdown of a rock into MINERALS and chemical elements as a result of the effects of such factors as frost, solution, and biological activity

wetland a general term covering all shallow aquatic eco-systems (freshwater and marine) together with marshes, swamps, fens, and bogs

xeromorphic structural adaptations in plants associated with drought resistance

zonation the banding of vegetation along an environmental gradient, as in the transition from submerged and floating aquatic, emergent aquatics, reed bed, and swamp, around a shallow water body

Further Reading

GENERAL ENVIRONMENTAL REFERENCE

Archibold, O. W. *Ecology of World Vegetation.* New York: Chapman and Hall, 1995. A broad and useful introduction to all the major biomes of the world.

Bradbury, Ian K. *The Biosphere.* New York: Wiley, 2nd ed., 1998. An introduction to global processes that link the various biomes and the human population of the planet.

Brown, J. H., and M. V. Lomolino. *Biogeography.* Sunderland, Mass.: Sinauer Associates, 3rd ed., 2006. A very extensive and exhaustive coverage of the scientific principles that unite biology and geography in the study of the living world.

Cox, C. B., and P. D. Moore. *Biogeography: An Ecological and Evolutionary Approach.* Oxford: Blackwell Publishing, 7th ed., 2005. An introductory text dealing with the historical and modern factors that determine species distributions on Earth.

Gaston, K. J., and J. I. Spicer. *Biodiversity: An Introduction.* Oxford: Blackwell Publishing, 2nd ed., 2004. An explanation of the concept of biodiversity, its meaning, and its importance in conservation.

Houghton, J. *Global Warming: The Complete Briefing.* Cambridge: Cambridge University Press, 3rd ed., 2004. The authoritative account of the most recent research by the Intergovernmental Panel on Climate Change.

GENERAL WETLANDS REFERENCE

Charman, D. *Peatlands and Environmental Change.* New York: Wiley, 2002. A full and detailed account of the ecology of peatland habitats.

Gore, A. J. P., ed. *Ecosystems of the World 4A and 4B Mires: Swamp, Bog, Fen and Moor.* 2 vols. Amsterdam: Elsevier, 1983. Two volumes that cover in great scientific detail the general functioning of wetland ecosystems and regional accounts of wetland types.

Keddy, Paul A. *Wetland Ecology: Principles and Conservation.* Cambridge: Cambridge University Press, 2000. An introduction to wetland ecology and an argument for wetland conservation.

Malanson, G. P. *Riparian Landscapes.* Cambridge: Cambridge University Press, 1993. A detailed academic volume concerning the wetland habitats associated with river environments.

Niering, William A. *Wetlands.* New York: Alfred A. Knopf, 1998. A well-illustrated field guide to the plants and animals of North American wetlands.

Rydin, H., and J. Jeglum. *The Biology of Peatlands.* Oxford: Oxford University Press, 2006. A readable account of the plants and animals that occupy wetland habitats and the ways they interact.

Williams, Michael, ed. *Wetlands: A Threatened Landscape.* Oxford: Blackwell, 1990. A collection of reviews of the state of wetlands in different parts of the world.

WETLAND CONSERVATION AND REHABILITATION

De Waal, Louise, Andrew R. G. Large, and Max P. Wade, eds. *Rehabilitation of Rivers.* New York: John Wiley and Sons, 1998. A series of case studies on the ways in which rivers and their associated wetlands have been improved and conserved.

Hook, D. D. *The Ecology and Management of Wetlands.* 2 vols. Portland, Ore.: Timber Press, 1988. A collection of detailed reports concerning wetland management mainly in North America.

Liddle, Michael. *Recreation Ecology.* New York: Chapman and Hall, 1997. A general account of the ways in which human recreation impacts upon natural environments, including wetlands.

Maltby, Edward. *Waterlogged Wealth*. Washington: Earthscan, 1986. An information-packed account of the contribution wetlands make to human welfare and the results of mismanagement.

Parkyn, L., R. E. Stoneman, and H. A. P. Ingram, eds. *Conserving Peatlands*. Wallingford, U.K.: CABI, 1997. A collection of case studies relating to conservation and management work, mainly European.

Perry, James, and Elizabeth Vanderklein. *Water Quality: Management of a Natural Resource*. Oxford: Blackwell Science, 1996. A book that stresses the importance of water and of wetlands as water resource.

Purseglove, Jeremy. *Taming the Flood: A History and Natural History of Rivers and Wetlands*. Oxford: Oxford University Press, 1988. A very readable account of the history of wetland management with an aim to avoid periodic floods.

Turner, Kerry, and Tom Jones. *Wetlands: Market and Intervention Failures, Four Case Studies*. London: Earthscan, 1991. The interaction between wetland management and economics.

Wheeler, Bryan D., Susan C. Shaw, Wanda J. Fojt, and R. Allan Robertson, eds. *Restoration of Temperate Wetlands*. New York: John Wiley and Sons, 1995. A series of case studies and general ideas relating to the reconstruction of damaged wetlands, mainly European in emphasis.

Whigham, D. F., R. E. Good, and J. Kvet, eds. *Wetland Ecology and Management: Case Studies*. Dordrecht, The Netherlands: Kluwer, 1990. More examples of the ways in which wetland restoration and conservation can be tackled.

HISTORY AND ARCHAEOLOGY OF WETLANDS

Brothwell, Don. *The Bog Man and the Archaeology of People*. London: British Museum Publications, 1986. A graphic and detailed account of the research conducted on the body of an Iron Age man found preserved in a British peat bog.

Coles, Bryony, and John Coles. *Sweet Track to Glastonbury: The Somerset Levels in Prehistory*. London: Thames and Hudson, 1986. The archaeology of ancient trackways preserved in wetlands in western Britain.

Delcourt, P. A., and H. R. Delcourt. *Prehistoric Native Americans and Ecological Change*. Cambridge: Cam-

bridge University Press, 2004. A broad and fascinating study of the ways in which the North American landscape has been affected by prehistoric human activity.

Darby, H. C. The *Medieval Fenland*. Newton Abbot, UK: David and Charles Ltd., 1974. A historical account of the draining of wetlands in eastern England.

Glob, P. V. *The Bog People: Iron Age Man Preserved*. London: Paladin, 1971. A general survey of the many corpses recovered from the bogs of Europe.

Godwin, Sir Harry. *Fenland: Its Ancient Past and Uncertain Future*. Cambridge: Cambridge University Press, 1978. An anecdotal and historical account of the development of lowland wetland studies in Britain.

———. *The Archives of the Peat Bogs*. Cambridge: Cambridge University Press, 1981. A companion volume to the last one covering bog habitats.

Moore, P. D., J. A. Webb, and M. E. Collinson. *Pollen Analysis*. Oxford: Blackwell Science, 1991. The standard account of this technique, containing an extensive collection of pollen photographs and an identification key.

Prince, Hugh. *Wetlands of the American Midwest: A Historical Geography of Changing Attitudes*. Chicago: University of Chicago Press, 1997. The history of human interaction with wetlands in the American Midwest.

Scott, Andrew C., ed. *Coal and Coal-bearing Strata: Recent Advances*. Oxford: Blackwell Science, 1987. A collection of research papers relating to the development and geology of coal.

Steinberg, Theodore. *Nature Incorporated: Industrialization and the Waters of New England*. Cambridge: Cambridge University Press, 1991. The problems of wetland and water management in the Northeastern United States.

Whitney, Gordon G. *From Coastal Wilderness to Fruited Plain: A History of Environmental Change in Temperate North America from 1500 to the Present*. Cambridge: Cambridge University Press, 1994. A very readable account of the consequences of European settlement along the eastern coast of America, including references to wetlands.

THE WETLAND ECOSYSTEM AND ITS INHABITANTS

Araujo-Lima, Carlos, and Michael Goulding. *So Fruitful a Fish: Ecology, Conservation, and Aquaculture of the Amazon's Tambaqui*. New York: Columbia University Press, 1997. The remarkable association of fish and tree seed dispersal in the Amazon Basin.

Chabot, B. F., and H. A. Mooney, eds. *Physiological Ecology of North American Plant Communities.* New York: Chapman and Hall, 1985. An ecological survey of North American vegetation with emphasis on the physiological properties of the plants; includes wetland habitats.

Crawford, R. M. M., ed. *Plant Life in Aquatic and Amphibious Habitats.* Oxford: Blackwell Science, 1987. A collection of research papers dealing with the ways in which plants have adapted to cope with waterlogged environments.

Rheinheimer, G. *Aquatic Microbiology.* 4th ed. New York: John Wiley and Sons, 1991. The microbes of wetlands, their variety and ecological importance.

Westlake, D. F., J. Kvet, and A. Szczepanski, eds. *The Production Ecology of Wetlands.* Cambridge: Cambridge University Press, 1998. A very detailed account of the trapping of solar energy by wetlands plants.

WETLANDS BY COUNTRY AND REGION

AFRICA

Allanson, B. R., and D. Baird, eds. *Estuaries of South Africa.* Cambridge: Cambridge University Press, 1999. Mainly coastal habitats.

Forrester, Bob, Mike Murray-Hudson, and Lance Cherry. *The Swamp Book: A View of the Okavango.* Johannesburg: Southern Publishers, 1989. A lavishly illustrated account of the wildlife of this African wetland.

ASIA

De Zylva, T. S. U. *Wings in the Wetlands.* Sri Lanka: Victor Hasselblad Wildlife Trust, 1996. The birds of wetlands in the Indian subcontinent.

Evans, Martin. *Bharatpur: Bird Paradise.* London: H. F. & G. Witherby, 1989. An illustrated account of one of the world's richest wetlands for bird life.

Knystautas, Algirdas. *The Natural History of the U.S.S.R.* London: Century, 1987. An account of the habitats and wildlife of Russia and its surrounding states, including some wetland accounts.

AUSTRALIA

Streever, B. *Bringing Back the Wetlands.* New South Wales: Australia, 1999. Wetland conservation studies in Australia.

COASTAL WETLANDS

Allen, J. R. L. and K. Pye. *Saltmarshes: Morphodynamics, Conservation and Engineering Significance.* Cambridge:

Cambridge University Press, 1992. A specialized text covering engineering aspects of coastal salt marsh development.

Carter, R. W. G. *Coastal Environments: An Introduction to the Physical, Ecological and Cultural Systems of Coastlines.* New York: Academic Press, 1988. A good general introduction to coastal habitats.

Packham, J. R., and A. J. Willis. *Ecology of Dunes, Salt Marsh and Shingle.* New York: Chapman and Hall, 1997. Detailed account of the development of coastal wetlands and dunes.

Viles, Heather, and Tom Spencer. *Coastal Problems: Geomorphology, Ecology and Society at the Coast.* London: Edward Arnold, 1995. The interaction between coastal environments, including wetlands, and people.

EUROPE

Bailey, R. G., P. V. Jose, and B. R. Sherwood. *United Kingdom Floodplains.* Otley, U.K.: Westbury, 1998. The specific problems associated with floodplains and problems in human housing and land use.

Bellamy, David. *The Wild Boglands: Bellamy's Ireland.* London: Christopher Helm, 1986. A readable and informative account of the bogs of Ireland.

Fernandez, Juan Antonio. *Doñana.* Sevilla: Editorial Olivo, 1974. A well-illustrated book describing the history and wildlife of one of Europe's richest wetlands.

Moore, Peter D., ed. *European Mires.* New York: Academic Press, 1984. A collection of reviews of the peatland habitats of Europe.

Mountfort, Guy. Portrait of a Wilderness: *The Story of the Coto Doñana Expeditions.* London: Hutchinsons, 1958. The original account of the expeditions that revealed the rich wildlife resources of this Spanish wetland.

Verhoeven, J. T. A. *Fens and Bogs in the Netherlands: Vegetation, History, Nutrient Dynamics and Conservation.* Dordrecht, The Netherlands: Kluwer, 1992. Problems associated with wetland management in The Netherlands.

NORTH AMERICA

Barbour, Michael G., and William Dwight Billings. *North American Terrestrial Vegetation.* Cambridge: Cambridge University Press, 1988. A very informative account of North America's vegetation.

Davis, Steven M., and John C. Ogden, eds. *Everglades: The Ecosystem and Its Restoration.* Delray Beach, Fla.: St. Lucie Press, 1994. A detailed account of the Everglades region and the history of its problems.

Martin, William H., Stephen G. Boyce, and Arthur C. Echternacht, eds. *Biodiversity of the Southeastern United States: Lowland Terrestrial Communities.* New York: John Wiley and Sons, 1993. Habitats of the Southestern region, including wetlands.

Myers, Ronald L., and John J. Ewel, eds. *Ecosystems of Florida.* Orlando: University of Central Florida Press, 1990. The wetlands and other ecosystems of this state.

Sage, Bryan. *The Arctic and Its Wildlife.* London: Croom Helm, 1986. A sound scientific coverage of the Arctic with information on its wetland and other habitats.

Schoenherr, Allan A. *A Natural History of California.* Berkeley: University of California Press, 1992. A very detailed coverage of California's ecosystems, their inhabitants, and their development.

SOUTH AMERICA

Araujo-Lima, Carlos, and Michael Goulding. *So Fruitful a Fish: Ecology, Conservation and Aquaculture of the Amazon's Tambaqui.* New York: Columbia University Press, 1997. The ecology of the fruit-eating fishes of the Amazon.

Goulding, Michael Smith, J. H. Nigel, and Dennis J. Mahar. *Floods of Fortune: Ecology and Economy along the Amazon.* New York: Columbia University Press, 1996. An analysis of the ecological significance of the Amazon floods.

TROPICAL WETLANDS

Rieley, J. O., and S. E. Page, eds. *Biodiversity and Sustainability of Tropical Peatlands.* Cardigan, Wales: Samara Publishing, 1997. One of the few books devoted to the peatlands of the Tropics, dealing with their ecology and exploitation.

Tomlinson, P. B. *The Botany of Mangroves.* Cambridge: Cambridge University Press, 1986. The biology and ecology of mangrove trees.

Web Sites

CONSERVATION INTERNATIONAL
URL: http://www.conservation.org
Accessed November 9, 2006. Particularly concerned with global biological conservation.

EARTHWATCH INSTITUTE
URL: http://www.earthwatch.org
Accessed November 9, 2006. General environmental problems worldwide.

ENVIRONMENTAL PROTECTION AGENCY
URL: http://www.epa.gov.owow/wetlands/
Accessed November 9, 2006. Wetland conservation problems and solutions in the United States.

INTERNATIONAL UNION FOR THE CONSERVATION OF NATURE
URL: http://www.redlist.org
Accessed November 9, 2006. Many links to other sources of information on particular species, especially those currently endangered.

NATIONAL PARKS SERVICE OF THE UNITED STATES
URL: http://www.nps.gov
Accessed November 9, 2006. Information on specific conservation problems facing the National Parks.

NATIONAL WETLAND RESEARCH CENTER
URL: http://nwrc.usgs.gov
Accessed November 9, 2006. Current wetland research in the United States.

NATIONAL WETLANDS INVENTORY
URL: http://www.fws.gov/nwi/
Accessed November 9, 2006. Information about the classification and survey of wetlands in the United States.

RAMSAR CONVENTION
URL: http://www.ramsar.org
Accessed November 9, 2006. The official website of the Ramsar organization, giving news of global wetland conservation.

SIERRA CLUB
URL: http://www.sierraclub.org
Accessed November 9, 2006. Covers general conservation issues in the United States and also covers issues relating to farming and land use.

U.S. FISH AND WILDLIFE SERVICE
URL: http://www.nwi.fws.gov
Accessed November 9, 2006. A valuable resource for information on wildlife conservation.

U.S. GEOLOGICAL SURVEY
URL: http://www.usgs.gov
Accessed November 9, 2006. Covers environmental problems affecting landscape conservation.

UNITED NATIONS ENVIRONMENTAL PROGRAM WORLD CONSERVATION MONITORING CENTER
URL: http://www.unep-wcmw.org
Accessed November 9, 2006. Good for global statistics on environmental problems.

Index